CAMBRIDGE LIBRARY COLLECTION

Books of enduring scholarly value

Darwin

Two hundred years after his birth and 150 years after the publication of 'On the Origin of Species', Charles Darwin and his theories are still the focus of worldwide attention. This series offers not only works by Darwin, but also the writings of his mentors in Cambridge and elsewhere, and a survey of the impassioned scientific, philosophical and theological debates sparked by his 'dangerous idea'.

William Bateson, Naturalist

William Bateson (1861-1926) began his academic career working on variation in animals in the light of evolutionary theory. He was inspired by the rediscovery of Gregor Mendel's work on plant hybridisation - which he translated into English - to pursue further experimental work in what he named 'genetics'. He realised that Mendel's results could help to solve difficult biological questions and controversies which others had glossed over, and to challenge assumptions underlying evolution as it was understood at the time. After two years as Professor of Biology at Cambridge he left in 1910 to become Director of the newly-founded John Innes Institute. Bateson's argumentative personality and unorthodox approach did not make him popular, and his reputation declined after his death. Was Bateson misunderstood? Was evolution misunderstood? This 1928 volume – including a substantial memoir by Bateson's wife – gives readers access to selected papers and addresses and allows them to consider him afresh.

Cambridge University Press has long been a pioneer in the reissuing of out-of-print titles from its own backlist, producing digital reprints of books that are still sought after by scholars and students but could not be reprinted economically using traditional technology. The Cambridge Library Collection extends this activity to a wider range of books which are still of importance to researchers and professionals, either for the source material they contain, or as landmarks in the history of their academic discipline.

Drawing from the world-renowned collections in the Cambridge University Library, and guided by the advice of experts in each subject area, Cambridge University Press is using state-of-the-art scanning machines in its own Printing House to capture the content of each book selected for inclusion. The files are processed to give a consistently clear, crisp image, and the books finished to the high quality standard for which the Press is recognised around the world. The latest print-on-demand technology ensures that the books will remain available indefinitely, and that orders for single or multiple copies can quickly be supplied.

The Cambridge Library Collection will bring back to life books of enduring scholarly value across a wide range of disciplines in the humanities and social sciences and in science and technology.

William Bateson, Naturalist

His Essays and Addresses together with a Short Account of his Life

BEATRICE BATESON

CAMBRIDGE
UNIVERSITY PRESS

CAMBRIDGE UNIVERSITY PRESS

Cambridge New York Melbourne Madrid Cape Town Singapore São Paolo Delhi

Published in the United States of America by Cambridge University Press, New York

www.cambridge.org
Information on this title: www.cambridge.org/9781108004343

© in this compilation Cambridge University Press 2009

This edition first published 1928
This digitally printed version 2009

ISBN 978-1-108-00434-3

WILLIAM BATESON

Cambridge University Press
Fetter Lane, London

New York
Bombay, Calcutta, Madras,
Toronto
Macmillan

Tokyo
Maruzen-Kabushiki-Kaisha

WILLIAM BATESON

Drawn by W. Arnold Forster *Dec. 1923*

WILLIAM BATESON, F.R.S.
Naturalist

HIS ESSAYS & ADDRESSES

together with

A SHORT ACCOUNT
OF HIS LIFE

by

BEATRICE BATESON

CAMBRIDGE
AT THE UNIVERSITY PRESS
1928

PREFACE

In the following pages I have attempted only to sketch a rare personality. Later, a more competent hand may, I hope, undertake a full biography and account of William Bateson's work.

For help and encouragement in compiling this short Memoir I am much indebted to my friends. The early letters I found amongst miscellaneous documents stored by his mother. I am especially grateful to my sister-in-law, Anna Bateson, for her splendid collection of her brother's letters; and for other letters which I have generous permission to use, I wish to thank Sir W. P. Herringham, M.D., Sir David Prain, Dr Ostenfeld of Copenhagen, and Professors R. C. Punnett, F.R.S., G. H. Hardy, F.R.S., G. C. M. Smith, and Major C. C. Hurst; and I thank Mr W. Arnold Forster and the Trustees of the National Portrait Gallery for kind permission to reproduce the frontispiece. I am very grateful to Miss Mary Fletcher for helping me with the proof-reading.

In 1920 my husband collected ten of his "lay" papers in the hope of finding a publisher. In the rough draft of a letter intended to accompany them he wrote:

"Here are ten papers. They are all more or less lawfully begotten by Mendelism out of Common Sense, *me obstetricante*. They may be classed as:

Digestible:
   (1) Herbert Spencer lecture (Biological fact and the structure of Society). 1912.
   (2) Science and Nationality. 1918.
   (3) Place of Science in Education. 1917.
   (4) Address to the Salt Schools. 1915.

For the *Eupeptic* only:
   (5), (6) Australian Addresses to the British Assoc. 1914.
   (7) Methods and Scope of Genetics. Inaugural Address. 1908.
   (8) Heredity and Variation. (*Darwin and Modern Science.*) 1909.

Indigestible:

(9) Gamete and Zygote. 1917.

(10) Address to Zoological Section, Brit. Assoc. 1904.

No. 9 has never been printed. No. 10 is a bit technical, and anyhow must be cut down. Nevertheless if such a volume appears I should like to put it in, because of its date. My friends are finding these things out for themselves now and I have a vain desire to tell them I was there first. All, except *Science in Education*, fell dead in the hour of birth. The Australian Addresses were delivered in the first week of the war....A title is wanted. A Scotch soldier, when I was lecturing in Y.M.C.A. huts, said: 'Sir, what ye're telling us is nothing but *Scientific Calvinism'*. Sometimes I think that would serve."

This was in 1920 before the Galton lecture (Common-sense in Racial Problems), which he intended to include. Since then there are additions to both classes, *digestible* and *indigestible*. Though some of the latter may intimidate, yet all the papers included in this volume were addressed to a lay public, i.e. hearers and readers not supposed to be versed in his studies. Had the volume been accepted for publication when he offered it, he would certainly have edited the papers freely, deleted repetitions, "cut them down" and "dressed them up", treatment that he felt necessary. To do this I have not ventured.

The papers are not in chronological order but are roughly classed. For their courteous permission to reprint them I have to thank: The Council of the British Association for the Advancement of Science; the Council of the Eugenics Society; the Council of the Royal Horticultural Society; the Delegates of the Clarendon Press; the Editor of *Brain*; the Editor of *Nature*, and Messrs Macmillan and Co., Ltd.; the Editor of *The Nation and the Athenaeum*; the Editor of *The Edinburgh Review*, and Messrs Longmans, Green and Co., Ltd.; and the Editor of *Science*, New York.

Beatrice Bateson

25 Bolton Gardens,
London, S.W. 5

CONTENTS

REVIEWS

MEMOIR

WILLIAM BATESON, the second child in a family of six, was born at Whitby on 8 August 1861. His father, William Henry, the fifth son of Richard Bateson, a prosperous merchant of Liverpool, was educated at Shrewsbury under Dr Samuel Butler, and at St John's College, Cambridge. He was a sound classical scholar. After reading for the Bar, he took orders, but eventually he settled in Cambridge. He was elected Public Orator in 1848, and Master of his College in 1857.

In the obituary notice by Professor Bonney, he is described as

especially distinguished by a clear logical intellect, by a singularly acute judgement and by a remarkable faculty for seeing the weak points of any scheme or argument. He was an excellent man of business, of great industry and patience, a first-rate chairman of a meeting, discerning its feeling with marvellous intuition.

An outward dignity of demeanour was combined with a real simplicity of character, and beneath a slight external coldness of manner lay a heart remarkably kind.

He married (1857) Anna, the elder daughter of James Aikin, a prominent citizen of Liverpool. The son of a Writer to the Signet in Dumfries, Aikin came to Liverpool at about the age of 15 (1806), and after some few years in a merchant's office set up business on his own account.

He made many very successful voyages, chiefly to the West Indies, and subsequently introduced a fleet of clipper schooners, which at that period were amongst the fastest vessels afloat. He settled in Liverpool as ship-broker, owner, and merchant. A staunch Liberal, he took a leading part in municipal affairs.

He was eminently practical as a public man, and though short in his observations, they were remarkably terse, and frequently

B

very epigrammatic.... There was no man who could so effectually silence an opponent by the vigour and readiness of his
repartee as James Aikin....

Of personal bitterness he never had more than was sufficient
to point the moral of perfidy, or to brand the baseness of a
public policy without nobility or ideas; but he had enough
plainness of speech to do this with effect, and if he made no
enemies there were not a few who found his short sword of
retort and epigram sharp and direct in its thrusts. (*Liverpool
Daily Post*, 8, 9 July 1878.)

His scholarly tastes and interest in art and drama are
noted.

Will's mother was also remarkable for her vigorous
vitality, lively spirit, and versatility. She was a very
beautiful woman.

My sister-in-law, Miss Anna Bateson, adds this note:

My grandfather, James Aikin, was an important figure in
our early days, an adorable companion to his grandchildren.
He was of very striking appearance, and extremely vigorous,
mentally and physically.

A successful man who had made his way unaided and had
preserved high ideals and simplicity of character, he was one
of those who inspire confidence at first sight. My brother
certainly inherited many of his characteristics, especially his
tireless physical energy and his power of endowing everything
he touched with a peculiar exciting thrill of interest.

In his prime he had been a man of indomitable determination and almost excessive severity. To the last his own children
regarded him with an awe which was unintelligible to us
grandchildren. He was a lively talker; where he was dulness
disappeared; his whole person radiated zest in life and vigorous
health. When we knew him he was an old man, but he preserved to the last the freshness of youth. It was characteristic
that he should die at the age of 86, from a chill caught sea-
bathing before breakfast, a habit he would never abandon.

•   •   •   •   •

Many a child shews curiosity and interest in natural knowledge, and it is but a phase in his development. With Will however the desire to know was not the normal, superficial curiosity of childhood, it was a passion that stayed by him through life to his last conscious days. He was always a student of nature—an enthusiastic student— and from quite early days commanded an extraordinary range of knowledge in his subject.

In an undated letter (about 1875) to a friend, Will's mother wrote: "Well, I am getting Willie a stunning little book by Sir John Lubbock on the Influence of Insects in the development of Flowers. I will see if he likes it. I expect him to burrow down in it like a bee in a flower himself".

Other small incidents are remembered which mark his very early interest in nature. As a little boy of seven or eight years he parried unfavourable comment on a somewhat disreputable-looking figure who haunted the ditches round Cambridge, exclaiming reverently, "That man is a naturalist".

His sister, Anna, tells me too of a school-room discussion of future careers in which he announced his intention of becoming a naturalist..."if I am good enough," he added doubtfully, "if not, I suppose I shall have to be a doctor". (About this time he procured by some means a copy of Gray's *Anatomy*, to which he was much attached.) His sister also recalls a children's party—occasions generally dreaded by the little Batesons—which Will enjoyed hugely. He spent the whole time talking to another boy. "That boy knows a lot about bones", he said, on his way home, in brief explanation of his unwonted pleasure.

At his Preparatory school (Mr Waterfield's, Temple Grove) all went well enough, and in 1875 he won a Scholarship at Rugby. Dr Jex-Blake, the Headmaster at that time, sent the Socratic Essay which Will wrote for this competition to Dr Bateson, describing it as "one of the best things done in the Examination".

Rugby Scholarship Essay, June 1875.

A. (Head of Eleven). Hope we win the match.

Socrates. How so? What good will it do you?

A.   Well, it won't do me any particular good, but it would be such an honour to the school.

Soc. Then you think it an honour to your school when you win a match, or, in other words, you think it is an honour to your school when 11 of your school can hit a ball better than 11 of S...?

A.   Well not exactly.

Soc. Well then, perhaps you think it an honour to your school that your Master and boys are richer and therefore better able to keep the field, etc., in good order than the Master and boys of S...?

A.   I can hardly say I do.

Soc. Well then, where is your honour? Perhaps you think it an honour when you have been employing 4 or 5 hours in practising cricket while the boys of S...were improving their minds; or rather, do you think it does you credit to have been wasting half the day in sport?

A.   No, I don't think it is.

Soc. Do you think it does honour to yourself or anybody else, when a lot of vagabonds have collected together to bet and get drunk, and all to see you and some others making a show of the way you run, etc.?

A.   No, I don't think that is an honour.

Soc. Well then, what do you consider the honour?

A.   I don't know exactly, now you ask me.

Soc. No more do I. It seems to me there is not much left. It is all very well to take exercise for your health, that everyone must admit, but it is another thing to think about nothing else but amusement.

But in spite of his very evident ability, Will was no success at school. Quarter after quarter his school reports express the dissatisfaction and disappointment of his masters, and his name figures ominously near the bottom of all his class lists. He was unpopular among the boys. Probably his intense and emotional sensitiveness, combined with an unusually alert critical faculty, made him an object of dislike to his school-fellows, and made his masters objects of dislike to him. Three letters written by him at Rugby shew clearly enough how sore was his need of friendly encouragement and sympathy in this dismal period.

10 March, 1878.

Dear Mamma,

Thank you for your note. I went to dinner with Hutchinson to-day, and to breakfast with Lee-Warner on Thursday. I happened to meet him yesterday when I was going a walk and we went together. It is delightful to have some one that you can talk to. He is so different from the other Masters and the fellows one sees. Sidgwick said the other day that I was doing better this term....He was not like a Master at all. I think I enjoyed myself more then than all the time I have been here.

I am going to begin Morley's *Life of Rousseau* which he recommended, next week. I have not been a walk of less than 6 miles any half-holiday since my headaches, and once 14, and twice 10. I have been all right since then. I have read a great deal this term, and have been happier than usual— *and I don't want you to tell this*—but somehow I never feel really happy,—I suppose I never shall, at least I don't think I ever did except when I got my scholarship; and when Sidg. says I am doing better. It never seems any better. Is anyone happy? I don't think I shall be. You will say, this is all morbid nonsense, but it is true. I never get on with anybody for long; at home even I am always in some scrape except when I am alone. And don't please write back that I am foolish and that, and then not tell me how to cure it.

Yours affectly.

W. Bateson.

A letter to his father was enclosed in the same envelope:

Dear Papa,

I am very sorry that I have spent so much, but I don't think I could have curtailed anything except pencils and the blotting book. With regard to the former I ordered the second lot because I had lost the first, which I have since found, and I sent a fag down for a blot book and he brought up that one, and I did not know the price until I had used it and it could not be sent back. We are supposed to use a separate note-book for each subject we do, but I do not near do that. The paper is the paper we use. I am sure I have always wanted a packet every three weeks before, but I will try and curtail it now. And I do so hope you won't despise me for what I have said to Mamma. I know I have no pluck. Fellows have kicked that out of me long ago. And I do mean it all. I could not tell you before, but now it would come.

Your affectionate son,

W. Bateson.

The third letter was written on the occasion of his grandfather's death.

Rugby,

July 8, 1878.

Dear Mamma,

When I saw another letter again so soon from you I feared what had happened.[1] Let us indeed be thankful for Grand-papa's long grand life! I felt so glad I saw him once more after all my bad reports. Don't you think he was more satisfied? We were all talking about the Certificate Exam. downstairs when I opened my letter. I am afraid I shall have to wait till next year. Shall you be disappointed? I had calculated the time to too great a nicety and made no allowance. You must have had a very sad Sunday. Is poor Mary very much cut up? She is so sensitive always. I have got a great deal to tell you, but I must wait till some other time.

---

[1] His grandfather Aikin died 5 July 1878.

I can leave from Wednesday morning till Friday night. I am
afraid of my things not being done in time. Thank you very
much for writing instead of telegram.

Ever your affectionate son,

W. Bateson.

Will's interest in Natural Knowledge never failed him;
it held him straight and steady through these discouraging
years; it armed him with the dogged endurance, which
seemed "self-satisfaction", offensive to his Headmaster,
but in later years stood him in good stead. If he referred
to these early days, the memory was either of his misery
at school, or of some boyish triumph of collecting or
observing in the field. He would dream through a summer
chapel service, watching the open windows. "Would a
Convolvulus hawk fly in? What should he do if one
actually did? Would they let him catch it?"

Under "Science", i.e. chemistry, the school reports
record that he is "attentive", and "really intelligent" and
his work is "V.G."

The indications seem clear; but Dr Bateson was a
classical scholar, and perhaps Natural Science did not
strike him as a possible "pursuit" in life. He wrote to his
friend (the Rev. Jermyn Cooper of Fylingdales)

Oct. 1877.

...I regard you as fortunate in one respect with regard to
your son Robert, that he has a distinct and decided inclination
for one pursuit in life. It is a great advantage where it exists
and it can appropriately be gratified. We have been wishing
that our eldest boy could manifest some special propension
but as yet there are no signs of any....

As a matter of fact, the boy knew very well what he
wanted to try to do, but his experiences at school had not
developed self-confidence. The remoteness of his interests
from his father's doubtless contributed to the illusion of

aimlessness. Parental misgivings cannot have been allayed
by the receipt of the following note from the Headmaster:

Rugby,

Dec. 16, 1878.

Dear Dr Bateson,

Your son's work is far inferior to its true level; and for
him to be 25th in Classics of Lower Bench, below Gedge,
Court, Cross—and below Downing too, a good mathematician,
but no Classic—is scandalous. His Divinity with me is very
poor; and in short the whole result is poor. Unless the two
next Terms are wholly different I cannot advise you to send
him back after the summer holidays; and it is very doubtful
whether so vague and aimless a boy will profit by University
life.

He is certainly capable, and certainly much to blame. He
ought to work during the holidays; and the Horace Prize will
give him a good target. While so exceedingly disappointed
in your son, let me say how glad I am to have seen his sister;
so cheerful a visitor and so nice in all the real points.

With kind regards to Mrs Bateson and regrets that I cannot
think well of your son while so self-satisfied, indolent and
useless, I am,

Yours sincerely,

T. W. Jex-Blake.

In July 1879 Dr Jex-Blake made some amends. He
writes:

I wish the report were better, though it is less bad than most
of late. You are right in removing him early, for he is very
self-satisfied and desultory—even indolent in most things here.
But he has vigour and character, though a little abnormal,
and I do not despair of seeing him make a mark in the world
yet. I enclose Mr M's report....

In after-life, addressing the Salt Schools (1915), Will
recalled his school-days:

I look back on my school education as a time of scarcely relieved weariness, mental starvation and despair. There came at last a moment when I was turned into a chemical laboratory and for the first time found there was such a thing as real knowledge which had a meaning and was not a mere exercise in pedantry. Our staple was of course Latin and Greek, of which I made nothing. Some emotional pleasure came towards the end of my school course from the Greek tragedies but otherwise those years were almost blank. Now what I, and thousands of other boys like me, discover in after-life is that by those very same materials, perhaps more than by any others, we might have been "waked to ecstasy" and to the joy of development.

Extremely sensitive, he was unable to assert himself in face of such discouragement. The same sensitiveness responded quickly enough to sympathy and interest, and this is evident in an immediate improvement in the school reports after the walk with Lee-Warner, also in his vivid happy memory of days when such encouragement came his way. Half-holidays spent with Mr Linnaeus Cumming (mathematical master and field-naturalist of distinction) were bright in his memory.

Later in middle life he came across an old lady, Miss Lakin, in Reading,[1] who had been governess in his family. With but little knowledge of Natural History she had shewn sympathy and interest in the boy's holiday pursuits, trying to help him collect his treasures on land and sea-shore. From this chance meeting he came home excited and moved by his memories of childish gratitude to her. Until her death he never failed to go and see her on the occasions of his visits to Reading; amusing her with accounts of his own boys, and helping her with small gifts of one sort or another. I have now her manuscript note-book with lists of "the seven wonders of the

[1] At that time Messrs Sutton and Sons generously allowed him a free run of their glass and gardens, giving him opportunity and material for study and research otherwise unobtainable, and he was often in Reading in consequence.

world", "The Labours of Hercules", the dates of the Kings of England...her stock-in-trade as governess—which he brought back in triumph one day after visiting her.

He left Rugby in 1879 and was entered at St John's College, Cambridge, of which his father then was Master.

But the grim cloud of school still overshadowed him. He was "ploughed" in mathematics in the "Little-go".

This difficulty he referred to many years later in a letter to *Nature*[1] on compulsory Greek:

Cambridge,

Feb. 1905.

The experiences of Mr Willis and others suggest that mine may be in point. Mr Willis was behind in Classics. He wasted $105\frac{1}{2}$ hours on Greek and passed. His present knowledge of Greek is *nil*. Mathematics were my difficulty. Being destined for Cambridge I was specially coached in mathematics at school. Arrived here, I was again coached, but failed. Coached once more I passed, having wasted, not one, but several hundred hours on that study. Needless to say, my knowledge of mathematics is *nil*. My case is that of hundreds. Why then are not compulsory mathematics to be reformed away?

Because they can be used in trades and professions for the making of money. But the things that put the touch of art in the life of a dull boy, that open his eyes for once to another world, where "utility" does not count—they, forsooth, must be dispensed with because in the market they have no value. And, verily, they are without price.

The teaching of mathematics since his school-days has been greatly improved. Often he envied his sons the easy grasp they had acquired of this subject, though none of them were "mathematical". When he needed help in his calculations he sought it frankly from a professional.

Once quit of school and "pedantry", free to choose

---

[1] *Nature*, LXXI, No. 1843, 390.

his own associates, sure of his high ambitions (if still
not sure of himself), his progress seems to have been un-
interrupted.

In 1882 he took a 1st in the Natural Sciences Tripos
and was elected to a College Scholarship; in the following
year he took a 1st again in Part II in Zoology, and
graduated.

Dr Bateson unfortunately, did not live to enjoy these
successes. He died in 1881. Many years later Will wrote
of his father:

I wish you had known my father—he died while I was an
undergraduate, before even I knew how rare such men as he
are in this world. He was the most unworldly of men, so
thoughtful and gentle, and yet a strong clever man too. I
know now that his ways must have been of the old *régime*,
courtly and delicate—and yet he was essentially a man of
action. Most of the changes made in Cambridge, bringing in
the things of the new knowledge, have been more or less
helped on by him. A very few weeks before he died he sat day
after day at the Arts School where the changes of the last
Commission were being debated, and stood up and tackled
the enemy on point after point, till at last no one could be got
up against him. I have heard many men say that it was a
regular rout; of course I knew nothing about these things then.

These College days and the next few years were packed
with work, and the pursuit of all sorts of interests and
delight. The opportunity of untrammelled freedom to
develop on his own lines in an environment of great
tradition and of great achievement had come to him, and
he made the most of it. Books, pictures, music, travel,
experiment, men and things—all passed under his critical
and appreciating review. With passionate depth of feeling
he began to know the masterpieces of art,[1] literature
and learning. His power of enjoyment, though very
discriminating, was intense; and it is this aesthetic and

[1] See p. 21. Letters about Dresden.

intellectual exaltation, "the awakening to ecstasy", that he urged again and again in his educational papers:

> We are all men born into a splendid and terrible world in which for a while our lot is to enjoy and to suffer. The one reasonable aim of man is that life shall be as happy as it can be made, with as much as possible of joy and as little as possible of pain. There is only one way of attaining that aim: the pursuit of natural knowledge.[1]

Nevertheless, certain aspects of life and certain fields of knowledge made no appeal to him. He would have liked to have acquired a more practical grasp of mathematics, but law, political history, geography and economy, he brushed aside at the outset of his career, as matters too narrow and too "ephemeral" (his epithet) to be accounted of interest.

He wrote to his sister, Anna (then studying botany), in a letter from Irghiz, June 1886:

> You are very thoughtful in thanking me for what little I have done, but in all seriousness, I often feel very heavy when I reflect that in all probability you would have done better, as regards class, if you had taken—History, *par exemple*, but I console myself when I reflect that perhaps you would have despised such a success. And don't think that I mean to throw mud at Mary—not at all—I am heartily glad she is doing so well and am proud of it too—but more because I don't think that she will make mere pot-boiling stuff of it, but also real work. But I believe that if you had been successful in such a subject, you would have ever regretted that your work did not lie a little nearer the origin of things, and was not, so to speak, "*purer*" than such work must be. And so I don't altogether regret for your sake, that I led you in at such an infernally strait gate.

He was a good linguist—one who enjoyed learning a language. Especially fond of French, he spared no pains to learn to speak and write it correctly. He had a wide

---

[1] See p. 409. Address to Salt Schools.

acquaintance with French novels in general, and was an ardent admirer of Balzac and of Voltaire in particular. I think he never travelled without a copy of *Candide* in his pocket.

Of his less complete mastery of German he was always ashamed; he had enough to read without difficulty, and as years passed and he saw more of his German friends and colleagues, he learned to converse freely. For many months—1901—we subscribed to the periodical *Die Fliegende Blätter*, which he read aloud "to improve his colloquial German, and to acquire a sense of their humour". He knew enough Spanish to read scientific papers in that tongue and to make his way, off the beaten track, through Spain. Italian I never heard him try to speak, but he could read it well enough to make out the gist of a paper, or treatise.

His critical and amused interest in the working of the human mind was probably never in abeyance. He watched his fellow-creatures keenly, noting every detail and commenting with abundant humour afterwards in his letters and conversation.

Amongst other matters ritual interested him intensely as a young man; he was extraordinarily well-informed about the conduct and faith of the different sects. He attended church and chapel and meeting-house assiduously.

Psychical Research also drew his interest. He watched, formed his conclusions and stood aside; but he became an expert "thought-reader" and enjoyed "performing".

He took part in a parliamentary election—but once only. On this occasion he threw himself with unstinted energy and enthusiasm into the Cambridge election of 1882, canvassing so hotly, and working so hard, that even the other day (1925), his son was asked by a Cambridge tradesman whether he were connected with "a Mr Bateson, Radical". He enjoyed the fun, but he never repeated the experience in connection with "Party" politics; indeed, it was much if he could even be brought to record his vote. Parliamentary politics, he declared, were without foundation in reality. Straining to wrest fundamental truths from the natural world, he paid but little heed to matters, the interest of which he judged to be "purely ephemeral".

Later on, of course, he took considerable part in University politics.

There is, however, not a little evidence in his earlier letters that he was as a youth intensely patriotic and proud to be an Englishman. He wrote to his mother from

Karkaralinsk, Siberia,
June 1887.

When I had seen even less of the world than now, I got somehow the idea that all men were equal and had equal rights. Hence it seemed to be clear that no one could be justified in appropriating his neighbour's goods or in controlling his neighbour's actions. A very slight experience suffices to shew the preposterous fallacy of this view. All men are no more equal than all animals and plants are equal. A Russian is no more the equal of an Englishman, and a negro is no more the equal of a white man than a Kirghiz pony is the equal of an English racer, or the Phylloxera the equal of the vine. If you think these things life stops short for you. Life without killing and without a struggle cannot go on. It is possible probably to increase or diminish the intensity of the struggle, but that is another thing. Now, one has made up one's mind that one is going to live life out and therefore one must give up the idea of treating all living things, and in particular, all men, as equal; therefore help some and fight the rest one must. Of course I know that there is no test of universal application by which a man's worth can be estimated: this cannot be helped. But there are a set of qualities which we, following our instinct for want of a better guide, regard as denoting superiority. Chief among these is the power of doing good work, and this, even you admit, is the prerogative of our race. Next comes the power of artistic production; and though we have plenty to be ashamed of in this respect, yet we have no more than our contemporaries; and if you take past time into account, with the Bible and Shakespeare we walk ahead of most. As regards physique we are miles ahead and also we are much cleaner. For all these things it seems to me that we are, as peoples go, well to the fore. It is no light thing that nearly all the great inventions are of English origin. ...For all this I believe we are, if not in the absolute a fine race, at least vastly finer than our compeers and therefore I think we are worth preserving not only for our own sakes but for the sake of the whole world too....And as I believe we are a fine race and wanted very badly on earth, so I believe we are worth preserving as a power for good and even if pipe-clay and Horse Guards are necessary to effect this, to that extent I go for pipe-clay and Horse Guards too; not because they are Philistine institutions, but precisely because they protect us from the Philistinism of the foreigner—to which ours is a joke. My feeling is quite clear that it is right to maintain these things at the highest point attainable without interfering with the handicrafts or hampering intellectual individuality. That is the point. If, for example we were one day to find that to keep our national independence the conscription was necessary as an institution, as I fear we shall, I could not vote for conscription because I believe that we should thereby lose the special characters for which it would be worth while to fight. But this is looking far enough ahead let us hope. So long as we can keep clear of the manufacture of shoddy, popular science, and other forms of Philistinism, we are safe enough to have a good place in the world....

His later knowledge of the Continent widened his sympathies, and in so much modified perhaps his ardour. But the great blow to his patriotism fell in 1899—with the Jameson raid and the consequent South African War. I remember how he came to my room,[1] white and shaken with emotion, to bring me the news. It was as though a well loved and trusted friend had been by him detected in some baseness or disgrace which he could not condone.

[1] Our second boy was a day or two old.

He lost then for ever all that pride in his country which constitutes patriotism; and he never found again confidence in the wisdom, nor respect for the activities of the politician, as a politician.

In 1901, presumably on the strength of his former interest, he was invited to join the Cambridge Association of Liberal Unionists. He replied:

Dear H.

No. I do not support this Government. Had I the voice of John Milton you should understand why not. I am not of those,

> *who rolled*
> *Mother with infant down the rocks,*

doing in the cool light of knowledge such things as the Holy Office did in the delirium of Faith, leaving a mark on the good fame of my country that centuries will not forget.

Yours truly,

W. Bateson.

He feared the oversetting of European equilibrium; and every small crisis—for instance the Fashoda affair or the strain at the time of the Agadir incident—caused him anguish of anxiety. Injustice, cruelty and bloodshed filled him with horror.

Religion, politics and law he mistrusted and disliked; to him they seemed systems of cumbersome intrigue menacing human progress and content.

As far as possible he kept his attention steadily on his scientific work and avoided the emotions which the public policy and events of the day stirred in him; indeed, for many years he ceased to read the newspaper regularly, buying one only when the housemaid demanded paper for her fires, or for amusement from matters such as a case in the Court of Arches, the doctrinal embarrassments of colonial Bishops, the Rougemont hoax, or the theft of a pearl.

In his day, the usual course for Cambridge Graduates in Zoology was to proceed to Naples and pursue some research, generally embryological, in the Marine Station there.

While working for the Natural Sciences Tripos Will's attention was caught by *Balanoglossus*, of which the life history was then unknown, and the possibility that this animal might be allied to the Vertebrates occurred to him. With help and encouragement from his friend, Mr W. F. R. Weldon,[1] he applied for permission to work in Professor W. K. Brooks' laboratory at Hampton, U.S.A.

Balanoglossus was to be found in Chesapeake Bay.

Professor W. K. Brooks gave the young man a kind welcome, and the summers of 1883 and 1884 were spent in the Southern States on this piece of work. He always congratulated himself on his "luck" in thus escaping the more ordinary and conventional training in embryological research work.

By his papers[2] on *Balanoglossus* he first made himself known to the biological world as an observer and a thinker. They are now a text-book classic.

A letter to his mother dated Hampton, Virginia, U.S.A., 20 July, 1883, gives an account of his first successful observations:

Yesterday I made a start, as I got some ova which began to segment. I fear however that something is still wrong, as the segmentation is terribly irregular. I have had to sit up all night (less two hours) watching for them to proceed, which however they do not do. I am in the greatest excitement about it, as no one, I believe, has got so far as this yet. The weather is mostly piping hot still, and, barring thunderstorms, is delicious. I never resided in such a mean boarding-house in my life. We fairly live on shore crabs and a little fish,

---

[1] W. F. R. Weldon, later Professor at University College and at Oxford, was then his most intimate friend. Extreme divergence of their views undermined this friendship which later dissolved in bitterness.

[2] *Quart. Journ. Micr. Sci.* xxiv, xxv, xxvi, N.S., 1884, 1885.

about the size of a Cam perch; Brooks and I are daily arranging a strike, but courage is still lacking....I am fairly tired out now and shall sleep an hour before the post goes.

Yours affectionately,

W. Bateson.

Pleasant and encouraging as the success of this research was, he valued more the opportunity the work had given him of intimacy with Professor W. K. Brooks.[1] He delighted in recalling the long hours of discussion (Brooks lying in his shirt-sleeves on his bed and Will siiting by), when problem and theory and practice passed in long review with ever-fresh interest. Professor Brooks, congenial and stimulating, was always remembered with affectionate respect. During the summer of 1884 Will fell dangerously ill; Professor Brooks and his students nursed him safely through with unflagging care and attention.

Soon after the publication of his papers on *Balanoglossus* he was elected Fellow of his College (1885).

But in two years he outgrew the *Balanoglossus* work and came even to regard it as trifling.

In the spring of 1886 he set out by himself to examine the fauna of the lakes and drying-up lake basins of Western Central Asia. This was his second plunge into independent scientific research. He had two objects in view; firstly, he hoped to find creatures still surviving from the days of an Asiatic Mediterranean; and secondly, he wanted to study the salt lakes in the Steppe and to learn how far the infinite variation of conditions (salinity, depth, etc.) of these waters affected and changed the forms of their fauna.

He was disappointed on both counts; his results were negative. But he must have learnt much, nevertheless.

[1] See reference to Brooks at the beginning of *Evolutionary Faith and Modern Doubts* (p. 389).

His field notes and exhaustive tabulations (which I still have) recording the measurements of the creatures under observation with locality, conditions, water densities, etc. in which they were taken, bear witness to his energy, enterprise and industry. He also acquired meanwhile colloquial fluency in the Russian and Kirghiz languages.

He travelled hard, he worked hard. The pleasures of travel could not entice him from its object, nor hardship make him shirk. He hoped to collect fauna from 600 different waters, and did collect from 500 or more and examined them in detail. At one time—in the second summer —he travelled by night in a "tarantass" from one locality to another; arriving at some settlement in the early morning he hired horses and rode some forty miles, examining and collecting material from four or five lakes a day, and then on again by night in his rough carriage to a new centre. Letter-writing was the solace of his long solitude. Many of his letters have been kept; apart from their lively record of his wandering life, they reveal his close attachment to his family, and interest in all home matters.[1]

In a letter from Kazalinsk (1886) he refers to his work on *Balanoglossus* (the subject is introduced by comment on the news that he has not been awarded the "Balfour" Studentship for which he had applied after setting out on this expedition):

Kazalinsk,

Nov. 22, 1886.

...About the "Balfour" business, I feel this way. Of course I should have been glad to get it; but looked at from an outsider's standpoint, I expect it is a good thing I haven't I have been much too successful pecuniarily and so on, lately, and I have really been getting more than I worked for, which is a bad thing. *Entre-nous*, the *Balanoglossus* business was a very

---

[1] Some passages of autobiographical interest I have quoted here, but the bulk of these letters relating to the journey and excluding personal and family allusions I hope to publish separately.

easy victory, and wasn't much work at all. The thing did itself. Of course, the *"kudos"* turned up most substantial trumps, but the thing isn't valuable really. Five years hence no one will think anything of that kind of work, which will be very properly despised. It hasn't any bearing whatever on the things we want to know. It came to me at a lucky moment and was sold at the top of the market; presently steam will be introduced into Biology and wooden ships of this class won't sell well. Pecuniarily I am well enough off; next year will be very cheap, really, all told, not dearer than living in Cambridge, if as dear....For next year I have no particular expectation of success, indeed, I doubt, as at present minded, if I shall compete. My work is of a kind with which the Cambridge people have little sympathy, and by next October, while the *kudos* got from *Balanoglossus* will be spent, any to be derived from my present occupation will be still in the future, and this is the kind of thing that the Committee, being human, will be guided by. My chance was far better this year than it will be for many years to come. But I don't repine. I went into this business with open eyes, knowing that chance of promotion from any quarter was thereby indefinitely deferred, and I am prepared to go on, on that understanding. How shockingly egotistical one does get in these monthly budgets! But when one hasn't said a word of one's affairs for Lord knows how long, perhaps it is excusable....By the way, whoever originated that ridiculous piece of bad logic about variations due to environmental change seeming not to be "permanent"? How the deuce should they be on any hypothesis, which supposes that they result from change, which when reversed, or withdrawn, leads naturally to a return to former state? If iron in soil make hydrangeas[1] blue, why is this to be regarded as a false variation? Because the same hydrangea without iron is *not* blue? By the way, also, round a small dry lake there are *generally* three rings of (?) *Salicornia,*

[1] The hydrangea stuck in his mind.
"We are told that salts of iron in the soil may turn a pink hydrangea blue. The iron cannot be passed on to the next generation. How can the iron multiply itself? The power to assimilate the iron is all that can be transmitted" (Melbourne Address, 1914, p. 290).

two green and the middle one red; isolated red and green ones being in all rings. The sediment in which they grow is very rusty, but I do not know that it is more so here than there.

And again in another letter:

I am very angry that — has not had my work on *Balano-glossus* reviewed as he promised....To be sure I don't think much of that work now, but one remains human all the same, and I should like to see my bantling prospering in life.

Winter-bound in Kazalinsk, after a lonely but strenuous summer in the open Steppe, he wrote the following letters to his sister Anna, who was visiting Dresden; they illustrate the second passionate interest of his life. It should be noted that Will had been once, in 1882, to Dresden,—five years before.

Kazalinsk, Turkestan,
22 Dec. 1886/3 Jan. 1887.

...But of course when you get this you will be far enough from Dresden. So you have heard "Tannhäuser" at last! I remember, too, thinking they did it poorly in Dresden. I saw it much better done at Covent Garden, but this is to be expected. Oh! when will you learn that we sit in the centre of the world!

I am glad that you are not overwhelmed by the Gallery. But I felt one could trust as good a Blake-ite as yourself with Rembrandt and Correggio. All the same I must say I have a weak place for the "Magdalene", which of course would have maddened Blake more than the rest even. I daresay it is old association, that makes me like it. By the way what has become of our copy of it? Was it College property? If you want to see why that picture is good work, go and look at Battoni's "Magdalene" which was obviously done to out-do the Correggio, and see what a mess a fool can make of precisely the same ideas. I fear you don't now remember the Battoni, but I was much amused to see how he tries to better Correggio's pose (which is just the beautiful part in C.'s picture), because, I take it, he thought, or his model thought, that she could not lie like Correggio's "Magdalene" for many minutes without getting the cramp, which is perhaps true; but, all the same, the thing is spoilt.

It is a happy chance that has put those two pictures in the same Collection. I wonder if you saw the Cranachs—some two hundred of them—perhaps more—in an off Gallery. You who are so "Germanly proclived" would have had a rude jolt, I expect. Of course Dresden isn't strong in 15th century but there are two lovely Francias, and a very fine Luca Signorelli there, which I remember impressed Margaret a good deal. Of course the *Johannäum* is a bore, but there is some magnificent oriental pottery somewhere. And by the way, let me thank you for the simile "as though it were prize wall-fruit!" which is the best remark I have heard for many a long day.... I do hope you will get to hear the "Dutchman". I fancy it was the first "Tone poem"! (*vide* St James' Hall programmes) that I really enjoyed.

...I hope on the whole you found something to like in Dresden. In my mind, Dresden and Antwerp are the two best places outside Italy (excepting Gt Britain and Ireland)!

It was in Dresden that I first saw pictures, first heard Wagner, and first made the acquaintance of Browning; so that I have some cause to remember it. And then, I was very fond of the "Terrasse", which is a kind of thing I delight in. But in winter, I fear, this kind of thing is in abeyance. However, I thought you would hardly have been through the Gallery and not even allude to the "San Sisto"—I can understand you not finding much besides to interest you, but I do believe the "San Sisto" is about the fairest thing on earth. Don't you even see that it is splendid *work*, apart from anything else? Did you see Sassoferrato's copy of the "Seggiola" in the room next to it? If you did, that should have suggested something by contrast. I know, I went to Dresden never having seen a good Raphael, and never having cared much about the engravings of the "San Sisto", and fancying that Raphael must be rot, and the "San Sisto" gave me a dreadful turn. I never shall forget turning the corner into the little room at the end, not knowing anything particular was there, and then coming

face to face with that picture. Those sort of things make one
feel awfully ashamed of oneself; and to my thinking, the "San
Sisto" more than all. Well! Lord spare you to see it again
and may it please Him to soften your hard hearts! You say
that your companion is as bad as you are—I hope that this is
a libel, but if it isn't, I think you ought both to have been put
out. I wonder if you discussed the price of beer, "*per litĕr*",
in the "San Sisto room", as I once heard two Yanks do!
I suppose you didn't notice A. del Sarto's "Abraham and
Isaac", which as a piece of sheer skill is very striking. How I
envy you! It does one good, in this desert, even to speak of
those things! By the way too, there is a good specimen of
L. di Credi at Dresden. Did you see it? I have grown out of
him now, but I remember liking it when I saw it first,—very
hard colour and so on. The "*Notte*"¹ I never cared for in the
least, though one or two of Browning's remarks sometimes
make one realise that one loses thereby; but all the Dresden
ones are very faded, and I suppose, in fact, I have only seen
one or two first-rate Correggios. Well, some day, I will go to
Parma and be converted. I grant you about the Titians, until
you begin to try and realise what such skill means. Of
course, I know that this is a totally different way of enjoying
things, and is not a proper one, but still to see anything in
perfection is a treat, even if it is only skill. Just try and think
what it means, to be able to paint a heap of clothes *absolutely*
correctly! The pitch of control that a man's muscles must have
reached, apart from the skill of getting colour, and seeing,
and interpreting the colour right, is perfectly appalling;
though I confess I never have been struck in any other way
by Titian, and I have seen some of the best now. I don't
remember any very first-rate one at Dresden. The Tintorettos,
and occasional Veroneses, I always have liked. The colours
they made aren't earthly, though I never saw any better than
our "St Helena" (?) in the National Gallery, for that matter.
I suppose you didn't go to see the Old Masters' Drawings at
Dresden? Perhaps there you get the very perfection of the
whole thing. There are some wonderful things there, as good

¹ Correggio.

as anything in Florence, I could almost say. When you see them, you see well enough why lots of things in pictures don't come up to scratch, simply because they were done for the world, while the man did the drawings for himself. All the drivel and half-heartedness of the pictures is away from the drawings. The fellows that did them, did them for downright love of them, and because they couldn't help it, and not for money and fame, as they did their pictures; and this one sees well enough in any of those drawings. And besides, too, only two or three people have ever been able to make paint shew exactly what they wanted, (and these persons have generally wanted nothing in particular!) while heaps of them could do this with chalk or bistre, and so you get to the man's real meaning at once in a drawing, while in a painting you may have to look for it through his mistakes, and often enough through a veil of tricks too, which poor folks were driven to, simply to make up for absence of skill, and not always for claptrap as the brutal critics always say. Well! my pen is broken and I am getting very crusty, so Goodnight!

I hope the Music will make amends for the deficiencies of the poor old Gallery!

Yours ever,

W. Bateson.

Kazalinsk, Turkestan,

5. 1./17. 1. 1887.

Have just got your long letter, dated 28 Dec. etc. for which accept heartfelt thanks. I am very glad my note reached you in time to arouse you from the shocking depravity you seem to have reached. Well! One must be thankful for small mercies and if you can't enjoy good work, it is at least a comfort that you don't run after bad—not that I ever feared you would, but still—I rejoice that your instinct warned you not to accept Battoni's huge blubbering female as an adequate representation of Mary Magdalene! But your equation about that stupid Holbein won't do. However you cultured folk may be agreed that the 7th Symphony isn't up to the 9th

(and you may perhaps be right) the 7th remains a splendid
thing, which the Holbein is not. Thank Goodness, I can like
Holbein as well as another, sometimes. It would be a great
loss if one didn't, as nobody else but Holbein and his set can
give one all those details and still keep their work as cool and
severe as a fresco; but in that big panel at Dresden there is no
detail at all, nothing but flatness. Of course the colours are
rather nice and soft. After all, it is only a copy. The original
I have never seen—isn't it at Basle? I forget. Now I am going
to return to the charge about the "San Sisto", so look out!
It seems to you ordinary—that is partly because you have
known always that there was such a picture, and the shape of
it, and have seen it employed as an article of furniture from
childhood. So it doesn't seem to you at all new or strange that
there should be at Dresden rather a bigger and brighter "San
Sisto" than at Harvey Road,[1] and I don't believe you looked
at it from any other point of view. You went into the room,
you two reprobates, probably laughing and exulting in your
folly and blindness, and you saw it again as an article of
furniture, and that was all. You weren't struck with the
surprising novelty and freshness of the thing, which is one of
its charms, and so you looked no further. I doubt indeed if
you honoured it as much as an American I heard speaking
there—he gaped at it and said to his friend, "See them angels
all around! Ain't there just lots! The longer yer looks, the
more yer see!" That fellow at all events saw and was struck
with about the most daring feat that has ever been done in
painting, which is more than you were, I take it. But that is
nothing. If you want to find out what a picture is good for, I
mean, what it is worth to you personally, and it does not
strike you at first glance, you should go by yourself and look
at it for a good while, and fancy that it is looking at you—and
then that the figure in the picture is there with you, and that
you are one of the company—and then, perhaps, you suddenly
become aware that the people in the picture are a new thing
to your conceptions of people and things, and have come down
from Heaven. It is monstrous that such a process should be

[1] The Batesons lived at No. 8 Harvey Road, Cambridge.

necessary with the "San Sisto", but, sometime, try it. Fancy you are where St Barbara is, and that that woman has just come into the room carrying that child—Would you think it an ordinary experience to be in the presence of such a person? What do you think you would say to her? You would then feel that you are in the presence of most surpassing beauty, such as you had never conceived. That is another great element to me, I know, in the delight of pictures—I mean the way the things drawn there transcend anything one can conceive. I know that careless people think it very easy to imagine a beautiful thing or person—Shut your eyes and try and you will find you can't. All that passes before you is nothing but a succession of vague reminiscences of what one has seen in the pictures. Of course only the very best pictures will come out of the canvas to you like that. You can't do that with most. With poor old Botticelli you certainly can't, except that wonderful picture in the National Gallery which is a thing apart, to me, at least. That is why Rossetti strikes home to so many people, in spite of the bad drawing; because his face comes right out and looks through and through you, and you feel that some new thing is happening to you, and that a sort of human thing is with you that is somehow better than the other humans you live with. I have never been much struck with any other Raphael. One sees that they are perfect but they aren't new to one's conception of people. But the "San Sisto" is new and is still perfect all the same. I know that St Barbara is rather a bore—that the angels are hanging on, on the *outside* of the window sill, which must be very uncomfortable, (especially as they have to look behind them). Also that that very handsome pair of curtains look as though they must have "cost money", that the Madonna is no Mother at all, but only a girl, and that she is entirely wrapt up in her own concerns and doesn't think about the child, and all the dozen other things that the damned critics have dared to discover, but she and that child remain the two loveliest things we shall ever see, and it is time you found it out.

This epistle has just been interrupted by the upsetting of a bottle of beer (*v. supra*)[1] which has also restored my ruffled composure, so I shall retire to bed and keep the rest for to-morrow. Oh! dear! I wish I had sheets on my bed! I think that will be the pleasantest novelty when one returns to civilisation.

After 18 months in the Steppe (wintering at Kazalinsk) he returned from Russia in the autumn of 1887, and within a few months was off again, this time to search the brackish waters of Northern Egypt. This was only a short expedition.

The outcome of these travels was the charming little paper on *Cardium edule*.[2] But he was baffled in his quest. He had no plain tale to tell. He always regarded these expeditions as failures and regretted that in his inexperience he had undertaken the investigation with too definite and narrow expectation, and had pursued the inquiry too closely to profit by the large opportunity of general observation.

The sense of failure however proved very stimulating: he had to make good: if he had followed a false clue, the greater the need to find the right one.

He was elected to the Balfour Studentship in November 1887.

The main preoccupation of the seven years following his return from the Steppe (1887) was the patient collection of facts bearing on Variation, their interpretation and assortment. This involved much correspondence; he travelled ceaselessly to see for himself alleged cases of abnormality and variation; he endeavoured, as far as possible, to examine every specimen, and verify every statement. These researches brought him in contact with the "practical man" in all his guises; his passionate interest communicated itself to breeder and collector,

[1] Ink run and the page half illegible.
[2] "On some Variations of *Cardium edule* apparently correlated to the Conditions of Life", *Phil. Trans. Roy. Soc. London*, CLXXX (B), 1889, 297–330.

amateur and professional. He had friendly acquaintance with "all sorts and conditions of men" all over the country; his enthusiasm made "the picking of their brains" a pleasant experience to them; and in their case, at least, his candid criticisms of their statements were not generally ill-received.

He ransacked museums, libraries, and private collections; he attended every sort of "Show," mixing freely with gardeners, shepherds and drovers, learning all they had to teach him. Thus in one way or another, he accumulated a vast store of facts of variation and of practical knowledge of animal and plant life; and to the end of his life, by the same means, he continued ever to add to this store.[1] To observe with him was an instinct, and to reflect and criticise, a life-long habit. Never for one moment, in the interest of detail did he lose sight of the central quest. For him no method, nor any branch of research, was ever an end in itself, or more than a useful tool to chip away with. He dreaded the pedantry and increasing narrowness of each branch of biological research almost more than he dreaded and despised the public ignorance.

Facts and alleged fact were weighed and judged; the "materials" for the study of Variation were piling up upon his study table, under his ever critical and thoughtful scrutiny.

For him there was no stopping-place. He put his question this way and that to Nature; persistent ingenuity and steady patience forced from this unwilling witness[2] evidence sufficient to keep him ever on the alert, fresh in hope, eager in observation, and—cautious in deduction.

But the straight answer was denied him.

---

[1] Three days before his death he was to have been with his students, and the Genetical Society, at the Cage-Bird Show at the Crystal Palace. On his study table lay the notes he had made in looking up the Canary literature, to refresh a memory always fresh, and make sure thus of missing no new points. His interest and enjoyment on such occasions never flagged; it was his aim, by sharing this pleasure and interest, to lead his "young people" to realise the wealth of opportunity about them for the study of the vital problems of biology.   [2] See *Method and Scope*, p. 323.

The literature of Evolution was at his finger-ends. This he supposed the normal outfit of his profession; but, on the other hand, the intimacy with *les belles lettres* and art that he found amongst his literary friends filled him with amazed and envious humility. "How do they manage it?" he would exclaim despairingly.

An exalted reverence for truth and beauty inspired him throughout his work. For Newton and Pasteur he had deep veneration, and for other giants in the annals of science too, but on the whole he rated scientific attainment lower in the scale of human achievement than that of the great Masters of art. This was a kind of intellectual modesty which was characteristic of him through life and was perhaps another aspect of his candour and unworldliness.

He wrote on an intimate occasion:

...Faith in great work is the nearest to religion that I have ever got, and it supplies what religious people get from superstition. There is also this difference, that the man of science very rarely hears the tempting voices and very seldom needs a stimulant at all, whereas the common man craves it all the time. Of course there is great work that is not science—great art, for instance, is perhaps greater still, but that is for the rarest, and is scarcely in the reach of people like ourselves. Science, I am certain, comes next and that is well within our reach....

Will would often hold back where others might have pushed forward, but his determination to serve science faithfully gave him confidence on occasion.

In 1890, in the midst of his active researches, and without interrupting them, he decided after much hesitation and anxious consultation with colleagues and friends to apply for the post of Deputy to the Linacre Professor at Oxford (to supplement Professor Moseley, then in failing health).

The application is long, but his enthusiasm permeates it.

He visualised clearly and definitely his aim and purpose, and set them forth with almost naïve candour.

To THE ELECTORS TO THE LINACRE PROFESSORSHIP.

St John's College, Cambridge,

June, 1890.

Gentlemen,

I beg leave to offer myself as a Candidate for the office of Deputy to the Linacre Professor of Comparative Anatomy according to the terms of the announcement in the Oxford University Gazette of May 28th.

I was educated at Rugby and at St John's College, Cambridge, and was placed in the First Class in Parts I and II of the Natural Sciences Tripos in 1882 and 1883. I was elected to a Fellowship at St John's College in 1885, and in 1887 the Studentship founded in memory of the late Professor F. M. Balfour was awarded to me. In 1888 I gained, together with Mr W. Gardiner, the Rolleston Prize in the University of Oxford.

Since taking my degree I have devoted myself to original research in Biology, and have published papers on some of the subjects investigated; of the chief of these, I submit copies herewith.

For the purposes of these investigations I have spent some time in foreign travel, having passed two successive summers in North Carolina and Virginia and a year and a half in Russian Central Asia and Western Siberia. I have also collected in Egypt and elsewhere. In the course of these journeys I have had opportunities of gaining a general knowledge of Natural History in addition to the training in Zoology and Embryology received in Cambridge.

As I have decided to offer myself as a Candidate, and as my views of the proper development of the study of Zoology differ somewhat from those held by others, I cannot avoid giving at least an outline of the method by which I think that development can best be brought about.

The first work which I undertook was a detailed investigation of the morphology and development of the *Enteropneusta*.

The papers which I published on this subject give a plain description of the anatomy and embryology of the group, and conclude with a discussion of the relations of *Balanoglossus* to other animals. The views there propounded are supported by the usual arguments which are admitted in modern morphological discussion, and they are applied in the ordinary way.

On finishing these investigations I became dissatisfied with this mode of attacking biological problems and resolved to seek a new field of inquiry. In pursuance of this resolution I have for the last five years been engaged in work of an entirely different kind. As experience has accumulated, I have become more and more convinced that this step was a right one. It is, therefore, from the methods I have since endeavoured to pursue that I now hope for the most legitimate and fruitful development of the Science of Zoology.

In order to make clear the nature of the work upon which I am now engaged and the results which may be expected from it, it is necessary to give some brief account of the relation of this branch of the science to what may be called the more orthodox methods of Zoology.

It is well known that since the promulgation of the doctrine of Descent the Science of Zoology has undergone a great change. The enthusiasm of zoologists has run altogether into different channels; a new class of facts is sought and the value of Zoological discoveries is judged by a new criterion.

The reason of all this is plain enough. When the theory of evolution first gained a hearing it was felt that it was of primary importance to know first, whether it was true that forms of life had been evolved from each other; and secondly, if evolved, on what lines had this been effected and what was the ancestry of each. All other problems sank into insignificance in comparison with these.

Now the readiest method of answering these questions seemed to be the embryological method. By using the so-called Law of von Baer, that the history of the individual is the history of the species, it should be possible to determine the pedigree of the form by examining the manner of its development.

Further, if stages in the development of different types

should be found common to several forms, this would be strong evidence of the common descent of these forms. To find such evidence has been the chief aim of zoological science since Darwin, and those researches have been held to be the most valuable which contributed to the solution of these questions.

For this reason the study of embryology superseded all others and elaborate deductions have been made from its results. The survey of the development of animals from this point of view is now complete for most forms of life and in all essential points. We are now, therefore, in a position to estimate its value.

The main object of such research was to discover the history of Evolution; but at the outset, embryological evidence as to the history of Evolution is open to the objection that it rests on an error in formal logic. For the stages through which a particular organism passes in the course of its development can be taken as indications of its pedigree only when it shall have been proved as a general truth that the development of in-dividuals does follow the lines on which the species was developed. But the proof of this proposition rests entirely on the observation that in the development of animals stages are passed through in which they resemble other forms: from this resemblance it is inferred that they are descended from those forms. Hence the truth of the general proposition is established by assuming it true in special cases; while its applicability to special cases rests on its being accepted as a general truth.

But even if it be generally true that the development of a form is a record of the history of its descent, it has never been suggested that this record is complete.

On the contrary, allowance must constantly be made for the omission of stages in development, for degeneration, for the presence of organs specially connected with larval or embryonic life, for the interference of yolk and so forth. But how much this allowance should be and in what cases it should be made has never been determined.

More than this: closely allied forms often develop on totally different plans; for example, *Balanoglossus Kowalevskii* has an

opaque larva which creeps in the sand, while the other species
have a transparent larva which swims at the surface of the
sea; the germinal layers of the guinea-pig, when compared
with those of the rabbit, are completely inverted, and so on.
When these things are so, who shall determine which developmental process shews the ancestral history and which is due
to secondary change? By what rules may secondary changes
be recognized as such? Does the transparent larva swimming
at the surface of the sea shew the ancestral type, or is the
opaque larva creeping in the mud the more primitive form?
Each investigator has answered these questions in the manner
which seemed best to himself. There is no rule to guide us in
these things and there is no canon by which we may judge the
worth of the evidence. It is perhaps not too much to say that
the main features of the development of nearly every type of
animal are now ascertained, and on this knowledge elaborate
and various tables of phylogeny have been constructed, each
differing from the rest and all plausible; but it would be difficult to name a single case in which the immediate pedigree of
a species is actually known. Embryology has provided us with
a magnificent body of facts, but the significance of the facts
is still to seek. Knowledge, however complete, of the anatomy
and embryology of animals can at best shew us what may be
called the formal relations of their structures to each other;
it can never shew us their genetic relations nor can it forecast
their future.

The question, then, which Zoology proposes to solve is this:
what have been the steps by which animals have acquired the
forms which they present, and what will be the future of their
development?

All naturalists are agreed that the process by which the
result has been achieved is one of Evolution, though they differ
as to the nature of the forces by which the forms have been
produced and fixed. Whether we believe with Lamarck that
adaptations are the direct result of environmental action, or
with Darwin that they have been brought about by natural
selection, it is admitted by all that the progression has come to
pass through the occurrence of variations. This is common

ground. Hence, if we seek to know the steps in the sequence of animal forms, we must seek by studying the variations which are now occurring in them, and by getting a knowledge of the modes of occurrence of those variations and, if possible, of the laws which limit them.

When we shall know the nature of the variations which are now occurring in animals and the steps by which they are now progressing before our eyes, we shall be in a position to surmise what their past has been; for we shall then know what changes are possible to them and what are not. Until the modes of Variation are known and classified no real advance can be made in the study of Zoology. In the absence of such knowledge, any person is at liberty to postulate the occurrence of variations on any lines which may suggest themselves to him; for there is no collected evidence to shew whether or not such variations do ever occur as a matter of fact.

We must, then, look to the Study of Variation for the further development of the Science. It will be found, I believe, that this branch of Zoology is the most fruitful and the most fascinating, as it certainly is also the most logical in its methods.

This study it is my ambition to pursue by the methods employed by Darwin himself; by studying the Variation of the animals and plants with which we are most familiar and especially by encouraging the study of domestic animals; by promoting experiments in breeding and cultivation; by the comparison of local varieties and generally, by amassing material which should serve to illustrate the nature and especially the modes of Variation.

Above all, I should strive to stimulate those whom I could influence to study animals *as they are*, in the field and on the sea, looked on as living things and not as diagrams and types. Nothing has been more unfortunate for Zoology than the separation between what is called its scientific side and the pursuit of Natural History; for this separation has gone far to destroy much that was beautiful in the study of life. It is only by the employment of both methods jointly, that artificiality of treatment and pedantry can be avoided. When the history of the study comes to be reviewed, it will seem a strange thing that the first result of the work of Darwin, who was above all men a field-naturalist, has been to bring the study of Natural History into contempt.

I think, then, that the proper way to develop Zoological study is to pursue Darwin's problems and to employ Darwin's methods. With this object in view I have now for some years been engaged in the collection of material bearing on the subjects mentioned above. In particular, I have brought together a collection of *Crustacea* and *Mollusca* from brackish and other waters in order to determine whether or not these diversities of environment produce specific alterations in structure. The result of this investigation goes to shew that, while such variations do occur in certain species, in the majority they do not. In a paper on *Cardium edule* I have described an instance of variations of this kind and an account of the *Branchipodidae* thus collected is now nearly ready for publication.

I have especially endeavoured to make myself familiar with the breeds and varieties of the domestic animals, and a good deal of evidence of importance has now been brought together illustrating the modes in which they vary.

For some time past I have been engaged in making a collection of the evidence bearing on certain special problems arising out of the Study of Variation. The chief part of the material thus accumulated relates to the following subjects:

(i) The Variation of Multiple Parts and of Symmetry;
(ii) Variations in the size, number, and fertility of animals;
(iii) Variations apparently due to change of Conditions;
(iv) Local Varieties.

A good deal of information has also been brought together relating to variations in sexual characters, the prepotency of special forms, the production of Hybrids, etc.

It is my purpose to publish these collections from time to time in the hope that they may serve as a nucleus for the further collection of facts of the same kind. Together with each of these collections I shall attempt to give an analysis of the evidence and to shew what inferences may be drawn from the facts given. The importance of these publications would, of course, turn chiefly on their value as collections of facts and not on the truth of the inferences. That such collections are urgently needed is apparent to any one acquainted with the progress of modern Zoology. Since Darwin's time no considerable collection of facts of this kind has been made. Each year sees the birth of new theories in this country and elsewhere, but each theory in turn is established by the facts of Darwin and Wallace.

The first of this proposed series of collections will deal with the Variation of Repeated Parts. It will be, as far as possible, a complete account of all observations bearing on the subject. This first part will, I hope, be published at the end of the present year. The importance of these facts lies in their value as evidence of the magnitude of the integral steps by which Variation proceeds and of the control which the symmetry of the body exercises over variations. It is impossible to give in abstract an intelligible account of the results to which these facts point, but taken together they will go far to suggest certain laws as to the variation of repeated parts: these laws, if established, will have an important bearing on the conception of the modes of Variation and would lead to a modification of the views now current as to the nature of Homology.

Similar collections of evidence on the other subjects named will be produced as opportunity may occur.

Of teaching I have had but little experience, having acted as Demonstrator in the Biological Classes in Cambridge for two seasons only. By the rules of the Balfour Studentship which I now hold, I have been precluded from undertaking regular teaching work; but during the Lent Term of this year I gave a course of lectures to advanced students on the immediate subject of my present work, and I hope that I succeeded in interesting some of those who attended the course.

Experience of the development of Science teaching in Cambridge has led me to doubt the possibility of both keeping up the enthusiasm which begets good original work and of maintaining at the same time that rigid system of scheduled instruction without which very large classes become unmanageable. I think it right, therefore, to say that I should not aim at indefinitely increasing the numbers of Zoological Students, but I should rather seek to stimulate a comparatively small number of students to produce work of a high class.

If the electors to the Linacre Professorship should think fit to nominate me as Deputy, I should strenuously endeavour both to continue my own work and to promote Zoological research in the University. In particular, I should try to stimulate others to pursue the study of Zoology in the manner which I have attempted to describe, and to investigate those problems which have become the most attractive to myself. I am convinced that for the Science thus pursued, a great future is in store.

Before concluding, I beg leave to make a personal statement.

I wish it to be distinctly understood that my Candidature is not in opposition to that of Professor Lankester. I recognise in the fullest manner the preeminence of Professor Lankester's claim. I have subscribed my name to the Memorial giving expression to this view and I do not retire from the position thus taken; for I hold that in the interests of Science and of the University the choice should fall on Professor Lankester and that the claims of no other candidate should be considered.

If, however, it be decided not to elect Professor Lankester, I do wish my application to be considered with that of others. The lines on which I desire to develop the Study of Zoology are distinct from those now generally followed and I believe that they are better. I decide therefore to submit them to the Electors.

I am,

Gentlemen,

Your obedient Servant,

William Bateson.

It is improbable that this candidature was ever for one moment considered. Professor Ray Lankester was of course elected.

Many years were yet to pass before Will was given the opportunity to carry out his ideals of teaching.

The quotation from his Central Asian letters (p. 19) shews that even as early as 1886 he felt that he had over-stepped the conventional boundaries hedging in biological inquiry, and that he foresaw the trouble and difficulty that awaited him—the bitterness with which his hetero-doxy would be opposed.

In one of his last letters from the Steppe before his home-coming he reviewed his immediate prospects and wrote to his sister Margaret (Sept. 1887):

...It is a small matter, and if I find that obvious expediency makes it necessary for me to teach there (Cambridge) a little, I shall pocket my pride and do it, but otherwise I shall look out for some other opening. The life of a beggar seems to me to offer this. I have my Fellowship, and a man could live handsomely as a beggar on this income. A beggar's life is a tonic and stimulating one, and is the only life that helps to the doing of first-rate work. Other careers are compatible with good work, but I believe the beggar's is a positive stimulus thereto; partly because the beggar is always angry and chafing and that righteous anger is the very salt of good work....

Fate took rather too literally this jesting declaration.

If in his school-days, without confidence in himself, school-fellows or masters, disapproval and failure acted on him as inhibitors, from now onward this was not to be the case.

Disappointment, disapproval, even failure stimulated him to more determined and more concentrated effort.

Difficulty excited him and he had to share his excite-ment.

If the right companion, one with whom he could talk freely of his work, were not at hand, he unburdened his brimming thought in letters. And so his breathless enjoy-ment as ideas came to him was often chronicled. For instance on 2 September 1888 he wrote to his sister Anna:

My brain boils with Evolution. It is becoming a perfect nightmare to me. I believe now that it is an axiomatic truth that no variation, however small, can occur in any part without other variation occurring in correlation to it in all other parts; or, rather, that no system, in which a variation of one part had occurred without such correlated variation in all other parts, could continue to be a system. This follows from what one knows of the nature of an "individual", whatever that may be. If then, it is true that no variation could occur if it were not arranged that other variations should occur in correlation with it, in all parts, all these correlated variations are dictated by the initial variation acting as an environmental change. Therefore the occurrence of any variation in a system is a proof that all parts have the power of changing with environmental change and must of necessity do so. Further any variation must always consist chiefly of the secondary correlated variations and to an infinitely small degree of an original primary variation. You will observe that if any variation occurring in one system is acting through the mechanism of correlation as a cause of further variation, it would then happen that on the occurrence of one variation, general variation must be expected, for if all the parts are to work in with the new variation, a long time must elapse before the whole organism is again a system. Hence the fact that if a particular variation is wanted, e.g. for selection, any variability may lead to it; also the numerous species "Variabilis" etc. I am sure that something would be gained if it were thus possible to separate any variation into its primary and secondary parts, or, in any case, keeping in mind the fact that any variation in a hitherto "fixed" form, must of necessity be made up of these two parts.

The accommodatory mechanism is the thing to go for. I don't believe it is generally recognised as existing, though when stated, it seems obvious.

Let me have a note.

Yours ever,

W. Bateson.

St John's College, Cambridge,

16 January 1890.

(To the same),

Thanks very much for your letter. I am glad to hear that you were so slightly "*grippé*" after all. I am working *à l'Africain* and have rarely slaved so hard. My lectures are to begin on Tuesday or Wednesday next.... (While I think of it, will you collect for me a sample, say 50–100 of the commonest water-snail you can find in any pool or stream at Ammanford? Put them alive into a box with cotton wool and send them to me here at your leisure).... If you have any ivies growing with you, look at the symmetry of the plants *as a whole* with a view to seeing whether the type of leaf does not change in a regular and symmetrical manner along the principal lines indicated, in those of course, in which there are different types of leaves. I think this rather striking, having seen it in two fancy ivies in the Botanical Garden to-day.

Is it that there is a mechanism co-ordinating the symmetry as a whole, or is it merely that the sprays are of different ages? It seems to me to be the former.

...A week ago I made a trip to Brighton and got a good many notes. Staying at the —— Hotel I dined at the *table d'hôte*. The people were as horrible a collection of broken crockery as I have seen. Damaged military men and almost everyone decrepit in mind or body....

In Bonafous' monograph of *Maïs*[1] is a remarkable figure[2] of a head of grain having grains in it of very many kinds—some of them identical with those which are normally found in other varieties (? due to pollen).[3] There is no description apparently of the figure.

---

[1] Bonafous, *Histoire naturelle du Maïs*, 1836.   [2] See Plate II, fig. 3 (p. 200).

[3] From breeding evidence only *Xenia* must be due to a second fertilising nucleus in ovule and pollen, making the endosperm. Experimental proof that the different coloured grains on a maize-cob are "due to pollen", at least in part, was lacking until 1898–9, when double fertilisation was discovered by Nawaschin (*Bot. Centralbl.* LXXVII, 62) and L. Guignard (*C. R. Acad. Sci. Paris*, CXXVIII, 869). A little later one of the first experiments demonstrating Mendelian inheritance was performed by Correns on the same material (*Ber. d. Deutsch. Bot. Ges.* XVII and XVIII).

I have written to Barr for some hybrid *Narcissi* (cut flowers
to shew to class), but have no answer. Do you think that any
other firms would send me some hybrid flowers together with
those from which they have been raised?

I am now a member of the Press Syndicate, and am about
to supervise the sale of the Scriptures and Liturgy with a view
to getting hold of some of the profits for the purposes of
Biological Research—otherwise they go to the building of
Labs. for Medical students, which is clearly inappropriate.

Now I am "gravelled for lack of matter!"

Yours as ever, W. Bateson.

26th March, 1890.

(To the same),

...Mary is going with me next week to a place called
Southwold in Suffolk. It has been reported to me that there
are some interesting Cockles there....

P.S. I had forgotten to say that I had a day at Barr's
nursery last week. It is a fine place near Surbiton, three great
fields but no glass except frames, I think. I got a good deal
interested in the doubling of *Narcissus*. Almost each additional
piece consists of one part like the outer perianth and one like
the corona in colour and texture. The corona-like piece is
often tubular; sometimes the whole flower is repeated (outer
perianth and all) nearly perfectly, inside the normal corona.
I was to have gone down there to do some cross-fertilising,
but it was so wet I came back here. Now it is fine again. Mr.
Barr may be useful, and his talk, though prolix, is interesting.

W. B.

I again start to record the fact that I spent a whole day at
the Thornelys', grafting apple and pear trees together and other
fanciful follies. Their gardener is a man of sense and some
experience. Weldon has a note in last week's *Nature* about
the turned-over Ivy, which I think I told you of. Our Fellows'
Garden is full of more or less reversed ivies, but I can't get
any understanding of them.   W. B.

St John's College, Cambridge,

19th July, 1890.

(To the same),

I am long owing you some letter. News is scarce. I am hard at work and am getting on fairly well. Macmillan has accepted my *WORK*[1] on Multiple parts at what is called "half profits" which is of course very pleasant, though there will never be any profits! On the other hand funds for future campaigning are likely to be deficient. Newton tells me that they don't want to renew the Balfour, and indeed I don't greatly wish it. He disapproves generally of renewing it, and doesn't think I have any special claim, which I admit. Sedgwick tells me that he would not wish me to have Weldon's lectureship if W. goes to University College. He says, as I expected, that I have gone too far afield, and that my things are a "fancy subject". On the other hand there is some chance of getting a renewal of Fellowship, though not a very good one. Newton told me he would try to get this done for me. He was very pleasant; Sedgwick was very *UN-*. So you see I have been "considering my position", like Sir Julian Twombley. N. thinks that I ought to be able to turn a considerable penny by regular literature—Magazine articles on Central Asia, the S. States, etc. It was his suggestion, and he urged me to try. Of course I should do so only to maintain connexion between body and soul, but it must be thought of.

When are you coming to town? I daresay we might manage to meet.

Yours ever,

W. Bateson.

Cambridge,

14th Sept., 1891.

Did I tell you anything about my new *VIBRATORY THEORY of REPETITION of PARTS* in Animals and Plants? I have been turning it over again lately, and feel sure there is something in it. It is the best idea I ever had or

[1] *Materials for the Study of Variation* (Macmillan, 1894).

am likely to have—Do you see what I mean?—Divisions
between segments, petals etc. are *internodal* lines like those in
sand figures made by sound, i.e. lines of maximum vibratory
strain, while the mid-segmental lines and the petals, etc. are
the *nodal* lines, or places of minimum movement. Hence all
the *patterns* and *recurrence of patterns* in animals and plants—
hence the perfection of symmetry—hence bilaterally symmetrical variation, and the *completeness* of repetition whether
of a part repeated in a radial or linear series etc. etc. I am, as
you see, in a great fluster. I have been talking to F. D.[1] about
it—and he thinks it "Really is very neat, upon my word",
also, "Oh!" he says.

How are you getting on? Send a line.

Yours ever, W. B.

I will tell it you more coherently when we meet.   W. B.

P.S. Of course Heredity becomes quite a simple phenomenon in the light of this.

And, scribbled on the envelope:

You see, an eight-petalled form stands to a four-petalled
form as a note does to the lower octave.   W. B.

Cambridge,
20th Sept., 1891.

Dear A.,

How are you getting on? I am still in great excitement
about my "Undulatory Hypothesis". I am sure it will work
and whither it will lead, Goodness knows. It makes one giddy
to think of the things that may come out of it! I have talked
it over with both Gaskell and Sedgwick and both of them take
it kindly. I hope to get it into a preliminary stage of writing
soon. I stopped a night with —— the other day. They are
rather nice people, Quakers you know. At dinner we sat
down without having any standing grace, so I supposed it was
all clear and took out my bread. I found, however, that
everyone was bowing over their napkins, so I did the same.

[1] Mr Francis Darwin, F.R.S., afterwards Sir Francis Darwin.

Nothing, however, was said, which somehow took me aback, as it was so unexpected, and I fear I sniggered slightly out of "nervousness". I do wish people would conform.

Main and I are the only two gathered together in Hall, but Lister is expected to-morrow. Send a line at your leisure.

Yours,

W. Bateson.

Lankester[1] is coming up for a day or two this week and I dare say he will liven us up a bit.

St John's College, Cambridge,

Sept. 26, 1891.

(To the same),

I am tremendously pleased with the *IDEA*. F. D. came round to-day and advises me to make a feature of it for the book. I have also had some talk with Lankester about it and he gave it a very decent *accueil*. You'll see—it will be a common-place of Education, like the Multiplication table or Shakespere, before long!

Yours ever, W. Bateson.

Sand-patterns and ripple-marks, the octave and wavelength, recur again and again in his letters and papers in illustration of his theory of repetition.

In 1910 from his favourite haunt, Happisburgh, he wrote:

I had a good look at ripples, both water and wind-formed. Also at the way the backwash, after a wave, breaks up segmentally as the water pours down the shelving beach. I am not sure that a pretty stiff, viscous fluid pouring in a thin layer down a plate would not imitate our segment as well as any other models! I don't see why the backwash does break segmentally, but it certainly does. That surely cannot be due to vortex formation which the authorities think the ripples are. I never noticed it before.

And in a letter two days later (27. 3. 1910), answering some comment of mine:

---

[1] Professor Sir E. Ray Lankester, K.C.B., F.R.S.

My wave idea is not difficult, for so far, it is a mere idea for
an idea. It has no body or anything to take hold of—but as
an idea for an idea, I think it is a good one.

These observations he developed at some length in the
second and third chapters of *Problems of Genetics*.

He wrote to Professor G. H. Hardy, 24 May 1924:

We have had some absurd attempts,—mostly from biometricians,—to apply mathematics to biology, but as I said
my hope is still that I may live to see mathematics applied
to biology properly. The most promising place for a beginning,
I believe, is the mechanism of pattern.

And to me: *Copenhagen*,

                                              5. VIII. 1924.

Streaming the streets I noticed bunches of cat skins in a
leather shop. I took the opportunity of buying two, illustrating
the octave idea of striping—not very cheap, about 5/6d. each.
Most had the simplicity of pattern disturbed in various ways,
but I got a pretty good pair.

And again in a letter to Issajev,[1] sent on receipt of his
paper on *Hydra* grafts for the *Journal of Genetics*.

                                              7. 11. 24.

...I think you have found a subject of great interest. The
regulation in the number of tentacles is the point which
strikes me as of importance. For me such organs repeated in
series are strongly suggestive of wave motion, and if 12 tentacles are gradually transformed into 6, I see this as a halving
of wave length, giving the kind of result which one would
obtain by suitably stopping the vibrations of a Chladni plate.
I once amused myself obliterating the sand ripples formed on
a dry beach under wind. I smoothed down the ripples and
watched them reform. At first they came double the original
number; that is to say, with the wave length halved. Then as
the wind went on blowing, each two waves were gradually

---

[1] V. Issajev, "Researches on Animal Chimaeras", *Journ. Gen.* xv, 1924,
273–354.

replaced by one, very much as the tentacles are in your Chimaeras. Of course animal material is not homogeneous, and consequently the figures are less regular. I strongly suspect that an "organisation centre" in Spemann's sense, must be a focus from which some force is emanating capable of setting up repetitions at rhythmical intervals, which present true analogies with mechanical vibrations.

And again in 1925:

We may go even further, and applying the analogy of wavelength we may speak of the pattern of Grevy's Zebra (*Equus Grevyi*) as approximately the upper octave of that of *Equus zebra*. Such a terminology, by familiarising the mind with the probable nature of the process of these segmentations, will prepare the way for a correct analysis, though admittedly open to abuse....

In attempting to form a conception of the way in which rhythmical banding develops and spreads over the body of an organism, or rhythmical barring over an organ such as a feather, the analogy with the propagation of wave-motion must in part, at least, be a true guide. ("On Certain Aberrations of the Red-legged Partridge", *Journ. Gen.* XVI, 1 Nov. 1925.)

This "idea for an idea" was always at the back of his mind, and was one of the causes of his hesitation to accept the view that "factors" may be, in any literal sense, transmitted as material particles.[1]

His arduous study of Variation in the field, show-pen, stall and green-house by no means exhausted his energies. He took an active part and interest in the affairs of the University and his College.

After an interval of eleven years the question whether to maintain Greek as a compulsory subject for the Previous Examination was again before the Senate. The Greek test

---

[1] See also p. 292. "That which is conferred in variation must rather itself be a change, not of material, but of arrangement, or of motion".

was held by some to bar otherwise promising men from the advantages of an University education. It was a matter on which Will felt very strongly. He was no classical scholar, but he had a deep respect for such scholarship and a firm belief in the value of general culture.

Tradition of learning gave the Universities of Oxford and Cambridge their position as centres of the intellectual life of the country, eminence which he thought highly important to the student as an individual and to the University as a body. His interest reached beyond education as a general equipment for the business of life; professional efficiency he thought of as the by-product of a higher aim.

An introduction to Greek literature, he maintained, gave to the student a chance of visualising higher ideals than any other study. It is true he would have revolutionised the teaching of Greek, but he would also have kept the broadened study of it essential for the admission to the privilege of being a member of the University.

He felt that lowered standards of ability and attainment were implied in the proposal to abolish compulsory Greek, and with heat and thoroughness he organised the party for the defence.

He wrote this fly-sheet:

FOR GREEK

The letter of "A Country Schoolmaster" tells the truth at last. As Mr Kitchener plainly says, if the Universities no longer demand Greek, the Classical System will pass away, and "a sound basis of Technical Education" will be laid in its stead. The real choice is between the Classical System and such Technical Education.

Most boys, we are told, have no literary aptitude; to them the Classical System is hateful and absurd; for them it is said to fail. It is as one of those for whom the Classical System may thus be said to have failed that I now speak in its behalf.

For many reasons which have often been put, the System is

good. It is both rigid and subtle. It is liked by few, thus few escape its discipline. It is foreign to ordinary life. It is "useless"; and from grim analogies in Nature it must be feared that it is in just this "uselessness" that the unique virtue of the System lies. If this were all, it would be perhaps enough.

But there is something more than this. The change is asked for on behalf of common men who are going to lead common lives. It has been asked for especially on behalf of Natural Science men. Now it is exactly for common men in general and for Natural Science men in particular that the System should be kept.

To common men a Classical Education gives the single glimpse of the side of life which is not common. It is in the Classics and especially in the Greek Classics which he is forced to read, that such a man is for once brought into the presence of the things which are beautiful and have no "use". He will not meet them again, but it is good that he should see them if only for a moment. He does not understand these things; they mean nothing to him. But sooner or later there comes a time when he looks back at these things and remembers, and he knows then that there is something which he has not got, which he does not understand, which is not for him. He is then afraid of that other side of life, if only a little, and when power comes to him he will perhaps not use it to destroy. In the arid mind of many a common man there is an oasis of reverence which would not have been there if he had never read Greek. For Society it would be dangerous, and for the common man it would be hard, if he had never stood thus once in the presence of noble and beautiful things. Some one may say that he may meet things as noble and as beautiful in the literature of England; that is true, but the common man does not read the literature of England. If with "A Country School-master" he lays the "basis of a Technical Education", be it never so "sound", he will begin his life where he must needs end it, in the Black Country of the commonplace.

Of all others it is the Natural Science man who most needs the things which Classics give, if only that he may know the greatness of his own calling. He, forsooth, will read the riddle of Nature. In the fulness of time he has set himself up to solve the old problems, and the answer that he will give is to be final. It is right, then, that he should know that his problems are those which the poets have put. If there had been no poets there would have been no problems, for surely the unlettered scientist of to-day would never have found them. To him it is easier to solve a difficulty than to feel it. It is good, besides, that the Science man should be made to know that there was a people as sharp as he is, who saw the same Nature that he sees, and who read it otherwise with no less confidence than he.

If Compulsory Greek is abolished it will be done by men of two classes. The one has culture by instinct; to them it is inconceivable that any should be really without it, or that to any it can be taught. The other class by instinct and training is savage and would fain destroy what it cannot understand. It is unfortunate that these two classes cannot become acquainted with each other, for perhaps both would then vote for Greek.[1]

William Bateson.

St John's College,
    23rd October, 1891.

Thirteen years later the matter was again under discussion, and again Will took an active part. He maintained that there were only two things of value in the world, Art and Truth. Education should bring the student in touch with noble and lofty thought. Incidentally he pointed out that education in the end was a selection, and that a change of subject meant selection of a different class of men, those, in this case, whose educational aims were so utilitarian as to be properly placed outside the University pale.

[1] The Grace was rejected:
"Every one was prepared for the announcement that the Grace to appoint the Syndicate was rejected, but when the Senior Proctor announced the numbers—Placet 185; Non-placet 525—hearty cheers were raised. The decisive majority was hardly expected. The fact that 710 members of the Senate recorded their votes shows the great interest taken in the question, and it will probably be many years before it will again be raised" (*The Times*, Friday, 30 October 1891).

The Syndicate appointed to consider "what changes, if any, were desirable in the studies, teaching and examinations" sent in a Report to the Senate, which was rejected (1904). The Syndicate was then re-formed and a second Report in 1905 was also rejected by the Senate. Will was one of the Syndics. He signed neither Report.

And in 1906 he opposed the Grace for the new titles of degrees, differentiating between students in Letters and students in Science, an innovation by which the compulsory study of Greek was doomed.

Will was intellectually an aristocrat.

Warm-hearted and quick in sympathy, nevertheless his pleasure and interest in persons as in matters were intellectual.

His practice, on the other hand, or rather, perhaps, in consequence was democratic. He would take endless pains to teach a garden boy *if* the boy shewed any sign of intelligent interest; he was at as much pains to prepare a discourse for the garden staff at Merton as, on more important occasions, for a distinguished audience at the Royal Institution.

But with stupidity he had no patience. "There is only one Deadly Sin", he would say, "and that is stupidity".

This activity in the larger affairs of the University did not preclude attention to the interests and dignity of his College. Lately I found the draft of the following note jammed behind a drawer of his writing-table. (It seems that in 1890 or 1891 a red pillar-box was placed near to the College gates.)

St John's College, Cambridge,
27 February, 1891.

Dear Master,

We have learnt with disappointment that the Council have not already agreed to take steps to remove the pillar-box, and we venture to repeat our sincere hope that they will do so without delay. We feel that the presence of the pillar-box in any form must take away the charm and the stateliness of the gateway. The beauty of our ancient buildings is still the great and singular glory of the College, and it is the duty of the College to protect them from such injury as this. We may perhaps be allowed to add that we are sensible of the difficulties; but we are confident that if the circumstances are represented to the Post Office, no objection to the removal will be raised.

From the Master's answer we learn that 15 Fellows were persuaded to sign this document, which was immediately brought to the notice of the College Council, and the offending red pillar-box was shortly removed.

Later, in 1892, he was elected to the Stewardship of his College, an appointment which he filled with zeal and zest. The question how to carry on his researches had begun to worry him already in these early days. For the next fifteen years the Stewardship together with his Fellowship relieved him of personal, financial anxiety, and just made the difference between freedom to work, and constraint to earn a living by other means. In other ways too, the appointment was valuable to him.

As Steward he had charge of a College farm for a short period, a responsibility from which with his quick and alert powers of observation he gained much experience and knowledge useful to him in his many subsequent meetings with the agricultural world. The College kitchen garden, also, was in his charge, and this was a source of great interest and pleasure; he could, for instance, keep and study bees there. He devoted the same care and energy to the discharge of the domestic duties of his office as he gave to pursuits of his own choice. When in 1902 the scandals of College kitchens were entertaining every household in Cambridge, it was a matter of self-congratulation to him that he had steered the Johnian kitchen staff safely through those seas of temptation wherein so many at that time foundered.

4-2

Meanwhile his book, *The Materials for the Study of Variation*, was steadily taking shape.

S. J. C.

22 July 1893.

Dear Anna,

How are you making out? I have not heard of or from you for ever so long. Here each day is as eventless as the last. Your wild is scarcely less exciting. A poor little hunchback, Darbishire by name, who was lately elected a Fellow of St. John's, died suddenly last Monday. It was very sad as he was a cheerful little person and generally liked....You should see Mrs. Horace's carnations—masses and masses of gorgeous colour. She gave me a wasp's nest yesterday. The larvae all sit in rows with their faces to the centre, like a school.

We have at last had two good soaks of rain. I hope to cut some hay in September.

About 370 pp. of my book are now in print and the publication is no longer so remote.

I rather want to put a little verse of some sort at the head. I know your severer taste, but for all that a little Grace of poesy is rather nice in a philosophical Work, so I hold.

I hesitate between several.

I forget if I have told you that 1 Cor. xv. 39 is set at the head of the Introduction. It always seems to me the best statement of the Problems of the Discontinuity of Species that I know. On an accompanying sheet are several alternatives. I bid you choose *ONE*. They are all tags that have often run in my head during this job. I incline most to putting the bits marked 2 and 3 together; but 1 is such a lovely verse I should like to use it, and the "phantom Caravan" gives the idea of the procession of living things so splendidly.[1] I shouldn't mind letting 3 go in by itself. 4 is a very fine couplet, but I don't set great store by Schiller, and only 1 in 1000 would be able to construe

---

[1] See Introduction to *Materials for the Study of Variation*: "On the hypothesis of Common Descent, the forms of living things are succeeding each other, passing across the stage of the earth in a constant procession. To find the laws of the succession it will be best for us to stand as it were aside and to watch the procession as it passes by".

it. 5 is a nice little bit, but it is rather too light in tone—I want something to strike more solemn. If I used 3 I should stop at "it". The Master wouldn't quite do—at least it might mislead!

If your sterner judgment says put nothing but "printed by C. J. Clay & Sons, etc." please let it speak—but I own to a weakness towards the versicle.[1]

Yours ever,
W. Bateson

1. A Moment's Halt—a momentary taste
   Of Being from the Well amid the Waste—
   And Lo!—the phantom Caravan has reacht
   The NOTHING it set out from—Oh, make haste!

2. Would you that spangle of Existence spend
   About THE SECRET—quick about it, Friend!
   A Hair perhaps divides the False and True—
   And upon what, prithee, does Life depend?

3. A Hair perhaps divides the False and True
   Yes; and a single Alif were the clue—
   Could you but find it—to the Treasure-house,
   And peradventure to THE MASTER too;

Omar Khayyám, XLVIII, XLIX, L.

4. Aus der Wahrheit Feuerspiegel
   Lächelt sie den Forscher an!

Schiller (An die Freude).

5. And being now at some pause, looking back into that I have passed through, this writing seemeth to me, "si nunquam fallit imago" (as far as a man can judge of his own work), not much better than that noise or sound which musicians make while they are tuning their instruments; which is nothing pleasant to hear, but yet is a cause why the music is sweeter afterwards: so have I been content to tune the instruments of the Muses, that they may play that have better hands.

Bacon (Advancement of Learning).

---

[1] None were used.

St John's College, Cambridge,
30 July 1893.

Dear A.,

I am glad you think some part of the Omar will pass. As to all you say, I quite agree. I feel and had felt the "Carlyle" difficulty as to the Schiller couplet. If I put the Omar (2) I should like to put (3) as well. "The single Alif" is a feeling I have every day, and altogether it seems to me to have the true spirit of "Research". I shouldn't mind (3) without (2) at all. Do you mind stopping in the middle of the quatrain? Such a proceeding is most unoriental.

Last week I had a night at Herne Bay, whither I went at short notice to get lady-birds, mostly failing in this quest, as they had gone with the rain. It was a novelty for me, for I never saw the true cockney pleasure-ground. Everyone was bawling "*Daiiiisy, Daiiiisy*" fit to split, likewise, "I 'ear the girls declare, 'e must be a millionaire" etc. Several bands ditto. Three decayed gentlemen, (perhaps undergraduates), earning coppers by singing "The Old Brigade" etc. in evening dress and black masks; and a vague youth telling us of the "Lamb of Gawd that taketh away the sins of the world", a burden to which Herne Bay was making a not inconsiderable addition. How strange it is to reflect on the past of such phrases, "*Qui tollis peccata mundi*"!

I quite agree that 1 Cor. xv, 41 is an immeasurably finer verse than *ibid.* 39, but I didn't see how to drag it in, though I had the thought before me.

Every day almost makes me misgive more and more about this book. Several of the sections of evidence seem very weak now they are actually "floated and rendered" as the builders say, and half of the best things seem never to have been got in at all. I am afraid it doesn't deserve a very fine "motter"....

His book, *Materials for the Study of Variation*, was published by Macmillan in 1894.

Will always maintained that Darwin's greatest contribution to science was his splendid collection of facts;

he admired this achievement wholeheartedly; and he had set to work to study Variation by like methods.

His book opens with a statement of his aim, followed by a tribute to Darwin:

> To solve the problem of the forms of living things is the aim with which the naturalist of to-day comes to his work. How have living things become what they are, and what are the laws which govern their forms? These are the questions which the naturalist has set himself to answer.
>
> It is more than thirty years since the *Origin of Species* was written, but for many these questions are in no sense answered yet. In owning that it is so, we shall not honour Darwin's memory the less; for whatever may be the part which shall be finally assigned to Natural Selection, it will always be remembered that it was through Darwin's work that men saw for the first time that the problem is one which man may reasonably hope to solve. If Darwin did not solve the problem himself, he first gave us the hope of a solution, perhaps a greater thing. How great a feat this was, we who have heard it all from childhood can scarcely know.
>
> In the present work an attempt is made to find a way of attacking parts of the problem afresh....

He was not satisfied with Darwin's (still less with Lamarck's) solution; he was, besides, impatient of the dull finality with which Darwin's work was blindly accepted, confining its scope to a narrow orthodoxy as rigid as that of the Churches. Will had seen too much and observed too closely to subscribe to such an orthodoxy.

In the way of both solutions (Darwin's and Lamarck's) there is one cardinal difficulty which in its most general form may be thus expressed. According to both theories, specific diversity of form is consequent upon diversity of environment, and diversity of environment is thus the ultimate measure of diversity of specific form. Here then we meet the difficulty that diverse environments often shade into each other insensibly and form a continuous series, whereas the Specific Forms of life which are subject to them on the whole form a Discontinuous Series. The immense significance of this difficulty will be made more apparent in the course of this work. The difficulty is here put generally. Particular instances have been repeatedly set forth. Temperature, altitude, depth of water, salinity, in fact most of the elements which make up the physical environment are continuous in their gradations, while, as a rule, the forms of life are discontinuous (*Materials for the Study of Variation*, Introduction, Section 1).

It was no doubt this difficulty that baffled him in his research in the Steppe. Since then he had found in the Discontinuity of Variation a new point of attack on the problem of Species. The preceding letters shew how in his study of Variation, he was led on from observation of Discontinuity to Meristics, Symmetry and the Repetition of Parts, and to tentative suggestions of Rhythm, which he never put aside from his considerations of the forms of life.

In the "Concluding Reflexions" (chap. xxv) he marks out his future work.

But however this may be, whatever may be the meaning of alternative inheritance and the physical facts from which it results, and though it may not be possible to find general expressions to distinguish characters so inherited from characters that may blend, it is quite certain that the distinctness and Discontinuity of many characters is in some unknown way a part of their nature, and is not directly dependent upon Natural Selection at all.

The belief that all distinctness is due to Natural Selection, and the expectation that apart from Natural Selection there would be a general level of confusion, agrees ill with the facts of Variation. We may doubt indeed whether the ideas associated with that flower of speech, "Panmixia", are not as false to the laws of life as the word to the laws of language.

But beyond general impression, in this, the most fascinating part of the whole problem, there is still no guide. The only way in which we may hope to get at the truth is by the organisa-

tion of systematic experiments in breeding, a class of research
that calls perhaps for more patience and more resources than
any other form of biological inquiry. Sooner or later such
investigation will be undertaken and then we shall begin to
know.

Comment and criticism followed; the very idea of
"Discontinuity" was so foreign to the biologists of 1894
that he was moved in answer to one critic to reply:

Part of the difficulty, that will I fear stand in the way of
any general acceptance of the suggestion of Discontinuity, lies
in the impossibility of at once giving it a definite expression.
I am unwilling to attempt this until there is much more
evidence, but I am afraid that as it stands the suggestion is not
in a very intelligible form. On the other hand, Continuity, as
applicable to Descent, has clearly a single meaning; for I
suppose it is clear that if in a given case we believe that form
B has arisen from form *A* by a continuous process, we can only
mean that every condition intermediate between *A* and *B* has
not only occurred, but, that it has occurred in one line of
descent, and in the order of succession from *A* to *B*. (The idea
of continuous Descent as now received contains the additional
implication that every one of these intermediates has in turn,
and in the same order, been a *normal*, though this implication
is not, perhaps, a necessity.) Any mode of Descent other than
this would, I think, come within the term *Discontinuous*, but
some of these possible modes are so distinct that it would be
confusing to associate them under one term. As to the amount
of Discontinuity in any given case I can, of course, make no
suggestion, but I believe that there are numerous cases in
which it is practically impossible to suppose the course of
descent has been Continuous in the full sense of the term as
given above.

Will's misgivings as to the fate of his work were justified.
The book was not a success—the Professors and lecturers
of the day did not introduce their students to it. Perhaps,
that they should was not to be expected. For a few years

the annual arrival of the publisher's account was a dismal event, and then the book was put "in remainder" and dropped out. The second volume as such was never written; but this was the result of the very exacting breeding work, in which he became so deeply interested, rather than an expression of discouragement. The spirit which inspired the great Introduction to his book never failed him in his work. It was of course very disappointing, and he must have felt disheartened, but he knew that the failure was not the fault of the book, and left it at that.

In this year (1894) he was elected Fellow of the Royal Society.

In June 1896 we were married, and I began to learn what life may be. His activity was unflagging. His many varied duties and responsibilities were severally discharged as though each one was his only care. His idea of rest was change of occupation, not a lessened expenditure of energy. The only reason why people "had no time" to do any given thing, he asserted, was that they did not really want to do it. And certainly he found time to do very much. Unhurried and critically careful, he seemed to get through many days' work in one, and yet have time for all social pleasures. He played as he worked, with a kind of rapid concentration of attention. His quickness made him impatient of slowness. He could not bear to wait.

Under a reserved and dignified manner lurked mirth and jollity. His intense and learned interest in life creamed over in gaiety. He was the best of companions.

Very absent-minded, his pencil, notebook, knife, forceps, scissors and even pipe (and in later life, spectacles) were perpetually mislaid. Of his clothes he was as reckless as a school-boy. He was capable of going up to London in old "garden" flannels, darned across the knee, or (in the other extreme), he might be found kneeling on the gritty garden path, in a brand-new "town" suit, recording some batch of seedlings. Such trivialities were not his affair, though on special occasions he was meticulously careful.

We settled in Cambridge, in a small house near the Botanic Garden.[1]

That summer he lectured to the University Extension school meeting in Cambridge. Then, as always, unless disallowed, he lectured extemporaneously. "With a *set piece*" he could not "keep his finger on the pulse of his audience". The preparation of a lecture was none the less arduous. In those still youthful days of physical activity, he liked to take a long walk with a companion (silent whilst he "chewed the cud"), on whom he would suddenly try a passage of his thought.

As about every detail, except dress, he was fastidious also about his performance as a lecturer. His voice, his gesture, his manner, his matter—each of these he considered and subjected to his own self-criticism. Of a big Hall strange to him he would always contrive to try the acoustics before the hour of his lecture.

He was very fond of teaching and enjoyed lecturing, but before starting out to his class or lecture-hall he was subject to sudden misgivings—"he had not enough to tell them"; as a matter of fact he never used up all the material briefly noted on his slip of paper, and never was "gravelled for lack of matter". This was the only outward sign of nervousness he ever shewed; it was noticeable just as much before his regular class-work as before more sensational occasions.

Towards the end of the year 1895, after an interval of nearly ten years, the question whether women should be admitted to Degrees came again to the fore in Cambridge University, and in May 1897 was put to the vote.

Will was born and bred in a home environment where the intellectual aptness of women for academical training was obvious and accepted as a matter of course. His

[1] Norwich House, at the corner of Norwich and Panton Streets.

parents had taken an active part in the first establishment of women students in Cambridge, and those early efforts had gathered together a remarkable group of women of high ability. More than this, his sisters shewed promise and brilliance that alone were enough to convince him that the University should admit women to the degree. He thought this desirable, not because the degree would serve as a reward of work done, or as a credential useful to women on entering professions, but because he wanted every facility for original work and research to be freely accessible to all genuine students, men or women. The University should lose no chance of securing the best original workers, whatever their sex. He flung himself enthusiastically into the contest, and worked with as much energy and resourcefulness in organizing his party as he had on the occasion of the *Compulsory Greek* vote. But this time without success. A bolder and more frank programme on the women's part, he thought, would have had better chance of success; but in this cause he submitted to the leadership of others.

He enjoyed such contests heartily.

In 1894, at Mr Francis Galton's suggestion, the Royal Society had appointed a Committee for conducting Statistical inquiries into the measurable characteristics of Plants and Animals. Early in 1897 Will was invited to become a member of this Committee. Though at the time there was a proposal to establish a biological farm in connection with this Committee (a scheme extremely tempting to him) he characteristically jibbed at the limitations laid down in the terms of the Committee's title, and declined. A little later this Committee was reconstituted as the *Evolution Committee*, and the scope of its researches was widened. Will became a member, and ultimately the Secretary of this Committee, active in work and enthusiastic in drawing up schemes of inquiry and interrogatories to be sent to breeders and growers.

His own work in the next years, and that of his fellow-workers and pupils, was partly carried out with the help of small grants from this Committee, and records of most of this early work were published by it.

In 1897, aided by such a grant, we hired an allotment (behind, and belonging to the Cambridge Botanic Garden) and the experiments in Poultry- and Plant-breeding were begun, which he continued to the end of his life, with constant interest, at times with excitement.

In 1898 our first child[1] was born, an event commemorated by the purchase of Blake's *Book of Job*.

In the next year we were already in need of more space for the poultry. We hired another allotment but very soon began to look for ground outside Cambridge. It was on the occasion of a house-hunting expedition that we came upon a house sale where Old Master drawings were being offered. For only a few shillings more than we could afford these coveted treasures were to be had. That was a most exciting day. Of all pastimes except that of painting, the hunting and the acquisition of Old Master drawings was his favourite. He had an extraordinarily sure "flair" for a good thing, and once started, his collection grew and paid its own way. Any few savings—but saving is not very compatible with research work—were invested in Old Master drawings. Compared with the gamble or backing one's own judgment in the "Art market", race meetings and Stock Exchange were "poor doings—dull waste of time and money" he would say; and he never touched either. Later on he was attracted by Japanese prints, and Chinese works of art, and these too he collected well; but the Old Master drawings were never degraded from the first place in his affection. All this however was holiday amusement. The real work for many years to come was very exacting.

His "flair", perhaps, stood us in good stead in the

[1] John, field-naturalist, of uncommonly ripe judgment and good promise of scientific ability. Killed in the war, 13 October 1918.

search for house and grounds. Merton house, Grantchester, was most admirably adapted to his needs. It was within University bounds, it had a good paddock, protected by a belt of shrubbery, good out-houses, and a very good garden and orchard. For the time being space sufficed.

We moved to this house in the autumn of 1899, after the birth of our second son.[1]

In this year too, Will was appointed Deputy to Professor Newton (Professor of Zoology in the University), who since early days had been his staunch and loyal friend.

This appointment pleased Will very much; it gave him the opportunity of regular teaching which he had long desired. A group of eager students soon gathered round him.

We had now ground; we had poultry-pens well stocked; we had row upon row of peas, poppies, lychnis; the garden was full of big and little experiments—some, tentative trials of subjects; some, serious undertakings. Our "gardener", Blogg, was a capital man and a delightful person, but by profession a coachman and already ripe in years. (He made all the proper stable noises with his mouth whilst cleaning the hen-houses.) Besides him we had our own hands. When lecturing and College work were done, the rest of the day was spent in hard manual labour; from the merest menial drudgery to high flights of scientific speculation, hand and brain were hard at work. There was all the sorting, sowing and gathering of seed to be done personally; the fertilising and recording; most of the digging, hoeing, weeding, staking and watering; the five incubators, each 100 egg power, and as many rearers (all run with oil lamps); the tiny chicks; and at times hundreds of larvae to be attended to. All writing (not reckoning the ordinary post, which was often heavy) was done at night.

Intimate letters he wrote freely and fluently; in all other forms of composition he wrote slowly and with difficulty, never content until he had found "*le mot juste*".

---

[1] Martin, scholar of Rugby School and of St John's College, Cambridge; of brilliant promise and great personal charm. He shot himself in 1922.

If he was critical of others' contributions he was doubly critical of his own.

His first manuscript would be one word superscribed over another, all "inserts" and deletions; fair copy succeeded fair copy in like case; often until the small hours of the morning we sat up, he writing, I copying, until at last he was satisfied that he had found the one way to say exactly and indubitably what he meant to say.

During these years, from early spring to late autumn we never left home together, except on "Flower Show"[1] day. Then we rose early, "did" the incubators, and bicycled to the station for an 8 o'clock morning train. We had time to see the Royal Academy summer show, and a Bond Street Gallery or two, before the Flower Show opened. We made a complete tour of that, noting any novelty or variety that was new to him, and then away to hunt up a few references in one library or another, or to make a hurried inspection of Christie's, Sotheby's, Robinson and Fisher's, or Foster's sale rooms; a few moments in the Temple Church, or a visit to confirm some impression of a picture in the National Gallery, and away back to the Flower Show, by this time less crowded so that the men in charge of the exhibits were freer to answer the questions suggested to him by the morning visit. We tried once or twice to finish with a theatre but the midnight ride back to Grantchester from Cambridge station—the eggs still to be turned, the lamps adjusted—taught us to be content without this extra pleasure.

Here are extracts from letters whilst I was holiday-making and he keeping guard at home:

Norwich House, Cambridge,

19. 7. 98.

...I hope you and yours are rested and content.

For myself I am thankful for a dull day. The emergences[2] are simply overwhelming, but the feeding is much lightened.

[1] Of the Roy. Hort. Soc. in Temple Gardens.
[2] *Pieris napi* × *Bryoniae; Pararge egeria.*

New troubles. The disease among the chickens has broken out with fresh virulence. After you went yesterday, I killed 2 as hopeless. One more died this morning. Two middle-sized ones are very shaky, and some four more small ones are in a doubtful state. It is heartbreaking. With the young ones it is the "tucked up behind" form. One of the larger ones has the bleeding well-developed.

The piano-tuner has been tuning your long mute instrument. A new barrel organ has come, or haply an old one with new tunes. I almost gave it a penny for the relief. Two others— old ones—have also been. Never more will you see my cane hat. It blew off yesterday and a ruffian lad drove his wheel over it on purpose. I was loud in language and demeanour towards him—but of no avail.

The arrivals of Mrs B., Master B. and nurse are by now, I suppose gazetted through Much Hadham. Probably J. B.'s weight to-day, and at starting, are household jottings throughout the parish.

"Pall-Mall" Mum?..."'Ome Notes" Mum?...It didn't take that lad long to get your measure. The Saturday match is postponed. I wish you would help eat things here. Yrs. w. b.

Norwich House, Cambridge,
22 July 1898.

...The rush is awful. Butterflies coming out in the pots, in spite of Frederick boxing off. The 2nd boy is useless and only makes F. waste time. [We used some of the grant in boys' wages, during my temporary disablement.] The last of the young chicks died to-day. It is disappointing. I had quite thought it saved. To-morrow afternoon if I can, I will come over.

Since writing I have made up some lee-way, and unless it is wet or in some other way unsuitable, I will come.

The middle of this day devoted in part to judging beers[1]— 7 in number, submitted by Buol, for the Refreshment tent. 1 is fair, rest bad, at least all I tried. After 5 bottles I gave up.

[1] This was done on behalf of the members of the International Zoological Congress which met that year in Cambridge.

24. 7. 98.

Am getting dreadfully behind. Is there any chance you could conveniently come back here from town and give me a day's work on Tuesday? Don't unless you feel thoroughly easy about leaving the cottage. If I could get a good lot of boxing off and setting done I should be clearer. w. b.

Norwich House, Cambridge,
28. VII. 1898.

...My own parent will have a good deal of criticism and remark to offer regarding J. B. when she sees him. It is a pity he can't keep asleep between 4–7 a.m.! That is beginning early indeed. Splendid rain this afternoon. To-morrow I am compelled to take part in the High Table match. It is very stupid, as the Museums match is on Saturday, and I can ill spare the time. I have bargained to be let off at 3 o'clock. If I rise with the cock I shall pull through. I set 60 today, so the tide has scarcely abated yet.

Sorry about the hat. Perhaps if you shew you have an empty head your ladies will not observe the want of fashion in your hat. Mrs Ward and friend called yesterday, and we had a nice chat. She wanted to see the infant. I said there was one handy in the street she could look at—which indeed there was—but she insisted on the veritable J. B.

Read Hooley's evidence. Fancy it taking £20,000 to buy a peer. I am amazed at the price. £100 seems ample.

Yrs. w. b.

Merton House, Grantchester, Cambridge,
27. 8. 1901.

John has had a stomach-ache all day and consequently been moping. We think the cause is probably no further than the mulberry-bush. I don't suppose there is much wrong. He is now asleep and seems comfortable.

I have had a thorough clean of all flues, incubators and

rearer. The latter had a leaky and empty boiler, now I trust repaired, as the night is cold. My hands reached a level of uniform blackness never, I believe, before endured. They were one clear negro tint, smoothly spread, and for days they will be imperfectly clean. Martin[1] was afraid to pass me in the passage and shrank in physical fear. He felt as the children of Israel did when they found that Moses had married a "woman of the Ethiopians". I have been thinking it was small wonder they "murmured" against him. Think how we should feel if the Archbishop of Canterbury did the same.

There is a slackening of the household tension which shews me that it is you who are supposed to be the critic by the kitchen. Yesterday I dined off faded hashed mutton and Sunday's pudding, but was borne up by thoughts of to-night's curry. This would have been admirable had the rice been cooked. Soup and potatoes were stopped. Salad is stopped. Veg. marrow has succeeded it, for good and all. *But* this morning I had the prethought to order turnips, which were yellow and excellent. Also we had the long hoped for raisin dumpling, but to one gorged with curry this was only occasion of regret for wasted opportunity.

I gathered the plums, three baskets full and they have been some jammed and some bottled. I have bought nothing but a paper all day.

Seeing that Moses had very little influence with the people at the best of times he ought to have thought twice before marrying a negress. And if any one else had done it he would have been the first to call out. 28. 8. Yours just come. John much better, though the cause is not yet removed.

Wilks writes asking whether I will go to the Hybridisation Conference in New York, September 1902, as one representative of the Horticultural Society. They offer £75 towards expenses. I am rather tempted but shall not write for a few days....   w. b.

Return 1st to N.Y. costs about £25 I believe.

[1] About 2½ years old.

Merton House,
Grantchester, Cambridge,
28. 4. 02.

Bitter frost and wind to-night. I do begin to feel the hardness of your lot. Some lamps will have to do their best on once filling in two days, I think.

John and I went cowslipping for an hour this afternoon. He asked me: Was I busy? and as I said, "not very", he asked, "I thought it was to-day you was going to take me to get cowslips?", which I had incautiously promised and forgotten.

...As your hands soften, mine harden. The eggs stump me utterly....

Besides definitely planned and carefully contrived experiment he liked to have what he called "a fool's experiment" in hand. These were either a kind of empirical short cut, or just a sketch of possible future undertakings. He wanted to know very much more than he had time to find out.

My first experience of such "fanciful follies" (see p. 41) nearly cost us dear. Right- and left-handedness continually intrigued his curiosity; the day would come when something would be known of the nature of this phenomenon, which might indeed bear upon the problem of species. The simplicity of the domestic mincer (to which he had just been introduced) inspired him with the desire to mince young conifers, and extract from them resins, right- and left-handed. A list of suitable conifers was drawn up and despatched (I forget to what firm). The first parcel of seedling or yearling trees arrived before the invoice. We set at once to work. Happily conifers are harder than beef, and our enthusiastic efforts were but poorly rewarded. We desisted to devise a better plan. Next morning brought a staggering bill and we lost all further interest in the venture. I do not now remember what these conifers cost, but the nursery-man very generously accepted Will's explanation and apology, and took back the surviving plants.

This phenomenon of right- and left-handedness was one of the many pre-occupations of his mind.

Some years later he heard of a pond not very far from the Hartlepools, where right- and left-handed snails (*Limnaea p.*) were said to be found together. At the first opportunity he was up and away to get these snails. He found them and brought home samples; but we had already too much on hand and to his great disappointment did not succeed in breeding them. Recently Professor Boycott and Captain Diver have carried out research work on this subject.[1]

About 1916 he thought he had found right and left swimming *Paramoecium*. He took endless pains to arrange the glass slides containing them, and watched with absorbed interest. So sure was he of his observation that he sent a short note to the Linnean Society for publication. Suddenly he became convinced that his observation was merely an optical delusion; he rushed up to town in feverish anxiety just in time to withdraw his paper.

He also constantly criticised with disapproval the artificial waving and curling of women's hair, mocking the ignorance thereby displayed of this right- and left-handed phenomenon in nature.[2]

Of attempted hybridising and grafting there were always numerous examples. Our little John, about four, trotted after him in the garden, full of curiosity and admiration. One day we saw him on the forbidden flower-bed, rubbing one flower on to another. "Hullo! John! what are you doing there?" "I'm on'y vaccinaten of them", he beamed back. The reproof for the trespass was never spoken; the business of the moment was postponed and John received his first simple lesson in the structure of a flower, with breathless interest.

Working as he did almost entirely at home, the children saw much of their father, and he very soon established a

[1] "The inheritance of Inverse Symmetry in *Limnaea peregra*", *Journ. Gen.* xv, 1925, 113–200.
See also *Problems of Genetics*, chap. ii, p. 44.

delightful *camaraderie* with them. He was keenly interested in watching their development, and enjoyed helping them. He was determined that their education should be based on literature. Every morning for many years he read to them after breakfast, generally from the Old Testament, but sometimes from Bunyan or other fine prose, or even from Shakespeare. If sometimes he rose from the table, absent-minded, one of them would say: "We haven't had our chapter, father", and back he came.

Meanwhile a permanent basis for his breeding work was urgently necessary. The Evolution Committee of the Royal Society alone took part at first in promoting the researches, and then only on a limited scale. Grateful as this help was, it lacked all promise of permanence. To make real advance in knowledge (Will's one great over-mastering ambition), prolonged experiment in breeding systematically must be made possible. The work must go forward steadily, and independently of the vicissitudes of human life.

Paddock and garden were soon as full as could be. But though thus cramped and hindered, we could ourselves, with health and strength, manage happily enough and get through a good deal of work. This did not suffice. Again and again he suffered vexation and disappointment in having to relinquish some pet scheme, or to cut down to narrow limits the seed-sowing—waiting for a second year to have full numbers.

Besides the work in our garden and barns, enthusiastic pupils had begun to experiment. Miss E. R. Saunders continued the plant-breeding begun in our former allotment garden; Miss Sollas reared guinea-pigs in a field behind Newnham College; Miss Killby goats. Miss Wheldale worked on flower-colours in Antirrhinum, Miss Marryat grew *Mirabilis Jalapa*; Miss Durham hybridised mice in a kind of attic over the Museums. In Oxfordshire Mr Staples-Browne bred pigeons; Major Hurst was busy with poultry and rabbits; Mr Doncaster crossed varieties of moth (*Abraxas grossulariata* × var. *lacticolor*); Miss Darwin began on trimorphic forms, with *Oxalis*.

Will was in touch with each experiment, and his interest and energy kept the work at a high level of accuracy, in spite of narrow means.

From time to time a welcome grant from the Evolution Committee or from the British Association helped in part either to defray an initial outlay, or some of the current expenses of these experiments; but the whole fabric, save for the enthusiasm which inspired the work, was grievously insecure.

Many investigations seemed hardly worth initiating without some guarantee of continuance in the future; and all the work as yet started was on too small a scale to satisfy Will's eagerness and ambition. In some respects, of course, the smallness of scale was wholesome, but it delayed results, and made advance very slow.

Will strove in vain to get this research endowed.

Probably the chilling reception of his doctrine of Discontinuity hindered his endeavour, and when in 1900 Mendel's papers were rediscovered by de Vries and Correns (and hailed with a kind of triumphant gladness by Will), the possibility of endowment seemed to grow more and more remote.

Mendel's work fitted in with Will's with extraordinary nicety—perhaps too well; at any rate at that time the English school of Biology had little appreciation of, and no use for it.

Those who accepted Mendelism could not reject Discontinuity. Will's intimate study of variational phenomena made him realise at once how very far these Mendelian clues could be followed. He was over the stepping-stones and away, scrambling up the further bank whilst the Biometricians, chiding, were still negotiating the difficulties of the first step.

The importance of the systematic analysis of individuals was not yet understood. Long and tiresome was the obstruction. Nevertheless, in spite of growing anxiety as to the maintenance and continuation of the experiments, those struggling years through which he led his little troop were full of healthy and very happy activity.

In the early days of its existence the Evolution Committee of the Royal Society had considered a project of acquiring, by subscription, Darwin's old house at Down, with the idea of instituting a station where work of this kind could be carried on. Partly because of lack of means, partly on account of inaccessibility—remote from other laboratories, and (worse still in Will's opinion) remote from libraries—the scheme fell through. In 1899 however there was again this suggestion of establishing a Biological Farm or Research Station at Down. Sir George Darwin[1] wrote to consult Will. I have a rough draft of his reply.

Norwich House, Cambridge,

21 July 1899.

If Down were preserved as a biological station it would, I take it, be devoted to work having a direct bearing on Evolution. The primary object of such an institution should be the maintenance of investigations which require to be continued for long periods of time. Of these the most important would certainly be an attempt to determine accurately by experimental breeding the laws of inheritance in animals and plants. Such experiments should be so designed as to throw light on many points some of which may be roughly grouped thus:

1. The magnitude of variations.

2. Modes by which variations may be perpetuated; the degree to which different varietal characters are, or are not capable of blending on intercrossing.

3. The influence of intercrossing on the fertility of races—fertility being used in a wide sense.

[1] Sir George Darwin, F.R.S., Plumian Professor of Astronomy, Cambridge.

4. The conditions of transmissibility; the nature of prepotency; the possibility of altering the power of transmission by selective mating—notably by consanguineous
breeding.

Work of the kind contemplated should of course be begun
simultaneously in both animals and plants. To have permanent value however such experiments must be performed
on a scale sufficient for the application of Statistical methods.
By this condition the larger animals are necessarily excluded,
but in addition to the Botanical work much might be done
with several of the smaller animals, especially the fixed breeds of
poultry, pigeons and perhaps canaries, several species and fixed
vars. of insects especially Lepidoptera, and perhaps others.

The beginning should be made on a very small scale, but
the work which could be carried on is limited almost entirely
by the amount available for labour, as simple appliances
would suffice. Both in the case of animals and plants the chief
expense would be under this head.

The scientific importance of this class of work cannot be
over-estimated. The field is almost untouched. It is for want
of continued observations on inheritance in animals and
plants that so little progress has been made in the science of
Evolution since Darwin's work.

From the practical side also, though the gain may be remote,
it cannot be doubted that sooner or later very considerable
results will be reached. What the precise form of these results
may be cannot be foretold, but it is certain that if only an
outline of the laws of inheritance could be attained the influence of such knowledge on the art of the breeder must be
immediately felt.   W. B.

Down however he considered impracticable for the
modern requirements and methods of research.

Without a well-stocked library he could not imagine
any Scientific Institute thriving. It was indeed one of his
aims to familiarise his students with the use of books and
to teach them to know the work of predecessors and
contemporaries at first hand.

Also, he held it to be of the first importance to be within touch of the students and teaching of other branches of science. That a zoologist should come into his work knowing nothing of botany, or a botanist knowing still less of zoology, and neither knowing anything of physiology, nor much of chemistry, grieved him sorely. He was convinced that the problems of Evolution, to yield their secret, must be attacked by Natural Science as a whole, not by a few specialised units. And so in those pre-motor-transport days he deprecated the isolation of Down, and the cost of installation of even elementary equipment for the work outlined again put an end to this suggested scheme.

This letter, dated some six months before the rediscovery of Mendel's papers, shews clearly how nearly he had approached to Mendel's methods of work, and to his solutions of the preliminary problems of Heredity; and this is also well shewn in two little papers which he delivered to the Royal Horticultural Society. The first, "Hybridisation and Cross-breeding as a method of Scientific Investigation" (p. 161), was read before the Society on 11 July 1899 (published 1900); the second, "Problems of Heredity as a subject for Horticultural Investigation" (p. 171), was given on 8 May 1900. He had already prepared this paper, but in the train on his way to town to deliver it, he read Mendel's actual paper on peas for the first time. As a lecturer he was always cautious, suggesting rather than affirming his own convictions. So ready was he however for the simple Mendelian law that he at once incorporated it into his lecture.

His delight and pleasure on his first introduction to Mendel's work were greater than I can describe; as when with a very long line to hoe, one suddenly finds a great part of it already done by someone else and one is unexpectedly free to get on with other jobs. He was fortified with renewed faith in the largeness of his research; he found in it new interest, new possibilities, and drew from it new inspiration. How warm his enthusiasm and how vigorous his confidence may be gathered from his "Defence", in the little volume (1902) *Mendel's Principles of Heredity.*

The "Defence" is fierce, and in some quarters this fierceness has been deplored and, probably, thought unreasonable. That is because its occasion is already past history. Mendelism now has a place in general biological instruction. Weldon's[1] insidious and bitter attack has fallen into oblivion, leaving the present-day reader wondering why the case should be defended at all. In its day the outspoken hardness and clarity of the "Defence" were not idle.

In his writing-table drawer lay a big envelope, labelled by him "Begging letters", containing drafts of letters in which at different times and seeming opportunities, to awaken the much-needed interest, he applied to various societies and individuals, setting forth the claims of the study of Heredity to rank at once as pure science and as of immeasurable practical importance. Alas, in vain.

In after years we often laughed at our eager expectation of the answers to these epistles: one disappointment followed another.

In 1900 Mr Carnegie, moved by a suggestion of Mr Herbert Spencer's that the question of the inheritance of acquired characters could be determined by experiment, wrote tentatively to Mr Francis Darwin to ask his opinion. This letter was passed on to Will. He made the following answer:

Andrew Carnegie, Esq.

Merton House,
Grantchester, Cambridge,
21 Nov. 1900.

Dear Sir,

Mr Francis Darwin has put into my hands, letters lately written by yourself and others[2] as to a possible station for experiments relating to Evolution. From this correspondence I understand that you entertained the idea of contributing to the endowment of such work.

As I think you are aware, there is in existence a Committee

[1] *Biometrika,* 1, pt. 2, 228–254.　　　　　[2] Herbert Spencer.

of the Royal Society having for its object the furtherance of investigations of this kind, and of that Committee I am at present Secretary. I ought to say at once that in my opinion and in that of others practically acquainted with these subjects, the problem of the inheritance of acquired characters is not a favourable one for experimental investigation as yet. To be effective such an inquiry must be preceded by a thorough study of the normal laws of heredity. The lines on which this work should be pursued are now clear and there can be no doubt that with proper equipment results of first-rate importance are attainable. With the aid of grants from the Royal Society I made a beginning on these lines here three years ago, in association with Miss E. R. Saunders[1] who is most especially concerned with the botanical side; but with greater resources it would be possible to attack these problems in a far more effective way.

If you are still favourably disposed to these objects I should be glad of an opportunity of laying before you suggestions for carrying them into effect under conditions which should ensure proper control.

I may add that in this matter I have no idea of getting anything for myself beyond increased opportunities of pursuing these investigations.

In the event of a favourable answer I would ask you to be good enough to grant me a personal interview at which I could put details before you. I am yours faithfully,

W. Bateson.

The personal interview never took place; the matter was not pursued.

Encouraged by his success in America, he applied after his return to the Trustees of the Carnegie Institution.

APPLICATION FOR SUPPORT OF AN EXPERIMENTAL
INVESTIGATION OF MENDEL'S PRINCIPLES OF
HEREDITY IN ANIMALS AND PLANTS

By W. BATESON, M.A., F.R.S., Oct. 1902

The confirmation and extension of Mendel's discoveries constitute a new departure in the study of Heredity, in our

[1] Lecturer in Botany, Newnham College.

conceptions of Evolution, and in all Sciences which are concerned with the essential nature of living organisms. In addition to the high scientific importance of these principles, there can be no doubt of their economic value to the practical breeder of Animals and Plants. The operation of these principles has now been perceived in a considerable range of animal and plant species, but a precise determination of their scope and limitations is urgently needed. The science of Heredity is in a position not very different from that which Chemistry once occupied when the objects and methods of analysis were known, and the empirical study of the chemical properties of the various bodies was beginning.

The objects to be attained by the experimental study of Heredity are now definite, and the methods to be employed are perfectly clear. The number of forms which can be simultaneously studied, and the extent of the work which can be undertaken is thus merely a matter of expense.

For several years I have been engaged in this work and the results arrived at up to 1901, in association with Miss Saunders, are set forth in the Report to the Evolution Cttee. Royal Society, of which a copy is sent herewith. The work of 1902 will appear in a further Report.

These researches I hope in any case to continue, but from lack of means my operations have hitherto been confined to the barest limits. In order to carry on these experiments on an adequate scale, I require more land, more labour, and appliances, together with the assistance of a skilled observer, which, with my present resources, are unattainable. In illustration I may mention that I have begun an inquiry into the statistical relations of the Compound Allelomorphs of the Sweet Pea (see Report, p. 143), and have carried the experiments through two seasons. Unless however I am enabled to increase the scale of the work it will not be possible to make more than a sample testing of the next generation, whereas for a proper study of the problem the entire crop must be sown and recorded.

At the present time I hold appointments which suffice for my personal expenses, and have my time very largely at my own disposal. The whole of this leisure is occupied with these researches and I estimate that they cost even on the present scale from £100–£150 a year, an expenditure maintained with increasing difficulty.[1]

From time to time I have received grants from the Government Grant Fund of the Royal Society, and I may perhaps look forward to receiving occasional help from that Fund. Nevertheless no considerable sum can be expected from this source, and I do not know of any other available. Though by limiting the work to its present dimensions I may fairly hope to carry it on, I greatly desire to extend it, and with this object I now make application to the Carnegie Trustees.

It will be understood that for work of this nature the support, to be effectual, must be in the form of an annual subsidy. As the work is carried on in the neighbourhood of Cambridge, no expense is needed for Laboratory accommodation, the laboratories of the University being available.

I estimate the sum required for carrying on the work as contemplated at £600 a year. Of this, £200 would be used for the payment of a scientific assistant; £150 for the payment of labour, the remainder being spent on hire of land (4–6 acres), food of experimental animals, purchase and maintenance of appliances, such as greenhouse, incubators, rearers, sheds of a simple character, &c.

Accompanying this letter are sent:

(1) Report to the Evolution Committee of the Royal Soc. No. 1, by W. Bateson and Miss Saunders.

(2) Mendel's Principles of Heredity, by W. Bateson.

The Carnegie fund committee was "snowed under" with applications, and this was not "a lucky number".

At one time the Secretaryship of the Zoological Society seemed likely to offer a favourable opening. Even before our marriage his friend and "patron", as he liked to call

[1] Under-estimated, for he only counted current out-of-pocket expenses. We had put our small capital into the house and grounds, for the sake of the work.

Professor Newton, had the idea that this was the very post for his *protégé*. For some years we vaguely looked forward to such a solution of the problem of how to live and yet work. There was some suggestion that the Society might found an experimental farm, which would provide material, and ample scope for research work in Will's line. But it became obvious that this post involved far too much attention to business matters and organisation to allow time for his interests. When it fell vacant, in 1902, I was away from home. I received the following notes from him on this subject:

Re SECRETARYSHIP OF THE ZOOLOGICAL GARDENS

17. 11. 02.

I had begun to feel almost rid of this Zoo nightmare, and now your letter brings it back. Looking at the thing fairly I don't believe I could do good work and the Zoo too. As I told you, Lankester thought the Z.S. would expect their man to give his whole time, and I believe he would have to, if he was to be a success. *It would never do to grudge every day given to the Zoo*, and that is just what I should feel. These last few days I have been posting up chicken notes and I find I have really been neglecting them a good deal, and if I had stuck to them closer I should have got more out. For instance, we ought not to have killed that last S.G. [silver-grey] cock, he had a most valuable point—now buried past [? recovery] in the usual "sepulchre for fowl". I have a feeling we shall get the Carnegie money, and if not that some other, and day by day I feel clearer that my proper place is on the land. I feel sure we are on a splendid line and the next few years are *the* years. If we are to get the first crop off this work it must be done then.

So my feeling is, unless the Z.S. *offers*, which I don't think at all possible, I shall not go in.

Taking the most favourable view, it would be two years before I got a good hold of the Z.S. work. Then a year to prepare for the start of the Station—then the first year of work there—so it must be at least five years before we should begin to get results if we stopped our things off now. And, remember, even if we got our station all started, there would be no serious *income* to be got there and I should have to keep on with the Z.S. to earn my living.

All things considered, I am in an exceptionally good position now—enough to live on, and plenty of leisure—and there is always the chance of an endowment for the work.

As regards London, when my year on the R.S. Council is finished I needn't go up so much, for there would be no reason for me to go on the Z.S. again—at least, unless the Station scheme demanded it. In my present humour I wonder we have taken the question of the Zoo so seriously!

So there are my views!

W. B.

19. 11. 1902.

... I had a few words with old Sclater yesterday, and elicited that he considers that for six months, Jan. to June, the Secretary must practically live in London. He says he has generally had a furnished house in London. I feel less and less keen to succeed him.

I was talking to Sedgwick in the train yesterday about Foster's successor,[1] and *he* was good enough to say he thought me suitable. This is a strange coincidence. But I can't think the Society would ever take two secretaries from such identical *provenance*. As to land,[2] I am probably on the way to take a rough $\frac{1}{4}$ acre of Pearce's, which would go far to supply us this next year. You indeed seem to be taking it easy! But I confess, by some extraordinary accident, that I did not come down till 9.30 this morning myself. There's for you!

W. B.

The prejudice, that obstructed and retarded the free development of his work, affected his position personally very little. In University as in College affairs he made his

---

[1] Sir Michael Foster was then about to resign the Secretaryship of the Royal Society.

[2] We had not room that year in the garden to sow all our seeds.

mark. His opinion and criticism, expressed with uncompromising directness and faithfulness to his ideals, were never unconsidered. Appreciation of him was not wanting. Had he been so minded, he could assuredly have cut a figure in mundane affairs, for besides his high standard of probity and charity he had also great quickness and was very shrewd. But his interest in matters temporal was only incidental. His heart was set on science. For worldly distinctions he cared not at all, except in so far as they implied recognition of his work, and so, "by good luck", increased facilities.

Such an occasion was the invitation of the Royal Horticultural Society to him,[1] to represent the Society at the International Conference on Plant-breeding and Hybridisation, held in New York in the autumn of 1902. Though anxious about leaving the home-work he was greatly pleased. Ten days in the States were spent with his usual zest. He wrote in delight:

25. 9. 02.

The Brooklyn Institute warns me to expect an audience of 300–500, thoroughly accustomed to the highest class of lectures!...am in for an awful rush now, and probably shall have no time for writing. Saw *black* as well as grey squirrels in Central Park! The ethnological collection, the fossil vertebrates simply astounding. But perhaps the organisation of the Museum beats them even. The telephones, the typing, the files and general electrical "double million magnifying gas microscopes of Hextra power", with which everything is sorted and dealt with, leave one's poor head whirling. It seems to me if I lived this pace for a fortnight I should be like milk that has passed a separator. W. B.

And later:

3. 10. 02.

My own performances are over, and I believe passed off well....It has been a really profitable journey, and it is

[1] See letter 27 August 1901 (p. 66).

really rather foolish not to stop another week....At the train yesterday, many of the party arrived with their "Mendel's Principles" in their hands! It has been "Mendel, Mendel all the way", and I think a boom is beginning at last. There is talk of an International Association of Breeders of Plants and Animals, and I am glad to be right in the swim....

And in the last of these letters:

I am accounted somewhat here! whether on false pretences or not I don't clearly see. I am really rather taken with N.Y., and if they offered me a research farm and £200 a year, would you emigrate? I don't feel a plain "no" in my heart, but the offer, so far, has not been made!

Goodbye, w. b.

Soon after his return (October 1902) from this Conference the preliminary offer of a post, with a hearty invitation to settle in Brooklyn, came unofficially through Dr A. G. Mayer. For days we walked about weighing the *pros* and *cons*; whatever our occupation, thinking of nothing else. We decided to remain. He wrote:

Dear Dr Mayer,

I did not at once write in answer to your letter because I did not wish to say anything until I had thoroughly got used to the idea, and had looked at the proposition from many sides. The receipt of such a letter I need hardly say has been a cause of very special gratification to me. As you know, I was deeply impressed with the opportunities now being offered for scientific work in the United States, but I had no idea that such an opportunity might soon be so generously offered to myself.

Though I am pretty firmly rooted here, I think I am cosmopolitan enough not to decline a handsome offer from the mere inertia that most people call love of their country. But being here, in the midst of a going concern—if slowly going— I see all sorts of risks that would follow a great breach of continuity at my time of life, and before incurring those risks I am going to make a serious attempt to get, over here, the kind of

opportunity I want. Since my return I have spoken to several
people able to judge, and from what they say I am not altogether hopeless of starting an Experimental Station in England.
If the resources for such a scheme are not forthcoming, then
perhaps my thoughts may turn in earnest towards emigration !

You must understand and express also to Professor Hooper,
how greatly I am moved by the evidence of sympathy with
my work and objects, that your letter has given. It will not
be long, I trust, before I see Brooklyn again, if not as an
emigrant, at least as a visitor.

Yours truly,

W. Bateson.

In 1903 the garden at Wisley was left to the Royal
Horticultural Society and the Secretary, Mr Wilks, invited
Will to send up a scheme for making a scientific Depart-
ment of this splendid gift.

Merton House, Grantchester, Cambridge,

25. 9. 03.

Dear Mr. Wilks,

Thank you for asking me to write to you as to Wisley.
My hope, of course, is that this may be the long desired chance
of starting a properly equipped station for the study of breeders'
problems by scientific methods. None of the American or
Continental stations do what is wanted. They are all *bound*
to try for immediate results of utility, and the permanent
importance of their work suffers greatly thereby, as several of
them admit.

The kind of ground-knowledge which both the naturalists
and the practical plant- (or animal-) breeders need can only
be got by sticking to a few simple cases and following them
out minutely for a period of years.

Three or four species would be quite enough to begin with.
As soon as the heredity-rules had been thoroughly mastered
in a few cases, there would be material for the construction of
a real science of breeding. The extension to other cases is
comparatively simple.

There is not the smallest doubt that this can now be done. The methods of work are clear. The consequences to biological science, and I believe to all industries concerned with breeding would be incalculably great.

This is the one really large field of research ready to hand which is unexplored. No one has yet gone into it with both the proper training and adequate resources. We are trying to do it here, but we are obliged to work on a very small scale.

You will doubtless have many suggestions as to the utilisation of this great opportunity, but it seems to me that to lay the foundation of a comprehensive and precise knowledge of heredity and variation would be a work worthy of the Society, and in the end gain the approval both of naturalists and practical breeders of all nations.

You will know best what likelihood there is of any such suggestion being entertained, but if this is at all probable, I should like to go into details.

Meanwhile of course everything turns on the question of available funds, after the upkeep of the garden has been provided for—I assume that to be a first charge.

As the ground is free the chief expense would be for labour, and the payment of the experimenter, who for the purposes contemplated must be a scientific man. He must certainly live on the spot, and, I imagine, if he devoted himself to the work it would not be possible for him to earn anything considerable in addition to his stipend. It would not do for him to attend merely when the plants are in flower. In my own work I find it practically necessary to be almost always here and this would be still more essential in the case of the larger operations which an institution should undertake.

There must also be some green-house accommodation, and it would be desirable that one at least of the species investigated should be a winter-flowering form (such as *Primula sinensis*) so that the work could go on all the year round.

The work, if on a large scale, cannot in the busy times be done by one man alone. For instance, even here, in recording my peas and sweet-peas when in flower, I am obliged to get some help; also in harvesting the seed, or the work could not be done in time. In Cambridge one can get such help from students, but at Wisley it would have to be given by one of the more intelligent gardeners. I do not know if such a programme as this is at all feasible. If it is, let us consider the details. The expense seems the chief difficulty, but I am not at all sure that it would be impossible to collect the funds if the objects were understood.

Yours truly,
W. Bateson.

The Society at the time was building the new hall in Vincent Square, and was not prepared to embark on further expenses; and Wisley was not developed on the lines suggested.

Towards the end of 1903 we were strained to our utmost. From season to season the work had inevitably expanded, whilst our original confidence that help must come for work of such interest and importance was whittled away. We were launched in an endeavour that yearly became more expensive and exacting, and we had to face the fact that we could not "carry on".

He had now to reckon what parts of the experiments could best be curtailed, what discarded and what postponed; but he could not bring himself to the point of really drastic action.

It was throwing away too much hope and cancelling too much past work. Discouraged we were, and our thoughts turned again to emigration. But in the midst of our deliberations most generous and timely help came. On 17 December he received the following very charming and welcome note from our friend, Mrs Herringham:

40 Wimpole street, W.
Dec. 16, 1903.

Dear Mr. Bateson,

If I died and left you a legacy for your scientific work you would be very pleased—with the legacy. If a person doesn't intend to leave legacies, may that person give gifts for similar

sorts of purposes,—which means, could you allow your work
to be helped in this way? Possibly you know someone to whom
it would be a lift to take a salary to work for you. I have been
thinking of this for a long time, since I asked you last summer
what sort of help you wanted.

This is *private*.

It seems a pity that you should not be able to get on fast
with this work—and that you should hanker after America.

If you are not pleased with my effort at patriotism, just
forget. What I meant is about £100 a year for two or three
years. I know that isn't as much as you want. Could it help?

Yours very sincerely,

Christiana J. Herringham.

Of Will's answer I have not a copy, but of his gratitude
I have a very lively recollection. It was not a solution
of our difficulties, but it was a reprieve. We could not
only ask for help with the work, but we could pay for it;
in two years' time *something* might turn up, and after all,
it was too soon to begin to worry seriously about the boys'
education. An arrangement was made by which he could
draw £150 a year for two years and Mr R. C. Punnett
was written to within the week—a small salary offered.

Merton House, Grantchester,

25. 12. 03.

Dear Punnett,

I am writing to ask if you will entertain the idea of
coming into partnership in my breeding experiments. It has
long been clear to me that much more could be made of them
if another observer came in. They are seriously undermanned
at present. Only the barest recording is done and we have no
time for more. For example, I have to leave the whole question
of sex, as you know, and the fowls are admirable material for
that.

We seem to be interested in very much the same problems
and if you begin breeding, I am sure you will find, as I have,

that it is not a one-man job. We could do far more in combination than separately.

The matter has become urgent for two reasons:

(1), My wife has hitherto done a large part both of the recording and of the many menial operations that such work involves, but I am sorry to say she will be more or less incapacitated this next season, so that help in some form or other I must get.

(2), A benefactor, who wishes to be anonymous, has provided a material contribution towards the work—£150 for 1904, with a very good prospect that the same gift will be repeated for some years more. I cannot *guarantee* the continuance, but I have confidence that it will continue.

My expenses this year were about £90, and it was a cheap year. I made about £30 by sales and produce consumed, and I had a R.S. grant £36, so I only lost about £24.

For next year in addition to the £150 I have a British Ass. grant £38. As I shall not be able to spend anything substantial on the work in future I propose to use £70 of the £150 for expenses. This leaves £80 which would be offered to a partner, for he would have to forego some remunerative work, though if you came in you could of course retain Newton's demonstratorship.

The alternative would be to get a clerkly or gardening hand; but I would far rather bring in a scientific man with full partnership, though it should be understood that for both of us there will be a good deal of merely *menial* work to do. This menial work is nevertheless sufficiently responsible, and while a responsible paid assistant would cost too much, such work, on the other hand cannot be trusted to a lad.

I am not worried with thoughts about "priority"—and all that—nor, I imagine, are you; and I think we should not have any friction in working and publishing in common. There have been some very successful partnerships on these lines— e.g. Lawes and Gilbert—and I don't see why we should not do as well.

Difficulty of that sort might ensue if a young beginner came in. He must needs do independent work for Fellowship, etc., but with a man like yourself who has got his Fellowship and has already an independent reputation these questions would not arise. In future we might have to ask for grants. I think I mentioned the possibility that the Quick Bequest (at present £250 a year) *may* be got for our purposes. In that case we should start a small "station"—zoological and botanical—at the Botanic Garden as well. The chance of getting the Quick would be increased if anything so substantial as our partnership were already established. I believe it is not merely a sanguine temperament that makes me think that this will grow into a big thing.

I may add that I have talked the matter over with Sedgwick, and he approves my scheme with some enthusiasm.

If you are favourably inclined several points of detail will have to be considered, in particular the possibility of starting a new set of experiments. I am going abroad to-day but expect to return on Wednesday. Yours, W. Bateson.

By the New Year all was arranged. Mr Punnett joined with enthusiasm, and very generously refused the said salary. Village lads were engaged and taught the rougher "menial" work. So far from closing down, Will was at once busy devising new experiments and extending the scale of others.

This partnership was highly satisfactory, and lasted unbroken to the date of our removal from Cambridge (1910), when Will took up the Directorship of the newly founded *John Innes Horticultural Institution* at Merton.

For long he had missed community of interest in his work, which was still destined to be looked at askance, but now he had a fellow-worker. He felt again the delight of having "some one that you can talk to" about his experiments, which in spite of all extraneous worry proceeded merrily.

Our third son, Gregory, was born in May of this year 1904.

The Quick Bequest to which allusion is made in this last letter was a splendid legacy to the University of

88   MEMOIR

Cambridge "to promote the study and research in the sciences of vegetable and animal biology". This at last seemed to be the coveted opportunity. Will was invited to attend a meeting of the Special Board for Biology and Geology to consider the scheme drawn up by the Council[1] of the Senate for the administration of the Bequest. (The paragraph alluded to in the following letter was a suggestion that one half of the income in any year should be devoted to the furtherance of original Biological Investigations conducted under the control of the University by grants made to persons engaged in such investigations, by the provision of apparatus, or otherwise.)

This piece-meal use of the Bequest however was not at all in keeping with the Quick Trustees' ideal. They naturally desired that the bequest should be kept entire to found one Chair of real and undoubted importance. Of this, in writing the following letter, Will was not yet aware.

THE QUICK BEQUEST, Nov. 1903.

19. 11. 03.

Dear Professor Newton,

It is impossible for me to attend the meeting of the Board as my audit[2] takes place at the same time. I venture therefore to write to you on the subject of the Quick Bequest.

The income shortly to be received is, I understand, about £500. If par. 6, § (1), is agreed to, which I assume to be likely, there will be an income of about £250 available for research.

The proposal I wish the Board to consider is that this part of the endowment be used for the constitution of a small department or station for the study of problems involving breeding experiments with animals and plants. The problems I have in view are primarily those of Heredity and Variation, but there are, of course, other subjects of research demanding similar facilities. The methods to be followed in these researches

[1] See *Cambridge University Reporter*, 3 Feb. 1903, p. 426, and *ibid.* for the Special Board for Biology and Geology, 9 Nov. 1903.
[2] Of his College Stewardship.

are now well developed, and results branching out in many different directions are rapidly accruing.

A considerable number of students are now devoting themselves to these inquiries. Each year the number increases, and it is not doubtful that this branch of science must form a recognised part of the studies of a great biological school. Owing however to the special exigences of the work, it can, in the present circumstances, only be carried out with great difficulty. Land and labour have to be engaged, and consequently many who desire to take part are unable to do so. It is obvious that occasional grants are inadequate for these purposes, and that some permanent installation is absolutely necessary. Those who, like our students, are able to give a few years research after graduation, cannot take land or buildings for themselves. We require some continuous organisation which shall do for them what the laboratories do for other branches of study.

The pressing requirements are land, labour, one or more green-houses and a building or sheds of a simple character to serve as potting-shed, seed-store, etc., and also for the breeding of small animals such as rabbits, mice, lepidoptera, etc.

I believe that the most economical and efficient way in which a beginning can be made is by devoting some of the University Allotments (now let on annual tenancies) adjoining the Botanic Garden to this purpose. The Botanic Garden Syndicate, with great generosity, have already granted one such Allotment to Miss Saunders, rent-free, as an experimental garden, but they could not afford to dispense with the rent of a larger piece. For want of more land and green-house space, even experiments in progress have had to be seriously curtailed, and extension of the work is at present quite impossible.

A part of the endowment would therefore be required to pay rent for the ground to the Botanic Garden. Though this land is dear (£20 an acre) I do not think any position, so advantageous, can be got for much less. The gain of proximity to the Garden would be so great as fully to counterbalance the expense. Though financially and administratively independent of the Garden, it would be possible to work the experiment-station much more efficiently and easily if the two were in conjunction. For instance, a most pressing want is a green-house like the present cool-pit. Mr. Lynch tells me he has no doubt that if such a pit were built beside those now existing the one boiler would suffice. In this case the experiment-station could then be charged for its share of labour and coals, and no separate boiler or attendance would be needed. Experience too, has shown that the help derived from association with Mr. Lynch and his staff is quite invaluable.

Many of the experiments contemplated would in their later stages demand more space, and therefore, cheaper land. In illustration I may mention those of Mr. Biffen on wheats, and my own on poultry. These must of course be conducted elsewhere; but in each case the garden scale precedes the farm scale.

I anticipate no difficulty in arriving at an equitable distribution of the endowment among the various researches, which might be pursued in various places according to their several needs, though the maintenance of the central station would be the primary object. It may be pointed out that the work in question is almost unique, in so far as it pertains directly to all the great divisions of biology, not only to zoology and botany, but to physiology also, for inheritance is a physiological function as much as respiration or digestion. It affords moreover abundant scope for research by a great diversity of methods, microscopical and otherwise.

No station exactly comparable as yet exists. The experimental stations of the United States, Canada, Germany and Sweden are prevented from doing any considerable work by their direct attachment to commercial objects, but I believe that the need for such an institution as this is widely recognised, and I am hopeful that if a modest beginning were made, opportunities of extension would follow. The Quick Bequest, gives us, I think, the chance of taking the lead in an important departure in biological Science.

I am,
Yours truly,
W. Bateson.

To his surprise, immediately after the meeting this letter was handed back to him; he was subsequently told that no comment was made on his proposal, except for a suggestion (as an alternative) that the money should be used in grants for researchers, without specification of the work in Heredity and Variation.

This rebuff he swallowed as best he could. He decided to allow himself no further interest in the matter.

Though in later years he was often accounted a pessimist, he was sanguine by nature. Disappointment and some disillusionment, in the end, taught him to be wary in giving expression to his natural hopefulness, but in these earlier days he still could not believe that England would not maintain her lead in biological science. The claims of the study of Heredity and Variation seemed to him paramount—indisputable. Nevertheless in his optimistic faith that an Experimental Station for the study of "Genetics", somewhat on the lines he had sketched, would be founded, he was doomed to disappointment. He desired to extend the compass of the work, to carry out his investigations on broader lines and a much wider scale than his own modest means permitted; he craved to advance more rapidly in experience and knowledge. But greater still than this ambition was his anxiety to see the work adequately endowed and placed on a permanent basis. A lifetime was as nothing for the vast field of research that he was opening out. It was a kind of nightmare to him that, "if anything happened to him" the work would not be carried on. He saw in it the likeliest means of advancement of pure science, and in this aspect lay his interest; he saw also that such knowledge if pursued must ultimately lead to amelioration as yet undreamt of in the conditions of human life of every grade. But he himself was entirely engrossed in the purely scientific aspect and interest of the work; he was bored rather than otherwise by the practical application that already could be made of the knowledge that he was acquiring, and trying, in the face of bitter opposition, to teach. He feared that commercial interests in the practical application would bring into the research the wrong kind of worker, those, namely, whose motive was gain and profit instead of quest of knowledge.

The income of the Quick Bequest did not fall to the University until two years later, when its destination was again the subject of discussion. The bestowal of this great benefaction was a matter of keen interest to many branches of biological study.

Incapable of making any step that might be construed as self-interest, and still remembering the fate of his proposition in 1903, Will declined to serve on the Board appointed to decide the affair, subject to the Trustees' approval.

But in February 1905 he was asked by the Vice-Chancellor of the University to draw up a scheme for a "Quick Institute for the study of Heredity and Variation", an invitation which revived the hopefulness natural to him. He drew up a long, careful, and very modest scheme, embodying, with fuller detail, most of the ideas set out in the letters already quoted. Incidentally he made this forecast:

Before long, moreover, it is easy to see that there will be a demand for men skilled in this branch of knowledge and it is certain that there will be students wishing to participate in the work of the Institute for shorter periods during their University course.

His endeavour and his foresight passed unheeded.

The shortage to-day of well-trained men, equipped in sound scientific method, to fill the many Agricultural posts at home and abroad is an interesting commentary on the disregard and neglect which his efforts met with at this time.

All through these negotiations, when alluded to at all, his science was called that of "Heredity and Variation".

It was a tiresome mouthful, and in answer to some comment from Professor Adam Sedgwick, this note was written:

> Merton House, Grantchester,
> 18. 4. 05.[1]

Dear Sedgwick,

If the Quick Fund were used for the foundation of a Professorship relating to Heredity and Variation the best title would, I think, be "The Quick Professorship of the study of Heredity". No single word in common use quite gives this meaning. Such a word is badly wanted, and if it were desirable to coin one, "GENETICS" might do. Either expression clearly includes Variation and the cognate phenomena.

The conditions, which I understand the Trustees contemplated seem to me admirable.

If the Professorship came to me, I should naturally look forward to extending operations considerably and I am hopeful that means for doing this would be forthcoming, once a Professorship were instituted.

> Yours, W. Bateson.

Eventually, in 1906, the Quick Professorship in Protozoology was founded.

This was a bitter disappointment.

To a colleague Will wrote:

Heredity and Variation failed, as proposals generally do, not by reason of opposition, but for want of support. I am responsible in so far as my strength has been insufficient to bring about already the revolution in biological study that sooner or later must certainly come to pass. By this time I ought to have been able to make it obvious that such an endowment should be given to "*Genetics*". As I said, I have failed with my contemporaries; with posterity I hope to be more successful.

Meanwhile with the growth of work that followed the

[1] This letter dates the first use of the term Genetics, and defines exactly what he meant to express by it.

partnership formed with Mr Punnett, the need for green-house accommodation became acute.

The suggestion was made that he should apply to the Managers of the Balfour Fund for help in the matter. I have this draft of his application:

The propagating house is wanted primarily for a number of inquiries now in progress, undertaken by several persons in conjunction with me, though in great measure it would be available from time to time for other researches also, provided the space were sufficient.

The common object of all the sets of experiments with which I am especially concerned is the elucidation of the nature of differentiation between germ-cells, to detect such differentiation in as many cases as possible, and to work out the laws it obeys. The outlook on these problems has been so completely changed in consequence of Mendelian discoveries that it is most difficult to explain our present aims and methods without beginning at the very beginning, which, on this occasion is of course impossible. Taking the case of *Primula* as a type, I may however be able to illustrate the nature of our work, shewing what is already done, and what we may hope to do.

Whenever sexual organisms breeding together produce a mixture of forms, there is, in the light of what we now know, *prima facie* reason to suspect that the mixture is due to differentiation of germs. The most familiar case is that of sex itself. A population consisting of males and females has so many features in common with the differentiated offspring resulting from the segregation of characters among the germ-cells of cross-bred organisms that it is impossible to avoid the suspicion that the two phenomena are similar in causation. A categorical proof of this conclusion would make a remarkable advance in biology.

Various lines of experiment are begun which may or may not provide such a proof. Certain complications make it peculiarly difficult to close in on the problem of sex, and many happy accidents must occur if the work is to succeed. We have therefore meanwhile turned to the case of Heterostylism, a phenomenon notoriously akin to sex, where these especial difficulties do not arise.

Every one knows that in many ☿ plants there are two or more types of individuals each bearing flowers of a distinct form. The commonest examples are the *Primulas*, with their *pin* and *thrum*-eyed types of flower.

Darwin and Hildebrand, as is well-known, shewed that the number of good seeds is greater when the two types are crossed "legitimately", than when either is "selfed", or fertilised by its like, "illegitimately". In several species of *Oxalis*, *Linum*, etc. illegitimate fertilisation gives no seed at all.

There the matter rested. But in Mendelian lights the most significant discovery made by the earlier work is the fact that the pin type breeds *true* to *pin*. This was naturally treated as a subordinate observation by Darwin and Hildebrand. Nevertheless it is the clue to at least half the problem.

Pin, in fact, as we have been able to shew, is an ordinary Mendelian recessive to *thrum*, and the pollen and ova of any plant are differentiated as bearers of the one type or of the other. This is in accordance with horticultural evidence. It has been held that *pin* is the right florist's type for *Sinensis*, and being recessive that type was *at once* made pure by merely saving the pin plants. Thrums exist nowadays only in strains from cheap seed, where quantity is the main object.

But *Auricula* and *Polyanthus* for some 40 or more years have been prohibited in the *pin* form. No self-respecting florist keeps pin auriculas. Nevertheless, as far as I can learn, no such thing as a purely *thrum* strain of auricula exists, or ever has existed. This fact points to a very important conclusion. Though pin, being a pure recessive, comes true from the first selection, thrum never does. But since thrum and pin are differentiated in gametogenesis, it follows that thrums, in forming thrum germs, must also be forming pin germs. Pin is thus a by-product of thrum, standing to it in the relation which de Vries calls "Halb-rasse", (a phenomenon to which the case of sex probably is an exact parallel). Therefore we find that thrum "selfed" or × pin always gives a mixture of pin and thrum.

The next step is to determine the statistical ratios in which the two forms are produced, work with which we are now occupied.

To return to the question of the meaning of legitimate and illegitimate fertilisation. The fact that more germs are fertilised when contrary pollen is used means almost certainly that here another process of differentiation has taken place, segregating germs destined for contrary fertilisation from those capable of fertilisation by their like. If we succeed in demonstrating this the "advantages of cross-fertilisation" come to have a definite physiological meaning.

The question can be tested in certain ways which could only be made clear at some length. Briefly, we use a *double* fertilisation, applying simultaneously pollens from two plants pin, and thrum. The ♀ must be white, and *one* of the ♂s must be white, and the other red (pure). Red being dominant over white, the number of reds which come in the offspring will shew how many egg-cells are fertilised by the type of pollen borne by the red, and the difficulty that thrum is giving off both pin germs and thrum germs is thus eliminated, or at least minimised.

This method was only devised last year and we do not yet know the results with any certainty.

Moreover we have a set of experiments with the much simpler case of the homo- or equal-styled type, which is well known in *P. sinensis*. This is a pure recessive to both thrum and pin.

With the latter it Mendelises in the ordinary way, being true from whatever parentage. Even were other evidence wanting, this fact would go far to shew the real nature of the relation between thrum and pin.

This is the kind of analysis we are trying to carry out. Apart from plants like *sinensis* which must be worked in a green-house, we are using many others which cannot be grown properly without one, and it is now an actual necessity, if we are to go on efficiently.

I hope this account may give you some idea of our work. Of course the bearings of the facts on the big problems could only be shewn by a treatise.

13 May, 1904. W. Bateson.

It transpired that though money from this fund could be spent on any work done inside a building, it could not be used in the erection thereof. This was very disappointing. The green-house was a crying need; but for another two years he had to make shift as best he could with what space Mr Lynch (Curator of the Botanic Garden) could spare him.

In spite of all these difficulties caused by his lack of means, discouragement was rare and not allowed to appear. His interest was so concentrated that the outside world could not distract him at work, buzz as it might.

One could not imagine, whilst working with him, that he had any trouble or anxiety greater than the sterility of a pea, or the death of a valued chick. In the garden and the poultry-pen, vexation, worry and annoyance were all forgotten; he could not be ruffled by any outside passing disturbance. He might wish for a less tumble-down potting-bench, but he was far too deeply immersed in his problems and observations to bother about the inconveniences he needs must put up with.

In the same way he could completely absorb his attention in a game of chess or bridge, and thus absolutely rest from his work, or from the friction of the latest controversy.

For, all these years controversy constantly engaged him, over and above the manifold duties of the day.

Quite simply, without ulterior motive he longed to know the truth about the natural world, and every statement, or interpretation of fact that he came across was subjected to an earnest and critical scrutiny, before he either accepted or rejected it. This passion for the truth often led to controversy. Armed cap-à-pie with information, and certain of his point before plunging in, Will was

B 7

a dangerous opponent. He was candid as only those who have no thought of self can be. Like the Knights Errant of old he took to the field, full of chivalrous and honourable ardour to defend Truth as he knew her; possibly a disappointed expectation of the same ideal of conduct in others added some sting of contempt to the battery of knowledge which he brought to bear on his opponent. There is little doubt but his outspokenness in these contests injured him in the sight of the world, to which, in his zeal, he paid so little attention. Among his contemporaries his new doctrine suffered discredit perhaps by the mere fact of dispute, and their hostility or indifference caused a bitterness clouding again and again his almost triumphant joy in his work.

Before Mendel's papers came to light, his pen warred in defence of his own doctrine of "Discontinuity". The Mendelian controversies which followed, though broken by variety of topic, were of course all parts of one whole; and they culminated in 1904 at the meeting of the British Association in Cambridge.

At this meeting he was President of Section "D" (Zoology). Of this opportunity he took full advantage. Relative positions were changed; to him the attack, to his adversaries the defence. From the point of view of science he cleared his way. Open attacks on his work and on Mendelism hereafter dwindled to mere bickering, but that did not mean that hostilities were at an end. He did not succeed in getting his work endowed.[1]

The year 1904 brought, besides the Presidency over Section "D" of the Cambridge meeting of the British Ass., another personal distinction which pleased him very much. He was awarded the Darwin medal by the Royal Society.

The need of a green-house, which he had tried to get in the spring of 1904, became more and more urgent—the

---

[1] See Address, p. 233. A rough list of the published controversies is subjoined. See Appendix, p. 464.

impossibility of providing it more and more exasperating. Then once more help came from the inner circle of intimate friendship.

Early in 1906 he received the following note:

> 13 *Madingley Rd., Cambridge,*
> Mar. 11, 06.

Dear Bateson,

I am on the track of a small sum for your work. I don't know yet how much but I am afraid not more than £200.

What I want to say now is that I am inclined to think that it would be more ship-shape (supposing that this £200 should attract some more subscriptions), if there were a Committee or a couple of Trustees, or some machinery for receiving subscriptions. Of course any one who knows would prefer to hand over the sum to you personally. But as a matter of business I am not sure that it is the best plan. I should like to have a talk.

> Yours,
>
> F. Darwin.

> *Grantchester,*
> 1906.

Dear Darwin,

Your letter gives me an extraordinary feeling of gratitude, as you may imagine.

Certainly such money should be administered by a Committee. I have leave to acknowledge my recent benefaction in the next Report, so I may say that the £300 came from our friend Mrs. Herringham.

When it was offered I suggested that it should come through the Evolution Ctee. She deprecated this, but I have never felt it was quite satisfactory. If possible I should like to see you before Evolution Ctee. meeting.

For this year we are fairly well provided, and if any such sum as you mention comes, we should be right for two years more. Apart from working expenses Punnett is rather on my mind. Mrs. H. originally offered it to provide a paid assistant.

I then negotiated with Punnett on that basis. He however refused any pay, and so we started as partners. Soon after he got the Balfour Studentship, which has a year to run. I am rather troubled about his future when this expires. He has worked with the greatest devotion. Indeed the clearing up the colour-inheritance problem has been mainly his work.

In July Mr Darwin wrote to say he had put the fund collected by him (really subscribed by him and his brothers) into the Experiments fund at Mortlock's Bank.

It is perhaps characteristic of both that as soon as Will was better off he tried to pay back this sum which had been such a boon at the time of gift, and that the repayment was not accepted.

> *Merton,*
> April, 1914.

It has always seemed to me that sooner or later I ought to repay the sum that you and your brothers so generously subscribed for my experiments, or at least that part of it (£150) which I drew out. The payment was, as you will remember, made to the Botanic Garden Experiments Fund, and when I left Cambridge the balance remained in the Fund. This happens to be a suitable moment so I send cheque for £150 in liquidation of about the pleasantest debt I ever contracted.

> W. B.

> *Brookthorp, Gloucester,*
> Ap. 21, 1914.

Dear Bateson,

Your letter has just reached me here. It is very good of you to want to refund £150, but as far as I and my brothers are concerned you are in a state of virtuous delusion. It was never meant to be a loan, but as help to Genetics to be administered by you. So you must really let me return it; and you must deal with it as seems to you fairest.

> Yours sincerely,
> Francis Darwin.

P.S. Later.

The last sentence of my letter is a muddle—You have
already spent the money on Science or something connected
with it—and all you have to do is to burn the cheque.

Generous in giving, he was yet very careful in spending.
With part of the Darwin gift he built the extra green-house
in the Botanic Garden, which had for so long been an
urgent need. By and bye, Mrs Herringham found that of
the sum she had put aside for his use about a third remained intact. He had counted this help as for two years,
and when that time limit was reached, left the balance
untouched.

She wrote to him, and he answered:

Dear Mrs. Herringham,

I can only say how grateful I am. I take it you mean that
we may draw the balance of the "£300 in two years" originally mentioned. It has been a great disappointment that no
part of the Quick Fund has come to us. At one time it seemed
almost certain that we should get part, and I quite hoped to
have got the work on to a permanent footing but—that hope
is gone. When we publish our next Report some acknowledgment of what you have done for us must I think be made.
Some of the more tangled cases are at last beginning to clear
themselves up, and without undue pride I feel pretty sure it is
a creditable piece of work.

Yours,

W. Bateson.

At first Will's desire to make public acknowledgment
of her generosity was not acceded to by Mrs Herringham,
but in the third Report to the Evolution Committee he
was allowed to insert an expression of his gratitude.

Grants received from the Government Grant Fund, and
from the British Association, have been applied in part to the
cost of the experiments upon which we now report, and in part

to other researches which will be described hereafter. These sums would have been insufficient to enable us to carry on the work, and the scope of the investigation must have been greatly reduced had it not been for the generosity of Mrs. Herringham, who in 1904 placed a fund at our disposal for this purpose. We wish to record our deep sense of obligation to her for this assistance.

The very sudden and unexpected death of Professor Weldon occurred in the spring of this year, 1906. I was away from home and received this letter:

Merton House, Grantchester, Cambridge,

16. 4. 06.

You will probably have seen the news of Weldon's death in to-day's papers. At present I feel it as a shock—in the literal sense. It must make a great change in our outlook.

I feel sure that no conduct on my part would have averted our differences, or have diminished their bitterness.

I owe a great deal to him. It was through the chance of meeting him that I became a zoologist, and afterwards through him that I got my first start on Balanoglossus.

Until the time—about 16 years ago—when his mind began to embitter itself against me, I was more intimate with him than I have ever been with anyone but you.

I rather wish I could write to Mrs. Weldon, imagining ourselves back in earlier days. Perhaps I shall, but I do not know enough of her present state of mind to feel sure that it would be taken in good part. Perhaps not writing may be equally misunderstood.

How big a disturbance this will make in our area I hardly yet know. If any man ever set himself to destroy another man's work, that did he do to me—and now suddenly to have one of the chief preoccupations of one's mind withdrawn, leaves one rather "in irons", as sailors say.

...There is much to talk of, and I shall be glad to have you back.   w. b.

And in another letter at this time:

To Weldon I owe the chief awakening of my life. It was through him that I first learnt that there was work in the world which I could do. Failure and uselessness had been my accepted destiny before.

Such a debt is perhaps the greatest that one man can feel towards another; nor have I been backward in owning it. But this is the personal, private obligation of my own soul.

His sister, Mary,[1] died in November of this year 1906 suddenly after a few days' illness. This was a tragic blow. He was very much shaken.

Mary (b. 1865, d. 1906) had the same joyous spirit and serious purpose, the same zest for life that Will had, and the same gifts of vigour and ability. She had his quick industry and devoted herself in the same simple and single-hearted way to the pursuit of knowledge.

Professor Maitland[2] wrote to Will:

Downing College, Cambridge,
1 Dec. 1906.

My dear Bateson,

I hardly know whether I ought to intrude upon your sorrow. I don't think that I should do so, if, as is so often the case, there was nothing to be said that was at all worth saying. But I should like to tell you that your sister has not passed away without leaving behind her a memorial that is worthy of her. The book that she published this summer, the "Borough Customs", is a really great book, and I have not the slightest doubt that among the people who concern themselves with the history of institutions it will long rank as a classic. They have

---

[1] Mary Bateson, 1865–1906, Historian, Author of *Catalogue of the Library of Syon Monastery, Isleworth* (1898); *Records of Leicester* (1899, 1901, 1905); *The Laws of Breteuil* (English Historical Review); *Borough Customs* (Selden Society, 1904, 1906); *Mediaeval England* 1066–1350 (The Story of the Nations, 1903); and many papers.

[2] Professor Maitland (b. 1850, d. 1906), Downing Professor of the Laws of England.

not had time to read it yet, and, as you would guess, such a book does not find its place in a month or two, nor even in a year or two; but about the ultimate result I am quite sure. I have now seen through the Press one-and-twenty of the Selden Society's volumes and there is not one of them about the future of which I am so confident as I am in this case. I saw the book growing in all its stages and I could not overstate my admiration for the qualities of head and heart that were revealed to me. Of them you want nobody to tell you—and I should not venture to speak of them to you, at any rate at this moment; but you may like to know that the book will make a very deep mark and will be often quoted and often praised when you and I are gone. Needless to say that we hoped for much more, still there has been a real achievement of permanent value; and I can find—can you even now?—a little comfort in the thought that your sister was just able to finish a piece of work which will long preserve her name.

Do not write. Why should you?

Yours always very truly,

F. W. Maitland.

Merton House, Grantchester,

2. xii. 06.

Dear Maitland,

I cannot wait till calmer times to tell you of the pride with which we read your words to-day. I did hope you would write, but I had not expected such a letter as this.

They will sing for her on Tuesday, but to me your letter is her Requiem. That is what she would have chosen.

Of the importance of her work amongst other work I know of course nothing: still for many years I have known that she was a real scholar, sane, punctilious, and broad, working in the larger hope.

In all her later work you were the inspiring force. We seldom spoke of what she was doing without some reference being made to you. "What Maitland would think" was her standard. I well remember her joyous tone one day when she said to me about some proof or other, "He thinks my stuff will do".

She has gone in full work: So may we all.

We have not touched her papers yet. There may be something left that looks in a fit state for publication. If so, I know you will let me consult you.

Yours truly,
W. Bateson.

As soon as he was clear of the immediate consequent business affairs we hurried abroad, to Germany. A little alarmed, I think, at his overwhelming fatigue he took himself in hand with energy and purpose, and at the end of a week or ten days threw off the physical depression that had almost overmastered him. We walked in the snow of the Harz mountains, and afterwards wandered through the Museums and Galleries of Northern Germany. His attention was instinctively arrested by pictures, and without effort his interest soon concentrated on the galleries of Cassel, Brunswick and Dresden. The impersonality of the Arts at once soothed and delighted him.

He was, too, very much attracted by the carved wooden figures which were then being collected out of the churches in Germany. It was a form of art that he had not before looked at; and these wooden Madonnas and Saints excited his admiration and covetousness—a pleasurable emotion to him even when it could not be gratified. At the end of about a month he returned to his experiments, lectures and College affairs, refreshed and restored.

The death of Professor Newton (1907) left the Cambridge Chair of Zoology vacant. Will was a candidate for the appointment; his application shews perhaps a wider outlook than that of 1890 but his standpoint is unchanged.

St John's College, Cambridge,
July, 1907.

Dear Mr. Vice-Chancellor,

I beg leave to offer myself as a candidate for the Professorship of Zoology and Comparative Anatomy.

I do so hoping to find an opportunity of widening the scope of the science. Zoology as taught in the English Universities has become almost exclusively a morphological study. Having regard to the history of biology it is not unnatural that such a phase should be passed through, but the consequent isolation of academic Zoology from many cognate lines of inquiry is greatly to be regretted. If the subject is to continue to attract the more vigorous students some enlargement of scope is necessary.

The direction in which extension must be made is evident. The phenomena of Variation and Heredity—the ground facts of Evolution—are at last being studied with success, by methods which are both precise and comprehensive. The experimental discoveries thus already made have sufficed to change the focus of interest in all that concerns this part of knowledge. No one who is familiar with those results can doubt that the prosecution of such studies must become a chief aim of biological science. It is on these lines that I should look forward to the development of the study of Zoology.

An Address delivered to the Zoological Section of the British Association in 1904 will more fully illustrate the nature of this departure. Of that paper I enclose copies.

Since 1899 when I was first deputed to lecture for Professor Newton, I have constantly endeavoured to interest my class in Genetics—to use the term by which this branch of enquiry is now designated. I have found no lack of eagerness on the part of the younger generation to engage in such researches where progress is definite and rapid, but they do so under exceptional difficulty while these pursuits are not incorporated in the regular curriculum, and almost all facilities have to be provided by private effort.

If I came to occupy an official position such as that which the Professorship of Zoology would confer, I should at once endeavour to get proper equipment for these investigations. The establishment contemplated must begin on a modest scale, and in view of the extraordinary value of the results, practical as well as scientific, which might be obtained, I believe that means sufficient for this one object would be forthcoming.

In various other directions zoological science has been lately extended by the introduction of novel methods. I have especially in mind the researches into the mechanics—or preferably the physics—of development and regeneration now actively proceeding in Germany and the United States. It is very desirable that work of this kind should be instituted in England where as yet it is unrepresented. Cytological work also of a most suggestive character is in progress with the object of elucidating the nature of cell-division and the relation of these processes to the physiology of inheritance. Though I have not personally taken part in such investigations I fully appreciate their value and I should do all that I could to promote them in Cambridge.

It is a fortunate circumstance that to the solution of genetic problems all classes of biological evidence may alike contribute. Upon them the experience of physiologists, systematists, collectors, and breeders can equally be brought to bear, and I am confident that a school of Zoology, combining all these varied lines of interest, could be built up. Such a school may be expected to accomplish a considerable work in science and to become a source of strength to the University.

I am,

Mr. Vice-Chancellor,

Yours faithfully,

W. Bateson.

Mr Adam Sedgwick was elected. Will congratulated him:

Merton House, Grantchester,

23. 7. 07.

Dear Sedgwick,

I have just heard that you are elected Professor. You must believe me when I say that this is the result I wished for. As no doubt you know, I eventually decided to be a candidate. To have done otherwise was scarcely possible without treating the things I most care for with public disrespect.

You have received a distinction very fully earned, and I feel as I have done with regard to few appointments of late, that the right choice has been made.

Yours ever,

W. Bateson.

The first suggestion that he should give the Silliman course of lectures in 1907 at Yale came to him through Professor Osborn early in 1906. He was at the time busy with *Mendel's Principles*,[1] and almost declined the tempting and to him flattering invitation on the score of not having enough material to hand for two books.[2] He had an intense horror of repeating himself and a dread of having "nothing fresh" to contribute. I do not think that he ever accepted an invitation to lecture without surveying carefully what he had of new interest to say; and now he hesitated, though as pleased as a boy with his first prize to be asked.

He wrote:

Though the work is developing very fast, I doubt if—so to speak—the cistern would be ready for a really fresh discharge again so soon.

I expect you look on the book as the main thing, and the lectures as secondary. However were you to make any proposal for the giving of the lectures I would do my best to carry out this part of the plan. I am a very slow writer, and it would take me a long time to make the manuscript, even after the lectures had been fully prepared, for I never lecture from verbatim copy....

The matter was arranged and combined with the 7th International Zoological Congress, held in Boston, and with many other engagements to lecture.

[1] Cambridge University Press, 1909.
[2] According to the terms of the invitation the course of lectures was to be published in book form.

He left England at the end of July 1907, and made a most successful tour, returning in November.

Marine Biol. Laboratory. Woods Hole. Mass.

23. 8. 07.

These have been wonderful days, but so full that letter-writing has been impossible. I only retain faint control over my movements. I am everyone's prey, being torn to pieces by my admirers. No exaggeration, I assure you !

Not so hot here and mosquitoes comparatively few. My first lecture here was a sad fiasco. The lantern would not work properly, but as it did not actually fail, the fault seemed to be mine. The thing was got right before the next night, when all went splendidly. I had a record attendance, and great enthusiasm. Including students (now leaving), there were some 100 scientific people here, many of whom I know something of.

No drink may be sold in Woods Hole. I would give 2/6 for a bottle of beer. The tyranny of religion and temperance is constantly making itself felt in U.S.A.

Servants are rare. My shoes have not been touched by a brush since I left the ship.

Boston,

23. 8. 07.

My address just over. Really a big success. I had a great crowd and was quite overwhelmed with the enthusiasm.

What tricks conditions can play! I have said all that often to empty benches!...

En route. (Boston to New York.)

24. 8. 07.

At last Congress proper is over, and in this sumptuous train I sit down to write of our doings. From our point of view the meeting has been a stupendous success. Heredity, Cytology, and experimental Zoology have kept the whole Congress. Nothing else has had any hearing worth the name. I never before felt what an exhilarating thing it is to speak to a really large and enthusiastic audience. It is dreadfully intoxicating !

How I shall settle down to the comparative humdrum of Yale I don't know. I fear the anti-climax a good deal. The Americans are rather absurd in their hero-worship and one has continually to remember that they keep a constant procession of heroes on the march. But after the years of snubbing it is rather pleasant to get appreciation even though in an overdose. I go about like a queen-bee on a comb, and I should not be surprised to see some admirer backing out of my presence. . . .

He wrote home almost every day and described gaily all that he did, all that he saw, and all whom he met. The intervals between his lectures were crowded with seeing the country (chasing butterflies) and seeing the Museums.

Cold Spring Harbor. Long Island. N.Y.

Thursday. 8. 8. 07.

The faithful Mayer met me on the dock. After 4 hours in New York, the heat and row of which I shall never forget, he insisted on bringing me straight here. We reached the Railway station about 4. and then walked the three miles here—great heat, but walk delightful—through woods. *Danais plexippus*, the big brown milk-weed butterfly actually flew on board to welcome me in New York harbour. It gave me a jump, and I thought how delighted the boys would have been. On our walk we saw *Papilio Troilus*, a huge black Papilio and soon Mayer found the larva, in a folded leaf, with a great thick head, like a snake. As soon as I get an hour to myself I will go for butterflies.

He wrote a characteristic note to his son, John, who made his entry into school-life whilst his father was still abroad.

Middlebury, Vt.,

28. 9. 07.

Dear John,

I have just got letters from your mother, from which I hear that by now you must be a real school-boy! That is a great stage reached in life and I hope that you have made a good beginning and are working hard. The great thing is to go your own way and not to bother what other people think or say about you. If you are satisfied that what you do is right, you can't go very far wrong.

I am afraid butterflies are quite over—at least for the present. I saw a "tufted grouse" in the woods. It was close to me, such a fine big bird. I have seen three kinds of squirrels in U.S.A.

(1) The big grey squirrel, which is common in all the parks.

(2) The Chipmunk, which has longitudinal stripes.

(3) The red squirrel, very like our own.

There are bears in the woods near here and very rarely a wolf is seen, but they really wander in from Canada.

One hears a good deal of Canadian French here. It is funny stuff and absolutely unintelligible to me. I only made out with great difficulty that it was French at all. It has changed far more from French than English has from English.

> Yours ever,
>
> W. Bateson.

There was however no anti-climax when he came in October to give his lectures at Yale. He enjoyed this visit to New Haven perhaps more than all the rest of his lecturing tour.

The Silliman lectures were eventually published in 1913 in book form, under the title, *Problems of Genetics*.

The truth was that he had to attend to so much practical work with his experiments that he never was at his writing table for long at a time, and it was difficult after long days broken with every sort of interruption to settle down to the book. He also was always much more eager about what he saw there was to do, than about what was already done.

About the year 1906 we found at Happisburgh, on the Norfolk coast, a solitary inn, and thither he often went for a few days' quiet with his manuscript; but these little intervals of seclusion he felt to be stolen and he would only stay away for a few days at a time.

The anti-climax that he feared to meet at Yale awaited him at home. On his return to Cambridge he was offered a Readership in Zoology (not even in his own subject of Genetics), at £100 a year. His first impulse was to refuse; after much persuasion from Professor Sedgwick and others, he reluctantly accepted the appointment; humiliated, he resolved to leave Cambridge as soon as practicable.

But he held this Readership for a few months only.

Once more in the hour of need help was forthcoming.

On 24 February 1908 the Council of the Senate reported that they had under consideration

a generous offer made to the University by a Member of the University who wishes to remain anonymous. It has come to his knowledge that there is a desire on the part of the biologists of Cambridge to celebrate in 1909 the centenary of Darwin's birth and the jubilee of the publication of the *Origin of Species*, by founding a Chair of Biology, the occupant of which shall devote himself to those subjects which were the chief concern of Darwin's life-work.

Convinced of the great importance of the subjects with which such a Professorship would be concerned, and of the peremptory need that such subjects should obtain immediate recognition, the benefactor offers, under an arrangement approved by the Financial Board, to pay to the University £300 a year for five years, provided that the University establishes for that period and before June 30, 1908, a Professorship of Biology of the minimum annual value of £500. The donor also offers to increase the £300 to £400 for any portion of the five years during which the Professor may be holding a professorial fellowship. The further condition is made that it shall be the duty of the professor or professors elected during the period of five years above mentioned to teach and make researches in that branch of Biology now entitled Genetics (Heredity and Variation).[1]

I do not remember clearly when rumour of this anonymous benefaction first reached him (possibly he waited

[1] *Cambridge University Reporter*, 3 March 1908, 632.

for official confirmation before risking to disappoint me),
but the interval between dejection and gratification cannot
have been long.

Bad luck, or good luck (as he had it), might vex or
elate—neither could touch him at work.

I was away from home in January 1908. The children
had been ill and I had taken them to the sea-side for
the prescribed change of air. On 9th January in quick
succession I received the following telegram and notes
(the telegram was sent off at 7.29 a.m.; I received it
before 8 a.m.):

> Silky problem[1] almost certainly solved.
> Solution very exciting. Bateson.

> *Merton House, Grantchester, Cambridge,*
> 9. 1. 08.   6.45 a.m.

I have almost certainly solved the Silky problem, and a great
part of sex with it!

Last night I began to try to think over Sex for my "book".
I started by reading what I had written for *Progressus*. Coming
on Doncaster's moth story I felt it unsatisfactory, and then I
tried working it with the hypothesis that ♀s are all heterozygous
in sex, viz. ♀♂, ♂s being homozygous, ♂♂.

To my surprise I found it went quite smoothly and saves a
lot of assumption. So then I tried the Silkies with this clue,—
the exact contrary to what I had always assumed before
Christmas. I stuck to it all evening and by dodging it about in
various ways before mid-night I had got a scheme which *approximately fits all our facts*. It is so simple and on the whole fits
so well I feel sure it can't be far out.

I wrote it out for Punnett and went to bed at 1 a.m. but only
slept a little! I got up at 5.45 as I could not stay in bed, and
have been touching up my letter to Punnett. When it gets
light I shall run in and wire you.

[1] The "Silkies" are a breed of poultry with black pigmented flesh and
bones, and unbarbed feathers. The heredity of this black pigmentation
when these birds are crossed with light shanked fowls was the problem.
See "Heredity of Sex", *Science*, N.S. xxvii, 785–787.

I feel rather like I did on the morning of Jan. 11, 1889[1]—
Very pleased with myself—only perhaps a little more *certain*
I am on the right track. Also the risks incurred are not so great,
because hypothesis can be amended—Wives less easily.

W. B.

Merton House, Grantchester, Cambridge,
9. 1. 08.

Later

I went in early in blinding snow-storm and wired to you and
posted letter to Punnett. Then, as it was nearly 8 a.m. I thought
I would go and tell Florence.

Her maid came to the door (in her night-shirt I suspect)
"with her golden 'air a-'anging down 'er back" and said Miss
D. didn't breakfast till 9.30. However I communicated the
joyful intelligence in her chamber. Then I found I had lost my
ring! I think it must have fallen in the snow when I pulled
off my glove outside the Post Office. I have hunted everywhere
I can think of, and at last I have offered a reward through the
police. Thanks for the telegram.

You seem to be having a pretty poor time!

W. B.

Evidence of similar excitement runs through his letters,
as for instance, when he wrote home after sitting up all
night watching the segmentation of *Balanoglossus* ova; or
of his earliest perception of discontinuity (nameless as
yet to him); or of his first clues, later developed in his
theory of Repetition of parts.

Those who have worked with him will recall many
another instance.[2]

---

[1] The day of our engagement.

[2] E.g. His emotion when he first saw the *San Sisto* Madonna.
The finding of the two pollens in "Emily Henderson" (sweet-pea).
His first reception of the news of Mendel's papers.
His eager waiting for a bud to blossom and confirm his forecast.
His intense interest in his "Root-cuttings" and in somatic variation generally, and his delight when he hoped to have found right- and left-swimming *Paramoecium*.
Cp. also *Problems of Genetics* (Chap. III, p. 72): "Some years ago I made

He had to share his triumph and delight.

He was at one with "the watcher of the skies", and he knew how it was with Cortes.[1]

Research was one long delicious adventure to him. He was patient, painstaking and ingenious—the drudgery was nought compared to the exhilarating thrill of treasure-trove, which sure enough awaited him. And yet as he worked, in the white heat of excitement, judgment sat within him cool and critical. Emotion could not compel him to unwary haste.

Something of this pleasure of achievement, though perhaps in less degree and without the magnificence of exaltation, moved him at auction sales, when, "backing his judgment against the room" he snatched a bargain from under the dealers' noses.

He lectured in Newcastle in February 1908 and wrote home in high spirits, but critically of his performance:

17. 2. 08.

I am in great case and enjoying quite an after-glow of my American glories. There must have been close on 3000 people last night and I saw scarcely a vacant seat—in fact, none outside boxes I believe. I never spoke in an easier room, and Sarah's[2] vocal feats no longer surprise me. The lecture went pretty well. I was in good, dull, form—told a plain tale, but brought off no brilliance....

18. 2. 08.

My last night's performance was not so successful as Sunday's. I put in too much. Nothing went wrong, but I did

an examination of all the examples of such monstrosities to which access was to be obtained, and *it was with no ordinary feeling of excitement* that I found that these supernumerary structures were commonly disposed on a recognisable geometric plan, having definite spatial relations both to each other and to the normal limb from which they grew". (My italics, C.B.B.)

[1] He was particularly fond of Keats' sonnet "On first looking into Chapman's Homer".

[2] Sarah Bernhardt.

not get into such close contact with my house as in the Theatre.

I don't think it was what anyone could call a bad lecture. Still it was disappointing to fall back at all.

His formal election to the Professorship of Biology was announced to him on 8 June 1908 by the Vice-Chancellor, with many congratulations.

Will replied:

> Merton House, Grantchester, Cambridge,
> 9 June 1908.

Dear Mr. Vice-Chancellor,

Thank you for a most kind note and the good wishes it contains. I will attend on Thursday for admission.

The Readership of Zoology is of course now vacant. The identity of the founder of the Professorship of Biology is, I need scarcely say, quite unknown to me. His generosity has provided what I feel to be an extraordinary opportunity, and I should like to express through you my deep sense of obligation to him.

The work with which my name is connected has for some years past been, as you know, largely the combined production of several workers. I am sure that they also are proud of the institution of this Professorship, seeing in it evidence that our joint efforts have been appreciated.

So in their name and my own I ask you to thank our benefactor.

His Inaugural Address "The Methods and Scope of Genetics" he delivered in October 1908 (see p. 317).

When Will felt stale, nothing but a complete change of activity and interest could stay his restlessness. Sometimes a suddenly improvised expedition on our bicycles satisfied him, or an auction sale; sometimes he stole a real holiday. For instance:

> 3. 1. 09.

Very glad to hear you are so well placed.[1] I am heartily sick of this paper.[2] It is crawling to an end at last and I think I

---

[1] I was at Eastbourne with the children.
[2] "Heredity and Variation in modern Lights" (see p. 215).

must get a thorough change before next Term is on me. Yesterday afternoon I had some chess with Deighton, but I missed the most obvious things, and haven't played so badly for years. My thoughts turn towards Berlin *for one week only!* What do you think? It costs about £4-8-8, 2nd class return. I daresay it would be cheaper than Paris *in the end.* Of course I have thought of Eastbourne too,—but I seem to want something more distinct. I wish you poor old thing could come somewhere—but there it is.

Berlin or London is my choice.

I daresay London will have it.

This afternoon I am going to sit with F. Darwin a bit. He is just up.

W. B.

I have enjoyed my high teas and long evenings enormously, and shall greatly miss them....

If I do go to Berlin I think it must be *incog. of course, incog.*

Later. I am still no nearer deciding about a journey, but the paper is almost done.

Berlin,

6 Jan. 1909.

I expect I have had a better day than you! Museums till light began to fail, then a bit of shopping. The streets are very attractive—not at all behind Paris, so far as I see, except that there are no out-door Cafés in winter. Museums superb. Very few A1 double star gems, but masses and masses of splendid things. All morning in the Print Room. No Old Master drawings are exhibited, but they are willing and seemingly anxious to shew everything. I had a chair and a rack at a window and worked through 8 Solander cases of Dürer alone. There again they have nothing to equal the *Hauptstücke* of Paris and Vienna or even London—yet some amazingly fine drawings. One—a well known drawing—portrait of a man in a fur coat —(I forget who) made me shiver with joy. After Dürer I did 4 boxes of Jap prints. They were not much. Two shops have Jap prints. The prices are *much above* what we have been giving for things similar and as good, if not better. There is a *complete set* of the Utamaro silk series, 12 prints—very nice prints, but all more or less wormed—not conspicuously so, or beyond repair—but nevertheless wormed. They ask £180—I own I did just for a moment begin reckoning whether I would plunge—but I did not. There is a *lovely*, perfect Harunobu at £12. I feel this to be much cheaper than Kato, and am hesitating. I have bought 2 charming, and to me quite novel, pieces of white cambric drawn work (?) said to be Persian. They say they had a lot of them, and that these two alone remain. There is at the same shop a gorgeous piece in colour— I do so wish you were here. No doubt that would put prices up a little—but still. . . . For to-night I am told, I can get chess at a Café—and as there is nothing I know anything of at the theatres except *The Pillars of Society*, which I never read, that will suit me well. *Macbeth* to-morrow will be just right for me, as I know it almost off.

There is a whole room in the Museum with *nothing* but Greek and Roman helmets! Another all Tanagra figures. There are three Art Museums which I haven't touched, including the Schliemann Befund. The Hildesheimer things of course are here. I notice that the things in the perfect taste are all useful things—pots and pans—the ornamental objects are decadent.

There is an Etruscan tomb of a type I never saw. Very plain, painted with barbaric large animals—most stirring; and, of the ordinary type, any quantity, (but not so fine a set as the B.M.). Greek vases in thousands and thousands—but I can't look at them. A thing quite new to me is the three-headed monster from the top of the Capitol at Athens. The one here is a model, painted in bright colours. It is something betwixt the gargoyles of Notre Dame and a Japanese devil, a very, very oldworld invention.

Of pictures I walked past about ½ mile, but the light was too dull, and the best had mostly a copyist camped in front. Two Vermeers, but not up to the Brunswick. Franz Hals in quantity but none of the very best. I think it is right to come to Berlin last, but sure enough you must come.

The general gaiety of the place impresses me—quite as much

as in Vienna, I should say. These Prussians are turning over a new and rather doubtful leaf, I think.

Money shews extreme tendency to dribble out of my pockets.

I am hoping to see the Rembrandt drawings, wondering if they got ours at Amsterdam—very likely, I think.

Great deficiency of labels in Museums. You are evidently expected to buy catalogues and read as you go round in the true German fashion.

What do you think about that Harunobu? Address, *Haupt-Post*, as I am not sure if I shall stop in this Hotel.

W. B.

9. 1. 09. ...I sent a card this morning, and shall probably leave here to-morrow, for Osnabrück, and that on Monday, arriving home on Tuesday midday.

Macbeth was a most moving performance—too much shouting, of course, but the finer parts I thought magnificent. Many of the ideas were quite new to me. For example, Lady Macbeth, having got rid of her guests, in the Banquet scene falls in a heap on the steps of the throne. This is perhaps too much like a Balzac heroine (who was it? She was the mistress of Ajuda-Pinto). Still it gave me a tremendous thrill. Macbeth stood at one end of the second table defying Banquo's ghost at the other in a way I never saw. In the sleep-walking, Lady M. spoke in a forced undertone, which sounds absurd, but I think it was right. I don't think I ever enjoyed Macbeth more—and that is saying a good deal. We were spared nothing,—even the murder of the Macduffs, and the stupid scene between Malcolm and Macduff—the significance of which I never have grasped. I felt so dreadfully old in the foyer—eating sausage and beer, as in old times—and then to think that the young people were as likely as not in long clothes when we were doing that sort of thing! To-night *Siegfried* will make it still worse.

Last night, having supposed I had a ticket for *The Pillars of Society*, I found myself at a musical extravaganza by some mistake (I believe the theatre must be a sort of twin structure). It was called *Die Revolution in Krähwinkel*. Unfortunately, I got kept, finishing a chess game (which I lost after winning a piece and having a won game), and so missed the first Act. Perhaps in consequence I simply understood *nothing* of the purpose or story of the piece. It was nevertheless amusing and extremely well done. What it was about I suppose I shall never know. In many respects it is a piece of quite a fresh type. Crinoline costume, with a really wonderful *Kaffee-Klatsch*. About thirty ladies in crinolines and enormous floral designs of vivid red and green, ten or so on a central ottoman and the rest on a bench round the room. Wall-paper like the crinolines. Back of stage consisted apparently of immense sheets of red and green glass with palms and lights behind. Such colours as one could hardly believe it possible to make. One of the finest bits of stage nonsense I ever saw. Then a lady with no voice sang a song of the period with ridiculous bravouras, that she couldn't take, even after waiting for breath. Her Mamma sat in the front, and tapped approving time with her hand. That I did thoroughly enjoy.

I have bought very little, but how the money does go! I don't care so much for the *Vierte Gallerie* style as I used.

The costume Museum is so full one can hardly see anything, and I enjoyed Brunswick more. But that is not the sort of Museum to do alone.

No, I feel sure that my Essay is not one of my best things. It is all scraps, and I expect if you had been by you would have insisted on neutralising some of the more acid remarks. This may yet be done in proof.

W. B.

11. i. 09. After all I am stopping another day, meaning to get home on Wednesday morning in time to prepare for my harangue on Thursday evening in London. I got into Osnabrück last night and had a very dull evening after the distractions of Berlin.

Siegfried was superb. I never saw such calling before the curtain! It was like the encoring of Sarasate in Madrid. I daresay Brünnhilde was brought back 8 or 10 times. Unfortunately I was half-an-hour late through mistaking the time. *Fafner* had the mouth and the teeth of a hippopotamus, which surely is a great mistake. Generally only the head appears, but the whole animal came on at Berlin. The tail was enormous and waved about across the stage back. *Wotan* (or whoever it was chose *Fafner* as guardian), must have been ignorant of Natural History, for apart from his electric eyes and steam, he was plainly an incompetent beast....

At Osnabrück the fifteenth century (and later) wooden figures of saints and Madonnas in the shops collected from local churches roused further enthusiasm:

The figure of the Anne is lovely, face like Quentin Matsys—all the folds true, and not fussed up at all....Another, an old St. Anne, a stout old lady sitting with the Madonna on one knee, and the Christ on the other. The paint is nearly all gone and the roses are restored, with one arm of the Christ, but I like those jolly old St. Annes. She is to be had for £5—a very modest sum....We really must run over and have a day or two in these places before they are picked bare. I have however somewhat prejudiced further purchase by buying an *Ortus Sanitatis* (the mediaeval herbal), a rather late one (1508, I think) with two leaves damaged, otherwise a good copy, for £7. I couldn't resist it and thought I had a bargain, but from what I now see, I think I have paid an average dealer's price, about. It would of course fetch quite what I gave but not much more. I bought no Harunobu.

13. i. 09. Just got home and washed. ... The Press is poking me up for copy, so it is just as well I have got back....

To the Professorship he had felt in honour bound to sacrifice his College Stewardship (though the greater appointment was provisionally for five years only), and this involved also the loss of a pension-Fellowship, then almost due. The research work was still unendowed; that the University would be enabled to make the Professorship permanent could not be taken for granted;—his position had more of dignity than security. Our boys were growing into school-years. Therefore when the invitation came to assume the Directorship of the newly-founded John Innes Horticultural Institute at Merton, with its promise of unlimited facilities for experiment in plant-breeding, Will decided to resign the Chair of Biology which he had occupied a bare two years, and to accept the post offered to him by the John Innes Council. But he made one condition: namely, that in spite of the horticultural character of the Institute, he should be allowed space on the land to run his poultry and continue his breeding experiments with them. Were that point not conceded, he was quite clear that he could not accept the Directorship. Happily, permission was granted.

Whilst these changes were pending we took a short holiday in the spring of 1910 at Happisburgh. He had no manuscript with him, and except for bathing and eating the fine local shrimps we had no pastime in view. The attraction of that coast was, he said, its absolute nullity, and without lectures to prepare or even proofs to correct, we did not rightly know how we should bear this absoluteness.

En route came the inspiration—we would paint. We had to change trains at Norwich; we rushed into the town, found an artists' colour-man and fitted ourselves out with paint, boards and brushes. From that day nearly all holidays and many happy leisure hours at home were devoted to painting.

This was not his first attempt. In the summer or spring of 1885, after his *Balanoglossus* work he spent some weeks at Concarneau (still occupied, I believe, in research on *Balanoglossus*). In the inn there he found himself in the midst of a merry group of art-students who sent him forth with panel, brush, and paint as a sort of initiation to their society. He often recalled this incident with pleasure.

The University of Sheffield gave him an Honorary Degree, D.Sc., on the occasion of the British Association Meeting, and he was elected to an Honorary Fellowship at St John's College in 1910.

And in the autumn of this year we moved to Merton.

Though the work that could be undertaken there was in the main limited to Plant-breeding (and he would fain have had a station zoological, physiological and botanical, with room for chemical and physical experiments too), the Institute gave him more nearly the ideal opportunity than he by that time dared hope to obtain.

The Council and Trustees shewed him generous confidence, giving him a free hand to design and plan the gardens and laboratories, and to manage and arrange the whole work of the Institute.

He had felt the award of the special Professorship in Biology a great honour; he felt the Directorship of the John Innes Institute to be a great opportunity. He had no difficulty in choosing between recognition of work done and free scope for continued research. His native enthusiasm within was stimulated afresh by these encouragements from without.

Before we had been two years at Merton however the Chair of Biology at Cambridge was permanently endowed, and the descriptive title changed to that of "Genetics". In 1912 he was asked to reconsider his resignation and to return to Cambridge to occupy this new Chair. But when he left Cambridge he felt that he had cast his lot permanently with the new place and he unhesitatingly declined this tempting invitation.

The land which the experimental garden at Merton was to occupy was bare and, but for crops of mint for Mitcham factories, practically uncultivated; as flat as Cambridge itself.

For months—until he had established contact of a sort with the London schools—his exile from laboratory and library depressed him. He feared at first the isolation for his students. But otherwise the new conditions were congenial and he was determined to exact the uttermost from the freedom which their plenty gave.

He visualised clearly and definitely the establishment of

the Institute; step by step the whole was accomplished without hesitation or need of readjustment of his plans, and without forgetting the probable future need of expansion.

Always impatient himself of any arbitrary restraint, he set none upon his staff, assuming from the first that only those students who wanted to work would come. All the facilities for which he had longed in past years were placed without stint before them; his confidence in them and his interest in their work were sufficient to make rules and regulations superfluous. There were none.

He was very sensitive to personal impressions and summed up rapidly and instinctively all with whom he came in contact, seldom erring. He used to mock at me for wanting a written "character" before engaging a servant.

The very first thing that he did on his appointment to the Directorship was to secure the services of Mr Allard as garden-superintendent. Allard as garden-boy had worked for Will at the time of his earliest experimenting with *Primula sinensis*, carried on in odd corners of the Cambridge Botanic Garden green-houses. Will had noticed his interest and aptitude, and had long since decided, if ever in a position to do so, to enlist Allard's help. The choice was justified. Allard played his part loyally and well, keeping the garden in first-rate order, and the men well-disciplined. Will's edict that every plant should be well grown was attended to with exactness. On one occasion, after a party of gardeners had been shewn round and told something of the aims and the work of the Institute by Will, their spokesman made a little speech of thanks: "They do say, sir, that you are too scientific, but now that we have seen round the place we know that is not the case!"

He was now quit of the long hours of "menial" drudgery, and some increase of literary output resulted. But the claims on his time multiplied. As first charge on his day came the post (until within two years of his death he worked without the aid of a professional secretary); then he was off to the laboratories and green-houses, busy amidst his students with his own and with their experiments; full of resource and ever new suggestiveness; full of ardour and enthusiasm, he shared with boundless generosity his thoughts and ideas with all who cared to take part.

His standard of accuracy was rigid and high, and he exacted critical carefulness from his helpers, both students and garden staff, in all their operations and experiments. He himself with his fine, sensitive hands was a beautiful manipulator. To watch him at work, delicately dissecting some fragile blossom, was a splendid lesson. He seldom bungled.

His reputation as a lecturer spread; he was constantly asked for his "talks," by all sorts of scientific and educational bodies. These engagements took toll of hours that should perhaps have been leisure, but he was glad to find and encourage interest at last. Though of such informal lectures no record was kept, each one was meticulously thought out. He had, it is true, one scheme which he called "Mendelism without tears" for "lay" audiences, but the restatement of his facts, even on such occasions, was never mere routine work.

In 1911 he presided over the Agricultural Section of the Portsmouth meeting of the British Association (see p. 260).

In November 1911 the Royal Horticultural Society awarded him the Victoria Medal.

In 1912 he gave the Herbert Spencer lecture,[1] "Biological fact and the structure of Society" at Oxford. This lecture he always thought one of his best, but it was given to empty benches; the date fixed by the authorities clashed with the *Torpids*.

[1] See p. 334.

Brasenose College, Oxford,
28. 2. 1912.

My talk is done—with what success I scarcely know. A thin
house in a very large room rather damped me, and I had an
uneasy misgiving the paper was too long—but in reality it did not
prove seriously long. It seems the Torpids are on, and Oxford
takes those things much more to heart than Cambridge. My poor
host and hostess had been down *twice*, with a meeting between!

I gather I shall have set tongues wagging, but whether with
praise or blame will appear later....I got 10 minutes with
the drawings—saw some Raphaels which were new to me;
also two lovely Richard Wilsons.

In 1912 too, he was elected to the Fullerian Professorship at the Royal Institution, London. This is an appointment for three years involving three courses of lectures.
He spared no pains in the preparation of these lectures
and enjoyed them heartily, but the notes from which he
spoke are so scant as to be barely intelligible.

And all the while he was slowly and carefully working
over the Silliman lectures which he had given to Yale
(1907) and which were published under the title *Problems
of Genetics* in 1913 (Yale). This book is no mere treatise
redolent of laboratory and study; it is the tale of one side
of his life for the period which it covers. His absorbed
interest in his work glows in these pages; the facts and
illustrations to his argument are drawn from his daily life.
There is not a chapter that does not recall his stimulating
talk and his vigorous enjoyment of the little occasions that
served to inspire it. The labour of writing it was nevertheless very heavy.

From 1912 he was Chairman of the Development
Commission of the Board of Agriculture; there were also
endless other Committee meetings to be attended, which
seems to be the lot of the middle-aged.

He overstrained his magnificent physique. In 1913 he
broke down under some sort of sharp anginal attack.

Suddenly one day without any warning he fainted. He was ill and in pain for many weeks but made a good, apparent recovery. From that day however his life was changed for him. We were never quite free from anxiety; he felt pain lurking round the corner waiting to pounce upon him. The gay sense of physical well-being was lost.

In earlier days he walked with a long rapid stride, with which I could never keep pace; he once turned to me in the Strand exclaiming: "Really, Beatrice, you are the only man's wife I know who always walks two yards behind her husband", to which I made the obvious retort, and we both laughed and walked together for some way until he gradually slid ahead again. Now the time had come when I must slacken my pace for him; and it was grievous to wait whilst he fussed with his pipe knowing that it was not the pipe that was giving trouble. To little ailments, except toothache, he had been a stranger; from now onwards we had to reckon with them and fight to keep them in their place. However, though for a short while the possibility seemed doubtful, he was soon gingerly at work again, and in 1914 was, after all, well enough to undertake the journey to Australia and preside over the meeting of the British Association there[1] (see p. 276).

The interest and enthusiasm which he met with in Australia delighted him, and he made a great effort all the while to "play up", but he was never really free from pain.

On the top of this personal disaster came the international calamity of war. Anxiety and consternation hung in heavy clouds over all the meeting. Instead of relaxation from strain, the grim prospect of tragedy made new demands on his endurance and mental activity.

On our return to England, November 1914, the staff at the John Innes Institute had already dispersed—the young men gone. Makeshift was again his lot, but now his heart was heavy. Aching for the wilfully inflicted

[1] He had been chosen President of the Botany Section of the 1926 meeting at Oxford and thus, had he lived, would have been President of three separate sections as well as President of the whole Association (1914).

128   MEMOIR

suffering at large, he deliberately ignored patriotic passions and set himself quietly to maintain the continuity of his particular branch of work. (From about this time onward he was giving much attention to somatic segregation and much work was done on root-cuttings, and with variegated plants.)

To him the arts and sciences represented civilisation, and by civilisation he meant a state of high ideal, above and beyond the reach of commercial ambition and political cabal. The sustaining stimulus of national enthusiasms was not for him.

He was genuinely amazed and shocked to find the world of learning and the professional classes shaken by national prejudices. The savage sentiments overheard in English drawing-rooms appalled him no less than the newspaper reports of foreign brutalities.

Amidst much else the masking of truth in public announcements of "news" especially angered him. He thought this unworthy of any governing class and degrading.

His discouragement was profound; nor was he reassured by the political shapings of peace in 1919. There was no place for punishment in his philosophy.

In January 1915,[1] he wrote:

At first I did not suppose that the international scientific relations would be affected, but they are, and very seriously too....I regard the war as *almost* purely a commercial matter—not quite, for I know there is a strong element of national feeling mixed up with it—but a lot of that is commerce in disguise....British hypocrisy is, I quite admit, a real and astonishing phenomenon. It serves one good purpose, that it makes us shy of public and ostensible evil-doing: something, anyhow, to the good, if not much.

July 15, 1915. I should be most grateful if you could get me the numbers of Baur's *Zeitschrift* since Bd. xii, which I have complete. I used to get mine direct from Borntraeger, and I

---

[1] The following extracts are from letters to Dr Ostenfeld, Copenhagen, who kindly allows me to use them.

should think they might be told that the order is really from me, and I could pay you. I suppose this is *trading with the enemy*! But the *consideration received*, to use our legal phrase, is so much greater on our side than on theirs that I look on it as rather robbing them. We hear reliable rumours that dyes come pretty freely from Germany—especially *Khaki*!

We all live on hopes. I cannot say that mine are lively. In general this country was glowing with optimism till about May. Since then the truth has become more evident,—that to win the war in any comprehensive sense must be a terribly difficult business.

Personally I feel better about the prospect than I did in the winter, for I never dreamt that we should be able to hold them back so long. The only thing I have had from the "enemy" so far is the new part of Przibram's book, which he sent me through Davenport.

It is pleasant to remember that the sciences and the arts are going on all the time, but who is to follow us middle-aged people? The pitiful thing is the destruction of young men coming on. All the best are in it—the pick of the best breeds in the country. Inevitably the succession—in learning, I mean —must be broken throughout Europe, and I greatly fear that a more or less permanent state of war, for a generation at least, is the most probable outcome of it all.

16. vii. 16 (after thanking for journals received). Most people here are content to do without the German periodicals, a course that seems extraordinarily stupid.... We are dreadfully short-handed. We keep the peas going and a few of the more promising things, the rest has to be suspended.

The "succession of learning" he hoped could be partly sustained by possession of unbroken sets of all important scientific periodicals. When, after the Armistice, he found that the better known scientific reference libraries not only still failed to complete their broken sets of German Periodicals, but neglected to continue them, he was irate; indeed he threatened in consequence to withdraw from the Linnean Society. Eventually, however, "just to shew that the papers were easily procurable", he himself presented the missing volumes of one periodical to this Society.

For the suffering his pity overflowed.

In the winter 1916–7 he went out to the base camp at Rouen, under the auspices of the Y.M.C.A., to lecture to the men, and help as he might to ease their dreary lot.

Le Havre.[1]

19. xii. 1916.   ...Boat was crowded and a large party of French and Italian peasants and children got no berths in the third class. I felt rather sad about them, as it was a bitter night. In morning I found a travelling ship's crew—roughest sort of sailormen—had given up their berths to the children. The benefactors were simply delighted. They were telling everyone out of pure childish enjoyment—not vanity in the least—"We guv' up our berths to them childer!" "That's them what had mine" etc. and the last I saw of them they were giving pennies to their protégés. I am quite sure who got the most out of that charity!

20. xii. 1916.   My neighbour at dinner last night was an Australian Tommy—quite a rough chap, but not uninteresting. We went round to a *Café* after, where I ordered coffee and *deux fines*. The young lady was much amused at my *gaffe*—for it seems privates, however sober, are not "servable"! Dumb-animal fashion he asked for a dash out of my *petit verre*, which I gave—a bad beginning, to break rules so soon in my military career. The young lady however had conceived a more excellent way, for seeing my glass empty she rushed across to refill, but to that I did not consent. The poor Australian had never before met the combination of coffee with *fine*, and was delighted. I trust I have not facilitated the descent of a soul....

I must get you a photo of the high relief over one of the W. front doors representing St John Baptist scenes. Salome is

---

[1] These letters are as usual full of fun and matter. I have chosen only a few sentences to shew how he remained himself and yet made himself felt.

dancing *on her hands*. I must have seen this many times but had quite forgotten it. When last here St Ouen moved me most, now scarcely at all compared with Notre-Dame. These changes in oneself are curious and surprising.... In all probability I am to take both military and workmen's camps. The latter, with large numbers of mechanics, are likely to make my best audiences....

22. xii. 1916. Yesterday evening I began my ministrations with a lecture in R.E. hut. It was *not* a great success, I fear. Conditions dreadful. Long narrow room with no illumination but the lantern, so that I could never light up and see my audience.... Many crept away under cover of the darkness —a few no doubt interested.

The accepted cue is to talk down to them. They live chiefly on comic songs, sentiment, and religion in many forms. I wish I could write freely but——. I called on B——. They took me to a concert in the afternoon. Three girls sang the completest drivel with wonderful spirit. It would have been as easy to have given the men some better convention. It is all a pose and the English sentimental song is no more natural to these men than Wagner was to Germany. However it is decreed that what "the boys" want most is "Roses in my garden", "Ten little toes", "Jammy face" etc. I found it difficult to sit it out with patience,—and I suspect or at least hope there were others who felt as I did. If not, the Lord have mercy upon us....

The R.E. men asked me after lecture whether man could be descended from monkeys, if so how had he come by an immortal soul? Also Genesis said, "In the image of God created he him", and could we suppose God was like a monkey in appearance? These were men leaving for the front next day... and so...well...I temporised....

This morning I went to the top of Bon Secours and looked once more at the wonderful panorama, which struck me greatly years ago. I thought of "A king sat on a rocky brow". Unless I can find more to do, my days are likely to be dull. I don't get to work before 5, and end about 9, so from 9.30 p.m. to 5 p.m. I am idle.... Some really interested—most are not....

I don't feel a great success. No one here in the same rôle at present—and as usual, I believe I have tumbled into an eccentricity.

27. xii. 1916. I am finding my feet rather better. Certainly there is a fair proportion of interested hearers.

28. xii. 16. Getting really busy. This week I am to give the preliminary lecture seven times. *Heredity of Sex* twice and a lecturette each day as well. Lastly with much hesitation I have consented to discourse to the workers (largely ministers of various denominations) on how religion looks to the naturalist. Rather a doubtful move I fear, but they press me very hard to do it.

2 Jan. 1917. ...Rather sick of *Preliminary Mendel* and glad to have begun *Heredity of Sex* last night. We had an animated "Quiet Room" séance after. So much so that I was told the men had forgotten their suppers! At Y.M.C.A. H.Q. also the audience was so entranced that that night *Family Prayers* were suspended! My sermon growing in length and virulence. Just think what a chance. I shall feel some of the joy the mediaeval laity must have felt on All Fools' Day.

I took a "counter" for an hour last night. Sold tobacco, candles, soap, buttons, etc. It made me ache for these poor men. Fancy no soap or light provided! They pay $1\frac{1}{2}d$. for the thinnest "bar" of Nestlé's chocolate, so thin you can scarcely lift it without breaking.... We are getting the mud all right now. Some wear waders, mostly gaiters. Ladies all in oilies, which are really wanted. Last night coming back in the car with wind on port bow I got simply sluiced from hat downwards. The Art Gallery here has three P. de Chavannes frescoes, and I am satisfied he is a *nix*. Simply the good old allegoricals, Painting, Pottery, Science, etc. wandering in a green meadow with hermaphrodite modern clothes instead of the naked-to-the-waist, sitting-in-temples style. *Science* is a lady holding up a flower, *sp. incert.* while another lady draws it on a lyre-shaped board. I don't know why.

4. i. 17. A sad change to record in my fortunes. Have had incipient sore-throat a few days and on Tuesday it got worse, so I cried off lecturing, spent yesterday in bed with much throat pain. To-day saw a M.O. who has put me in Hospital, where he says I must be five days in bed. They say everyone is having these throats, but I fancy I am on the worse side. No ground for worrying, it is simply a nuisance.

7. i. 17. Thankful to say I seem to be through the worst. Have had a dreadful time, such aches in throat and lungs as I never remember. However temp. lower now and my advisers seem content. But to-day Sister told me she did not expect I should clear for 8–10 days, and that probably I should not then feel like work, so I suppose I may come home.

Will was still in the hospital on the 15th, and then being allowed to come out he made a dash for home, where he was laid up for another three weeks with bronchitis in bed.

Notwithstanding this illness he was not easy until able again to go out to France. He could not bear to be in safety and comfort without making some effort to alleviate, if only for the moment, the distresses suffered by the men, and in Jan. 1918 he went out again.

Étaples (I think).

14. i. 18. I am letting off my three "rounds" at each of two places with an extra turn thrown in to-night. Between 5 and 9.30 to-day I shall have done 3 lectures, one example of each turn in my repertoire—rather stiff! I am momentarily expecting Herringham,[1] or rather I should be, if we had not had a prodigious fall of snow in night which may have made roads almost impassable. Till I see him my plans are undecided. I am pressed to stay on, but incline to return this week at all events. A comfort to feel one is some use at last.... The queer half-nomadic life of the Y.M.C.A. just suits me. No one, or very few, keep up any kind of *gêne*; of course centres differ, some having more of the Y.M. ingredients, others partaking more of the C.A....

[1] His friend, Sir W. P. Herringham, M.D., C.B., K.C.M.G., Consulting Physician to the Forces in France, 1914–1919.

He was working very hard to keep things going at the Institute and regarded this very strenuous bout of lecturing as merely a holiday fling. He came home tired out.

In the spring of 1918 Will was asked to contribute an article on Genetics and Heredity to make one of a series on various branches of science which were to be translated into their several languages and distributed amongst the neutrals, to counteract an impression abroad that science was at a standstill in this country. Will frankly replied:

1918. Propaganda of the kind which you tell me is contemplated by the Ministry of Information seems to me objectionable, and incompatible with the spirit of science. I cannot accept the invitation to take part in it. My own branch of science has hitherto, I am happy to say, been free from Chauvinism. We acknowledge the contributions of the several nations and we know fairly well what each is doing at the present time. If, however, as you fear, the impression prevails abroad that scientific work is here at a standstill, the fact is not surprising. By closing the Museums[1] the Government gravely injured the cause of science and learning, and advertised to the world the contempt in which such Institutions are held in this country. Whatever propaganda may be disseminated by the Ministry of Information, that action is likely to be remembered abroad and its consequences persist.

Oddly enough this letter was not understood and a second bait was thrown to him to which he made answer:

If I found some other country, say the United States or Germany, holding forth in the various languages on what they had been doing in Genetics I should, without being unduly fastidious, think it not good scientific manners, to say the least. If the work is of value it generally makes its way, at all events in sciences having a proper international organisation as ours has. We care very little about the geographical provenance

[1] The British Museums, with the exception of the Reading Room and some of the students rooms, were closed to the public from 1 March 1916 onwards.

of discovery. I don't quite know how I should be able to meet my International Committee—as I still hope some day to do— if I had written what you ask for.

I can imagine circumstances in which it might be useful and proper to write an epitome of progress in Genetics made since 1914 in the world at large, but that is not what the Ministry wants. It would also have to record the unwelcome fact that the Prussian Government has, since the war began, started a new Institute at Dahlem for purely scientific research in Genetics, which was opened in May 1916.

They are not closing Museums in Germany, I believe. When we hear they have begun to do that we may get the fire-works ready.

My reference to the closing of Museums was not made on the ground that our Government had done a foolish thing, but as a commentary on your complaint that neutrals have got it into their heads that science is discouraged here. I was merely naming one of the many events which may be supposed to have contributed to that opinion.

The matter then dropped.

In the summer of 1918 his mother died, aged 89.

In October 1918, our son John, aged 20, was killed. Will wrote to our second boy, Martin:

I know what you must be feeling. John was a splendid creature; such breadth and balance are rare in men so young. He would have been a comrade for life to you and his mother. Like all of us I think he had no illusions. I never heard him say a foolish thing, even when quite little, and he knew how to bear trouble. . . .

There never was a nobler soul.

And to our friend, Professor G. C. M. Smith:

We knew your kind heart would be with us. John had no illusion either. He just knew his duty was to do his best. We never discussed the war with him. I think he felt caught in the wheels of a hideous destiny. To have numbered with brave men is much; but had he lived I think he might have done better than that. His good sense and balance were of a kind rare among quite young men. . . .

And in another letter he wrote:

27. xi. 18. Our eldest boy was killed at the very end—some 14 days before the Armistice was signed. He was a brave, good boy. He won the Military Cross for a piece of bravery. Like us also he had no illusion as to the war. The second has not been out. . . . We have had other sorrows. My superintendent-gardener, a most gifted and exceptional man, died of pneumonia, and now we have lost R. P. Gregory, also of pneumonia. He used to work here a great deal. It is difficult to keep up much spirit.

Characteristically putting aside his own unhappiness, Will set himself at once to help the widows of these two friends.

The address "Science and Nationality"[1] belongs to this dark period. He sent the proof of this paper to his son Martin, then stationed at Yarmouth, and inconsolable for the loss of his brother. To his boy's criticism and bitter questioning, Will made answer:

13. xii. 1918. . . . The question whether we were right or obliged to fight in 1914 can, I admit, be argued. But certain things seem to me clear. I take it that (1) a Government elected by a representative system is bound to do everything possible to defend the country from hostile pressure, direct or indirect. An autocrat, or an independent group of citizens, may on any occasion choose a course of action and follow it, whether it is approved by the country or not. Those who say, "Let us be *sans patrie*, like Jews", have much to be said for their opinion. But elected Governments are in honour bound to follow pretty closely in accepted courses. That is a principle of wide application and though at times it approaches sophistry, on the whole I think it a good guide. As a man I should take a certain line—but it does not follow that as an official I should take the

[1] See p. 356.

same line—especially if I were appointed to my office by people who do so in confidence that I shall act on lines accepted as reasonable.

(2) The real question therefore is whether we were in danger of actual military injury in 1914, and whether to join in was the safest way of meeting the danger. I can't see that either point is really in doubt. We are not concerned to inquire whether there was a wide-spread or general feeling against us in Germany, (such that to attack us after the conquest of France would have been there a natural or popular policy), but merely whether this policy would as a matter of fact have been adopted by the German Government. I can't doubt that it would. As to majorities one can't tell; but I believe after the Channel ports had been occupied a quarrel would soon have been picked with us which would have made war on us popular, at least with the commercial classes in Germany, and they certainly would then have been much better able to injure us so, than they could be if we went in before France was destroyed. On that point I think there must be general agreement.

The whole question therefore turns on the fundamental duties of Government. I certainly admit that the obligation to Belgium, though paraded as the main motive, was in fact little better than a useful pretext.

Looked at in the broader way I also think there is no serious doubt that Germany, as then led, would have set about destroying us at the first possible moment. To have avoided the war, our policy for the last 25 years would have had to be changed, as I said in my address.[1]

Where sophistry may more reasonably be suspected is in the principle, which many thoughtful and honourable people maintain, that in an event so critical as war the subject is bound to support the Government that has (we assume in good faith) decided to fight. One may no doubt argue that if enough subjects refused, the war would stop. But as a matter of observation, enough do not *at that stage* refuse; and all that happens through the refusal of some, is that the loyal have to take on more than their share of suffering and danger.

[1] "Science and Nationality."

The argument is not quite the vicious circle it at first seems to be, because in practice there are critical points and points less critical, and it is in the intervals that the objector may make his weight felt. A skipper *has* to put his ship's safety first, and during the critical moments the only chance in the long run is for the objector to subordinate his opinion and do what he is ordered to do. Sophistry and casuistical reasoning may be used in representing occasions as critical which are not critical, and *vice versa*, but in 1914 the moment, as I have said, was in my judgment most certainly critical in every sense of the word. . . .

The spirit in which we have carried on the war is enough to make anyone sick and ashamed of their country. But that I believe is the line we have usually taken in recent wars. It certainly was so in the Boer war, and I believe in the Crimea also.

It is always a prominent thought in my mind that strictly speaking such people as we are do not belong and are only here on sufferance. To think for oneself in most societies is a crime. But there it is! One has just to make the best of the situation and be thankful we are allowed our niche.

W. B.

Partly because he felt that they had been largely avoidable, the anxieties of peace depressed him as heavily almost as those of war. So long as material advantage and reprisal were the motives of public action he thought an abiding peace impossible to attain. A dread of continued ignoble jealousies and hate made him appear unduly pessimistic to many of his friends; probably the suspicion that he, for one, could not live long enough to see goodwill restored between the nations added to his unhappiness. But he never let his vexed and grieving spirit break the continuity of his scientific work; that went steadily on. His despondency shewed itself in a demeanour more grave and serious, his lessened bodily health somewhat undermined his confidence, but his interest and enthusiasm never flagged. As conditions became more normal the scope of the work at the John Innes Institute extended and increased; new laboratories and a library were added.

These years were the interval between "times critical
and less critical"; he tried "to make his weight felt" and
succeeded better than he dared hope.

In 1919 the Royal Society invited him to act as one of
the delegates at the meeting of the International Research
Council to be held at Brussels that summer.

The restoration of peaceful international relations in
the scientific world was what he most wished to see;
though he had some doubts as to the right of this Council
to use the word "international", he accepted and went.

Grand Hôtel Britannique, Bruxelles,
26. vii. 1919.

...The international position is even worse than I had supposed. Every scrap of common sense is gone. The French are
determined to turn the world upside down, and the rest
acquiesce. If I had understood the purposes of this meeting I
should not have come. It has certain real and useful objects in
view, but it is being largely made an occasion to exploit
science for chauvinistic purposes. I keep a quiet tongue and
have only intervened when it could be done without exciting
a row. Probably home on Wednesday. I have been trying and
trying to see Bruges....

(To Dr Ostenfeld)

23. x. 1919. I am greatly interested in what you tell me. Of
course these "unions" are not in any real sense international,
and are not meant to be so. In chemistry particularly the new
Union is really meant to fortify the chemists of the Allies
against the German chemists.... I fear all real international
development is over, at least for our time and probably for
much longer....

I regret that my name appears as Vice-President of the Biological Union established at Brussels. When I accepted I
thought that I was being appointed a V.P. for the *Brussels
meeting.* I refused to be president of two sections, but when I
understood my first mistake, I thought that to protest and

withdraw then would be too "fussy" as we say, and that I had better let things take their course.

And a year later (to the same)

27. xi. 1920. You will be interested to hear that the Ctee. appointed to consider the question of joining the Biological Union (under the *Conseil International des Recherches*) declined to join. We decided that the admissibility of the Central Empires was an indispensable preliminary to the constitution of an International Biological Union. Those were not the actual words—but that is approximately what we settled.

This result was to me quite unexpected. It has given me also very great pleasure. I had no idea that I should be able to bring about this result. We hope that we have heard the last of this Biological Union which nobody wants. It seems biologists have more sense than other people after all!

W. B.

(To the same)

7. xii. 1923. ...I am pleased to be able to say that the Biological "Union" is—so far as England is concerned—killed at last. This time I do not think the question will be raised again. I do not know who started it, but we had it once more before the Biological Committee of the Royal Society, and it was carried *nemine contradicente*, that we do not join until it is possible to admit the Germans, or words to that effect.

Nevertheless he was, in 1924, invited by *l'Union internationale des Sciences Biologiques* to preside over one section (biology and physiology) of that body. He replied:

The Manor House, Merton, London, S.W. 20,
21st January, 1924.

To receive such a letter as that which you have sent me in the name of the *Union Internationale des Sciences Biologiques* is an extraordinary honour, and I greatly value this sign of the confidence which my foreign colleagues are willing to extend to me.

The Presidency of the *Union Internationale* must be a position of distinction, and you will believe that it is not lightly or without serious consideration that I ask to be excused from acceptance of that office.

I have some doubt whether a general international organisation is in biology as yet a necessity, and whether the international organisations of the several branches of biology do not sufficiently supply what biologists require. Though doubtful on that point, I should nevertheless not regard this hesitation as ground sufficient to preclude me from accepting so flattering an invitation.

I have a much graver objection. The *Union Internationale des Sciences Biologiques*, as at present constituted, seems to me likely rather to restrict than to promote international scientific communication. This consideration has, as no doubt you know, weighed so much with us in England that the Royal Society has declined to join. I enclose a copy of the resolution which we lately adopted, though probably it has already reached you.[1]

For my own part, I deplore the introduction of the question of nationality into scientific affairs. The representatives of learning and the arts might provide a dominant influence in maintaining the sanity of the world. The force they might exert is prodigious. That they should promote further division is to me lamentable. The formation of the *Union Internationale* from which Germany is in effect excluded, can only increase the mischief. There is also to me something unreal and grotesque in the thought of a biological union containing no German names. After all, the developments of cellular biology have always from the beginning been very largely German. We gain nothing by refusing to recognise facts. One of the great deprivations of the war was the suspension of intercourse with our German colleagues. I cannot be a party to any measure tending still further to alienate us.

[1] "That the restoration of normal conditions of full and unrestricted scientific communication be an indispensable preliminary to the adherence of Great Britain to the International Union of Biological Sciences."
This recommendation was presented to the Royal Society on 30 Nov. 1923 and adopted.

Pray do not charge me with a want of sympathy. I can understand the feelings of resentment which have led to such proposals, but I mistrust the wisdom or utility of reprisal. If in the judgment of posterity we are found at this critical time ourselves to have done nothing unworthy, it is much, and with that ambition we may be content.

W. Bateson.

His interest in education never waned, and he was much pleased in 1919 to be invited to become a member of the University Grants Committee, but he served on this Committee only three years. He resigned in 1922 when elected Trustee of the British Museum.

In 1920 the Royal Society awarded the Royal Medal to him. This award pleased him. "Thanks", he wrote, "for kind words on my Royal Medal. I had supposed my time for that particular honour was long gone by. They gave me the Darwin medal years ago, and I did not expect another".

Another source of satisfaction to him this year was the invitation to give the "Croonian" lecture (see *Genetic Segregation, Scientific Papers*, vol. II); and close on this, in the summer, came an invitation from the American Association for the Advancement of Science to give an address to their Society at their meeting in Toronto in December, 1921. He accepted gladly.

The war had deprived him of association with his continental *confrères*; he had missed sorely the hearty welcome and the frank discussions and sympathetic exchange of ideas with his German, Dutch and Scandinavian friends. This was his first opportunity since 1914 of meeting again fellow-Geneticists outside the small home group of younger men and his own students. In England his professional life was always somewhat isolated; his work never found much favour or understanding amongst his contemporaries in biological studies.

He left England early in December 1921, and spent some days with Professor Morgan in New York before proceeding to Toronto. He saw Professor Morgan's laboratories and was there introduced to cytology as a living method of inquiry. Hitherto he had been inclined to hold aloof from cytology and to doubt the value of the American work on *Drosophila*; he thought it interesting, but a by-way, off the main road to knowledge of Evolution.

Messrs Bridges, Sturtevant and many others demonstrated their work to him and he was much impressed and interested. He wrote to me:

24. xii. 21. ...Part of each morning is devoted to chromosomes. I can see no escape from capitulation on the main point. The chromosomes must be in some way connected with our transferable characters. About linkage, and the great extensions I see little further than I did.... Cytology here is such a commonplace that every one is familiar with it. I wish it were so with us....

26. xii. 21. ...I am heartily glad I came. I was drifting into an untenable position which would soon have become ridiculous. The details of the linkage theory strike me still as improbable. Cytology, however, is a real thing—far more important and interesting than I had supposed—we must try to get a cytologist....

29. xii. 21. ...Only a line to say last night's address[1] went well. I had to announce my conversion on the main point, "that chromosomes are definitely associated with the transferable characters" is how I express it. Much enthusiasm over this of course, but as a candid man I don't see how any other view can now be maintained.... Evening after speech wonderful. I shook 60–70 hands.... Many old friends, Shulls, Davenport, Blakeslee, Ibsen, Nabours, Sewall Wright, L. J. Cole, all our household world. Enthusiasm intoxicating, and supplies place of drink!... Most of the American cards in our catalogue are now persons to me.

[1] *Evolutionary Faith and Modern Doubts* (see p. 389).

He also gave a short course of lectures in Toronto. These he described:

Those who got in and those who were shut out from my 1st. discourse—together with I know not how many of the relatives and friends assembled for the second lecture—quite 1100—and into them I hammered the principles of linkage and the significance of the X chromosome, bellowing it out in words of one syllable for 70 minutes. Did you ever hear of such a thing? In the end I warned them that I must get on faster and, though I was proud to see them, that I doubted whether they would be able to get much further.... This afternoon I shall see the result. I am put on for two extras already here. It is settled I am a great *popular* lecturer—That's *me*—take it or leave it, that reputation is what I have to carry.

He lectured also at many Universities in the United States, and in Philadelphia (University of Pennsylvania) he gave the Joseph Leidy Memorial lecture published *Journ. Gen.* XVI, 2, 1926, under the title "Segregation" (see *Scientific Papers*, vol. II).

The Toronto address, "Evolutionary Faith and Modern Doubts", is a candid and lucid statement of his creed; his frank handling of his past and present views of the chromosome theory are very characteristic of him. The address, however, apart from intrinsic and autobiographical interest, became almost notorious on another score. How or why it happened that it should have been seized upon as a weapon of attack on Evolution, is difficult to understand, but the fact remains that a sudden strange ebullition of back-country illiteracy brought it for a short time into journalistic prominence in America. He wrote to me:

1. i. 1922. Osborn very much perturbed because newspapers exaggerated my criticism of conventional Evolutionary views—scare headings "Darwinism disproved", or some such

words.... Of course I was most careful to declare that as against evolution and obscurantist conceptions the choice was obvious, and that we could comfortably believe evolution though we could not see the origin of species....

7. i. 1922. ...An auditor gave me these cuttings. They are among the milder that have appeared. I wish that I had got some of the others! I hear there have been some good answers, but personally I do not interfere....

(A sheaf of cuttings of ludicrous crudity was enclosed.)

This wave of "obscurantism" swept angrily over some of the States, challenging attention. In 1923 the Editor of *Nature* asked Will for a short article on the subject.

THE REVOLT AGAINST THE TEACHING OF EVOLUTION
IN THE UNITED STATES[1]

The movement in some of the Southern and Western United States to suppress the teaching of Evolution in schools and universities is an interesting and somewhat disconcerting phenomenon. As it was I who all unwittingly dropped the spark which started the fire, I welcome the invitation of the Editor of *Nature* to comment on the consequences.

First as to my personal share in the matter. At the Toronto meeting of the American Association I was addressing a scientific gathering, mainly professional. The opportunity was unique inasmuch as the audience included most of the American geneticists, a body several hundreds strong, who have advanced that science with such extraordinary success. I therefore took occasion to emphasise the fact that though no one doubts the truth of evolution, we have as yet no satisfactory account of that particular part of the theory which is concerned with the origin of *species* in the strict sense. The purpose of my address was to urge my colleagues to bear this part of the problem constantly in mind, for to them the best chances of a solution are likely to occur. This theme was of course highly academic and technical. Nevertheless, to guard against mis-

[1] *Nature*, CXII, 1 Sept. 1923.

representation, I added the following paragraph by the advice
of a friend whose judgment proved sound, though to me such
an addition looked superfluous. "I have put before you very
frankly the considerations which have made us agnostic as to
the actual mode and processes of evolution. When such confessions are made the enemies of science see their chance. If
we cannot declare here and now how species arose, they will
obligingly offer us the solutions with which obscurantism is
satisfied. Let us then proclaim in precise and unmistakable
language that our faith in evolution is unshaken. Every available line of argument converges on this inevitable conclusion.
The obscurantist has nothing to suggest which is worth a
moment's attention. The difficulties which weigh upon the
professional biologist need not trouble the layman. Our doubts
are not as to the reality or truth of evolution, but as to the
origin of *species*, a technical, almost domestic, problem. Any
day that mystery may be solved. The discoveries of the last
twenty-five years enable us for the first time to discuss these
questions intelligently and on a basis of fact. That synthesis will
follow on an analysis, we do not and cannot doubt."

The season must have been a dull one, for upon this rather
cold scent the more noisy newspapers went off full cry, with
scare-headings, "Darwin downed", and the like.

All this seemed foolish enough, and that practical consequences
would follow was not to be expected. Nevertheless, Mr William
Jennings Bryan, with a profound knowledge of the electoral
heart, saw that something could be made of it and introduced
the topic into his campaign, which, though so far harmless in
the great cities, has worked on the minds of simpler communities.
In Kentucky a bill for suppressing all evolutionary teaching
passed the House of Representatives, and was only rejected,
I believe, by one vote, in the Senate of that State. In Arkansas
the lower house passed a bill to the same effect almost without
opposition, but the Senate threw it out. Oklahoma followed
a similar course. In Florida, the House of Representatives has
passed, by a two-thirds vote, a resolution forbidding any instructor "to teach, or permit to be taught Atheism, agnosticism,
Darwinism or any other hypothesis that links man in blood-

relation to any form of life". This resolution was lately expected to pass the Senate. A melancholy case has been brought to my notice of a teacher in New Mexico who has been actually dismissed from his appointment for teaching evolution. This is said to have been done at the instigation of a revivalist who visited the district, selling Mr Bryan's book.

The chief interest of such proceedings lies in the indications they give of what is to be expected from a genuine democracy which has thrown off authority and has begun to judge for itself on questions beyond its mental range. Those who have the capacity, let alone the knowledge and the leisure, to form independent judgments on such subjects have never been more than a mere fraction of any population. We have been passing through a period in which, for reasons not altogether clear, this numerically insignificant fraction has been able to impose its authority on the primitive crowds by whom it is surrounded. There are signs that we may be soon about to see the consequences of the recognition of "equal rights", in a public recrudescence of earlier views. In Great Britain, for example, we may witness before long the results which overtake a democracy unable to tolerate the Vaccination Act, and protecting only some 38 per cent. of its children.

As men of science we are happily not concerned to consider whether a return to Nature, as a policy, will make for collective happiness or not. Nor is it, perhaps, of prime importance that the people of Kentucky or even of "Main Street" should be rightly instructed in evolutionary philosophy. Mr Bryan may have been quite right in telling them that it was better to know "Rock of Ages" than the ages of rocks. If we are allowed to gratify our abnormal instincts in the search for natural truth, we must be content, and we may be thankful if we are not all hanged like the Clerk of Chatham, with our ink-horns about our necks.

For the present we in Europe are fairly safe. A brief outbreak on the part of ecclesiastical authority did follow the publication of the "Origin of Species", but that is now perceived to have been a mistake. The convictions of the masses may be trusted to remain in essentials what they have always been; and

148   MEMOIR

I suppose that if science were to declare to-morrow that man descends from slugs or from centipedes, no episcopal lawn would be ruffled here. Unfortunately the American incidents suggest that our destinies may not much longer remain in the hands of that exalted tribunal, and that trouble may not be so far off as we have supposed.

In a letter written 10. viii. 1924 from Stockholm he alluded once more to this incident. He had heard

wonderful stories about Washington State University (away West). The more intelligent of the younger staff had secret "Sphinx" meetings at which Darwin and Spencer and similar disconcerting books were read and discussed. All were under oath not to divulge the nature and subjects of these meetings, for surely they would have been expelled, had the governing body learnt about them. That was in 1913, eleven years before I joined W. J. Bryan. So my responsibility is not grave.

W. B.

From this full and busy tour in Canada and the United States he returned home in February 1922, not in good health. His energy had again overtaxed his strength. He could not remember, when interested, that he must go carefully.

In April of that year we lost our second boy.

Months of despondency and dejection followed. Even the garden and laboratories almost failed to rouse him. But, happily, a new interest began at this time to push into his life.

In May 1922 he was unanimously elected to be a Trustee of the British Museum. This honour gave him extraordinary pleasure; it was an expression of confidence that he felt he must stir himself to justify. No foreign travel or medical *régime* could have helped him as did this unexpected appointment. Gradually, from fortnight to fortnight, his interest in his new responsibility grew, and with his interest grew his pleasure.

From boyhood he had been a devotee of Museums, whether of Natural History, objects of art, or pictures. He knew most of the collections in Europe well, both those of the great cities and the minor exhibits up and down the provinces.

Perhaps no distinction the world could have given would have gratified him more. His energy and his spirit grew again.

The honour of Knighthood also was offered to him in May 1922, but this he declined.

He recovered full interest in garden and green-house, and was again busy watching and devising new experiments. At this time the phenomenon of somatic segregation held his attention and occupied his thought.

His health mended. In August 1924 he made a tour round the Plant-breeding stations of Denmark and Sweden. This was a great success. He was roused to enthusiasm by the quantity and the quality of the Genetical work that he found going forward in these centres of research.

From Stockholm he wrote:

10. viii. 24. Three wonderful days—chock full—morning till 1 a.m. Endless things to see, to eat, to drink. Of all the Royal Progresses I have gone, this is the royallest. The chromosome cult here is after all only a half-hearted affair. Evidently misgiving has supervened and I hammered a few wedges into the cracks.... Stockholm seen only at sunset, a most lovely place....

12. viii. 24. A long solitary day. After so much talk not unpleasant. In the train I found myself next a young Alsatian, (French) but of the most cosmopolitan, and his society passed the time well. A gentleman in all senses. I did not discover his

occupation. The conversation touched on beer and I told him about Carlsberg, Hansen's pure yeasts etc. This led to inquiries as to how beer was made, and it transpired that he had not the faintest conception of what fermentation might be—the nature of yeast etc. etc. He asked, "Mais qu'est-ce que c'est, que la levure?" Astonished to learn it was a plant which grew, etc. So I think he must be in some high Government service, probably Diplomacy. A charming young man about 30, or rather less. Evening (presumably sunrise also) is Stockholm's hour and I was lucky to arrive at the right moment. Beautiful nevertheless all the time.... Pictures—the good ones some dozen—very fine. *Claudius civilis* (Rembrandt) superb. It has the grandeur and unexpectedness of the *Polish Rider*. Chardins—at least two of them delicious; I have worked straight through some 25 solanders of French drawings and hope to do Italy to-morrow....

The friendly welcome from his colleagues gladdened him; the food and the climate suited him, and he was very much interested and delighted to see so much promising work in hand. He came back in better health and spirits than for some time past.

He had an intense curiosity to see for himself what Russia was making of her huge and cruel experiment; deeper than curiosity lay the thought, or the wish, that a show of sympathy and interest might contribute to the restoration of civilisation and help to give security to intellectual life and scientific pursuits in that land of tyranny.

Warm friendship for his former pupil, Professor Vavilov, provided another motive, and also at the back of his mind lay a half dream reminiscence of his youthful expedition to the Steppe.

Negotiations for permission to travel in Russia in 1924 broke down—the permit came too late; but in 1925 on the occasion of the bi-centenary of the Russian Academy of Science he received an invitation to attend that function, and the way was made smooth for him.

From the moment of asking permission to enter Russia in 1924 he had characteristically set to work to learn Russian. He soon revived the knowledge that he previously had had of the language and added to it enough to be able to understand most of what was spoken in his presence. During his visit to Russia he was able to act as interpreter for other foreigners on several occasions. He saw much and was deeply interested, but he was also very much shocked; he returned safely. He sent a discreet account of his journey to *Nature*:

SCIENCE IN RUSSIA

By W. BATESON, F.R.S.

In consultation with Sir Henry Miers and Prof. D'Arcy Thompson, who were among the British delegates, I have prepared the following account of our recent visit to Russia to the celebrations of the two-hundredth anniversary of the Russian Academy of Science. The celebrations began on September 5. We were to have six days in Leningrad, full of every kind of function both solemn and festive, to be followed by four more, similarly crowded with events, in Moscow. We came determined to see and hear all that we possibly could in the time; and aware that the gathering had been organised largely with an eye to its propaganda-value, I suppose that most of us with our curiosity combined a considerable measure of scepticism as to the real value of anything we might be shewn.

Nothing could exceed the courtesy and hospitality with which we were received. Railway travelling in Russia was to be entirely free, and those who cared to do so were invited to visit any place in the country, Kieff, Odessa, etc., that could be seen before the end of September. At the frontier our baggage was passed in (and on the return journey passed out) unexamined, and we were authorised to send cables gratis. New motor omnibuses, specially sent from Moscow, met us at the station, and after we had been duly filmed, conveyed us to our hotel— a fine and well-appointed house. Rumour alleged that no other hotel could provide even reasonable comfort, but that

may have been untrue. Of the ordinary charges we paid about
half, but at Moscow, where we were distributed among several
hotels, no charge at all was made.

If these preliminary symptoms inclined us to complacency
we soon saw others of a different character. In the long drive
through the streets of Leningrad the evidence of the empty and
half-ruined buildings, the peeling walls, the dishevelled pave-
ments, and the broken roadways—full of dangerous and neglected pits, above all the clothes of the population, with rarest
exceptions, dingy and often improvised of coarse and unusual
materials, were indications of troubles which no camouflage
could conceal. Wherever we went in Leningrad, though in a
slightly less degree perhaps in Moscow, we had the experience—
to scientific persons novel and rather disconcerting—of finding
ourselves conspicuously well dressed. It was indeed a little
embarrassing to meet men of refinement and learning whose
trousers were eked out with large and unrelated fragments,
though almost worse to pass groups of artisans going home from
work in rags which suggested the casual ward of a workhouse
rather than decent employment. Readers of *Nature* will not
expect a report on social conditions, but some hint of the setting
in which we saw what we did see can scarcely be omitted.

In preliminary announcements the numbers of foreign
visitors expected had been given as about a hundred and fifty.
We were not furnished with any list of those who did attend,
but the highest estimate of our actual numbers which I heard
was ninety-six. Of these perhaps half were representatives of
mathematics and natural sciences in the stricter sense, the remainder being economists, historians, orientalists, and especially
students of Slav languages, etc.

The proceedings opened with an evening reception in the
rooms of the Academy. On this and on most other formal
occasions our arrival was attended by a guard of honour with
bayonets fixed, and we marched in under the lime-light of the
ubiquitous cinema men. Here we found ourselves in an immense gathering assembled from all parts of Russia, representative of every kind of learning, and if several of the eminent
foreigners whose presence had been promised failed to appear,

there were no defects on the part of our hosts, for practically every one with whom we had acquaintance, direct or indirect, was there.

On the following Sunday we assembled for the first of several public séances, which both in Leningrad and Moscow constituted the formal business of the congress. Proceedings began—as always—with the singing of the International, which was followed by discourses of various kinds. After an introduction from the president, Karpinski, the secretary of the Academy, Oldenburg, gave an interesting history of its origin and activities, reminding us of the many distinguished men who had worked under its auspices. Other members of the Academy, especially Steklov and Lazarev, followed him, but alternating with them were speeches by members of the Government, giving their views on what science had done and might do for the people. At various times we were thus addressed by Kalinin, president of the Soviet, Krassin, Kameneff, Lunacharsky, minister of education—on several occasions—and other well-known leaders. At one of these séances Zinovieff, president of the Leningrad Soviet, gave a long and rhetorical address, developing an analogy between the aims of science and those of the revolutionists, which was afterwards printed.

Among much in all these speeches which of course escaped us, one conclusion very plainly emerged, that the revolutionary government is perfectly sincere in its determination to promote and foster science on a very large scale. Signs were not wanting that science, especially perhaps in its applications, is regarded by the present governors of Russia as the best of all propaganda. It was interesting to hear the faith that the advancement of science is a first duty of the State proclaimed by professional politicians. We ought perhaps not to inquire too closely whether they and we mean the same thing by the term science. Zinovieff, for example, speaks in the same breath of the "discovery" of Karl Marx and the "discovery" of Charles Darwin. Each of these men valued the work of the other, he told us, in a very high degree, and in evidence of Darwin's admiration of Marx he spoke of a letter, sent in acknowledgment of the book on capital, which was to be published. In an impassioned passage, which was enthusiastically applauded, he drew a parallel—derived from Lenin—between what science had done for the abolition of syphilis, patiently trying and discarding 605 reagents before 606 was reached, with what revolutionists were attempting for the abolition of capitalism. Capitalism was a scourge worse than syphilis, and if need be, not 606 but 6006 remedies must be tried so that the planet may be delivered once for all from its "yoke". Judging, however, from the proportion of the population lost in the first experiments,[1] those who survive to benefit by the ultimate deliverance will be few.

Whether science will soon fulfil the many promises made in its name must be doubtful, yet not merely from what we heard, but also from what we saw, the failure we may be sure will not be for want of genuine effort. Next in importance after communism, the tenets of Leninism assert the doctrine that science is the basis of happiness. Religion is to be eradicated as a vice; science and the arts are to be promoted in its place. Those who regard this fervour for science as mere affectation for propaganda purposes miss an essential fact, and in our visits to laboratories and institutions, as related below, we found abundant and imposing evidence of the expression of this faith in material works. Of these new institutions we saw a great many, and though of course none of us could personally visit more than a part, we found on comparing notes that certain features were common to most of them. Palaces and great houses from which the owners have been dispossessed have been hastily adapted for the purposes of science. The effect was often incongruous. Though objects of value had been removed we saw laboratory benches improvised among the remains of Empire furniture and statuary, and under Boucher ceilings representing nymphs sporting with *amorini*. Formerly each institution would have the Tsar's portrait and probably an occasional ikon. The

---

[1] Though accurate figures are wanting, some estimates may be formed from the "Report on Epidemics in Russia since 1914", 2 Parts, by Prof. L. Tarassévitch, published at Geneva, 1922, for the League of Nations. On the basis of the statistics provided he estimated that "Russia may consider herself fortunate if she emerges from the present crisis with the loss of 20–25 per cent. of her population". Pt. 2, p. 44.

modern analogue of these is the "red room", arranged as a sort of shrine of communism with the bust of Lenin, surrounded with copious literature of propaganda. Signs also there were that, in other and less innocent ways, not only the universities but also the various scientific institutions were utilised as agencies for the dissemination of communistic ideas.

Doubts might be entertained as to whether the best atmosphere for research has been created in the institutions, but no praise is too high for the zeal and vigour with which work is being conducted in the new circumstances. As typical of them may be mentioned the Institute of Zoological and Botanical Research under Prof. Philiptschenko and Prof. Dogiel, which has been set up in the house and parks of the Leuchtenberg family at Peterhof. Besides the permanent staff, many hundreds of students are there accommodated in the summer months, partly for instruction and in part co-operating in the experiments we saw in progress. The whole gave the impression of a very active and well-organised school, which has already done excellent work both in fundamental and applied biology. At Moscow Prof. Nawaschin and a band of cytologists trained by him are housed in a fine building and equipped with good instruments. His researches are of course classical, and we saw with satisfaction that in this case at least, one of the "purest" of biological sciences was not suffering through the competition of the applied branches.

Another very large house, also in Moscow, has been assigned to Prof. Koltsov as an Institute of Experimental Biology. This includes numerous departments, especially a genetical station under Prof. Serebrovsky, work in experimental morphology, hydrobiology, etc. At the old Agricultural School at Petrovsky-Razumovsky, we saw more genetical work in application to agriculture and the long-established researches of Prof. Prianischnikov on agricultural chemistry. The Darwin Museum, created by Prof. Coats and his wife in illustration of the theory of evolution, was a remarkable and I believe unique curiosity. It contains a quantity of valuable and largely unpublished material bearing on the incidence of variation and a great number of other novelties. We were invited to inspect many more institutions, especially various establishments connected with public health, but all these I had to forgo.

Among the new organisations of a biological character the most extensive is the Institute of Applied Botany and Plant Breeding. The immediate object is to provide breeds of cereals and other agricultural plants for the various parts of Russia. The work is in the hands of Prof. Vavilov, who has already built up a great establishment for this purpose, employing 350 people, of whom some 200 are trained workers. In his travels through Turkestan, Afghanistan, and neighbouring countries, and by a vast correspondence, collections of seeds of wheat, barley, rye, millet, flax, etc., have been brought together on a great scale. The central office is in Leningrad and occupies a very large building, which is in great measure a living museum of economic plants as represented by their seeds. Of wheat alone some 13,000 forms are here collected. In various parts of the country are twelve subordinate stations, and by sowings made once in about three years it is proposed to perpetuate most of the collection alive. Besides a cytological department under Prof. Levitzky, there are special sections for meteorology, statistics, etc. Exceptional opportunities have been provided for investigating the geographical distribution of cultivated plants, especially in its bearing on problems of origin, which have resulted in several novel suggestions. At Dyetskoe (formerly Tsarskoe) Seló is the home breeding-station of this Institute. The chief building on it is a pleasant house, originally intended as a villa for Queen Victoria, and adjacent to this a number of laboratories and additional accommodation have been provided.

Of the old-established scientific institutions, the Museums of Zoology and of Palaeontology and the Botanic Garden were the only ones which I managed to visit. The fine and largely unique contents of the two museums are of course familiarly known. One recent acquisition I saw with mixed feelings—the almost complete skeleton of the giant Indricotherium, a rhinoceros-like creature, standing about twelve feet high at the shoulder. These bones, the collection of which has taken three years, were found exposed at Chalkar Tengiz, a remote and little known locality, but from the photograph of the site I saw that in 1886 I must have passed within a few yards of the very spot. The Botanic Garden alone still shewed signs of the hardships through which it had come.

Great provisions are being also made for the development of art, archaeology, and ethnology. Though unable to visit the Musée Russe, in which the collections illustrative of the history of the races included in the Soviet are preserved, I heard continually of its extraordinary excellence. Though it is the work of several decades, its recent extension has been much assisted under the new régime. Of the new developments at the Hermitage, Sir Martin Conway lately published an account. Among later novelties I was shewn the astonishing collections brought from East Turkestan by M. Kozlov's recent expeditions. From tombs originally constructed at a depth of twelve metres below the surface and permanently immersed in water, have been brought a vast series of textiles, both woollen quilts and silks in admirable preservation, together with ornaments of various kinds and domestic objects, the whole belonging to a civilisation previously unknown. The decorations representing stylised animals, especially reindeer (possibly elk) and oxen, are evidently Sassanian in affinity, as M. Orbeli demonstrated. Some of the designs in the silk fabrics are of incomparable beauty, and very striking also are fragments of a curious red and black lacquer shewing the utmost delicacy of taste and workmanship. These remains have not as yet been dated with certainty, but are presumed to cover a long period belonging chiefly to the sixth-seventh century A.D., but in part probably to a much earlier epoch.

Of all the things seen I suppose the interior of the Usspensky Sobor, the great Cathedral of the Coronations, exhibited the greatest break with past traditions. Until lately the whole was plastered with gold leaf and paintings rising little above the *objet de piété* type of art, giving an effect certainly solemn and gorgeous but tasteless and oppressive. The archaeologists have declared the whole of this display of orthodoxy to be modern and debased. The process of stripping has been begun and is steadily proceeding under scholarly and skilful guidance. The ancient pictures are being slowly recovered and will be seen before long as they were intended to be seen upon the original white of the walls. The most famous and venerable of all ikons, the Korsun Madonna, traditionally ascribed to St Luke, has already been cleaned. I remember it looking like a piece of black leather. Seven layers of paint have been removed and a most remarkable original picture has been displayed in its entirety. Terrible judgments were expected by the common people to follow these impieties and the opening of the reliquaries by which they were accompanied. Time alone can shew whether their anticipations will be justified, but meanwhile there is no gainsaying the fact that in the restorations which are to be extended throughout the Kremlin churches, and carried out upon every ikon of importance, the history of religious art in Russia will be revealed.

Of the banquets, lordly, if indeed the term princely were not more appropriate, graced, as that in Moscow at least was, with Imperial porcelain and madeira; of the operas, of the ballets and the sumptuous production at the Arts Theatre, where several of the courtly costumes were real brocades—lately somebody's "properties" in the fullest sense—one I verily believe lent from the Kremlin collection itself—this is not the place to speak at length.

We left with no clear conception of the principles or practice of communism; and in particular as we looked round at the 1200 persons assembled for the Moscow banquet, in a city teeming with beggars, we tried in vain to conjecture any system by which admission to the table, or exclusion from it, may have been determined. Of liberty we saw no sign. We here are accustomed to think of science and learning as flourishing best in quiet places, where they may come to slow perfection, under systems providing a reasonable measure of personal independence and security. Present conditions in Russia have brought about the very contrary, and among the grave indications of disharmony, which every visitor observes, the want of freedom is by far the most serious.

There remains but little to recount.

In August of 1925, he wrote to the Council of the John Innes Horticultural Institution to warn them that in the following year he would reach the age of 65, and offering to retire. Sir David Prain (Chairman) in answer wrote a delightful letter urging him to continue. Will was deeply moved by these expressions of confidence. He replied:

The Manor House, Merton, S.W. 20,

8 Jan. 1926.

Dear Sir David,

To receive such a letter as that which has just come from you is one of the rare pleasures of life. When I came here first I looked forward to doing much more than has been accomplished. Nor can I lay the blame for deficiencies wholly on the bad times through which we have passed. When the centre of chief interest in Genetics shifted away from work of my own type, to that of the American group, I was already too old and too much fixed in my ideas to become master of so very new and intricate a development. It has taken me years even to assimilate the new things and I recognise that the Institution has a right to a younger man in my place.

But another short-coming which may be even more serious is that I have not myself succeeded in bringing forward such a man, one I mean who might carry on the Institution somewhat on the lines we have hitherto followed. This anxiety has been continually in my thoughts and doubtless also in yours. The causes of the failure I need not now discuss. There are fashions in all things especially in research, and just now those who in our day would have been zoologists or botanists are bio-chemists, almost every one. Having been so long disappointed I am not sanguine that in the short time that remains I can succeed, but I shall make the question of a successor more and more my first preoccupation.

I accept the invitation to go on a bit longer with pride and gratitude. This is but a blunt answer to your charming and graceful intimation. The message of confidence which you have sent from the Council is an extraordinary gratification to me, and I should like on this occasion to thank them, not only for their kind words, but also for the encouragement they have so long given to us all of the Institution staff in freely and continuously facilitating our various undertakings.

Yours truly,

W. Bateson.

How short the time was to be was happily veiled from us all.

He died a month later, on 8 February 1926.

By his wish his body was cremated, and his ashes scattered, at Golders Green, 12 February 1926.

C. B. B.

HYBRIDISATION AND CROSS-BREEDING AS A METHOD OF SCIENTIFIC INVESTIGATION

Journ. Roy. Hort. Soc. Vol. xxiv, 1900

(Read 11 July, 1899)

It is with a special pleasure that I accepted the kind invitation of the Council to address this Conference of persons interested in hybridisation. Of all the methods which are open to us for investigating the facts of Natural History there is perhaps none which is more likely to bring forth results of first-rate importance. Not only is the field a vast one, but the work is ready to hand. Though the patience and labour needed are very great, the practical methods are simple, and can be in many cases carried out by any person who has leisure and is able to carry out anything accurately. Leisure, accuracy, and a garden of moderate extent are almost the only equipment necessary for such work. On the other hand, the scientific importance of the results to be obtained is transcendent.

It is perhaps simpler to follow the beaten track of classification or of comparative anatomy, or to make for the hundredth time collections of the plants and animals belonging to certain orders, or to compete in the production or cultivation of familiar forms of animals or plants. But all these pursuits demand great skill and unflagging attention. Any one of them may well take a man's whole life. If the work which is now being put into these occupations were devoted to the careful carrying out and recording of experiments of the kind we are contemplating, the result, it is not, I think, too much to say, would in some five-and-twenty years make a revolution in our ideas of species, inheritance, variation, and the other phenomena which go to make up the science of Natural History. We should, I believe, see a new Natural History created.

It seemed to me that I could not better make use of this opportunity than by indicating, as far as I can, some of the *aims* which I think a worker in this field should put before him, and the class of work which, as it seems to me, is most likely to prove fruitful in bringing about the result I have indicated.

The problem, it is assumed, on which all such work is to be brought to bear is the *problem of species.*

I must ask you for a moment to consider the present position of knowledge in regard to Evolution and the nature of Species—

B

for it is with a clear reference to the problem of species that breeding experiments, in the first instance, should, in my opinion, be undertaken. We see all living nature—animals and plants—divided into the groups which we call *species*, groups often so sharply marked off that there can be no doubt where they begin and end; groups often, on the other hand, so irregularly characterised that no two people would divide them alike. What are the causes that brought this about and keep it so? What are the facts underlying this phenomenon of *species*? For phenomenon it is; and, believe as we may that all these forms are related in descent, there they are now, grouped into species as we know. How did this come about?

We all know the accepted view. We start from the fact that, since of all forms of life many more are born than can possibly survive, some—indeed, nearly all—must perish and leave no descendants. Next we observe the fact of Variation—that even the offspring of the same parents are never precisely alike, but vary. Now, since all cannot survive, it is clear that different individuals have a different chance of survival and of being represented by descendants. For each individual this chance will depend on the degree to which its structure and aptitudes fit it to bear its part in the struggle to which it is exposed. Briefly, *on the whole* the fittest will survive and breed.

Lastly, as the places in life that the organisms fit are diverse, so the forms of the surviving organisms are diverse too.

Everyone who cares at all for Natural History knows this reasoning, and knows also the difficulties by which its application to the facts of Nature is beset—how simple the theory seems when thus stated in general terms, but how hard it is to apply it in detail to a particular case.

Of all these difficulties the most serious are two. The first is the difficulty which turns on the magnitude of the variations by which new forms arise. In all the older work on evolution it is assumed, if the assumption is not always expressly stated, that the variations by which species are thus built up are *small*. But if they are small, how can they be sufficiently useful to their possessors to give those individuals an advantage over their fellows? That is known as the difficulty of *small or initial variations*.

The second difficulty is somewhat similar. Granting that variations occur, and granting too that if they could persist and be perpetuated species might be built up of them, how *can* they be perpetuated? When the varying individuals breed with their non-varying fellows, will not these variations be obliterated? This second difficulty is known as that of the *swamping effect of intercrossing.* Now on each of these two points the work of the hybridist and the experimental breeder comes in exactly. It is he who can see the variations arise, and can note their size and find out exactly how large they are—whether they are great or small—whether offspring do really differ but little from their parents, or whether, in certain cases and in respect of certain characters, the differences in variation may not be very great and definite; whether, also, the supposed swamping effect is a real one or not, or to what extent it is real, and in regard to what characters.

I need not tell a body of persons, most of whom have themselves made experiments of this kind, that in numberless cases both great and thoroughly definite variations do occur. This much every practical man now recognises. But we are far from knowing which kinds of variations may thus be definite and palpable, and which are not. All we know is that both large variations and small variations occur, some in one character and others in other characters, and that characters which in one species may vary greatly and suddenly, in other species vary only slowly or hardly at all. All this is a matter which comes daily under the observation of the breeder—especially the cross-breeder of plants or of animals. It is to him that we look for first-hand evidence as to the *magnitude of variations.*

At this point a word of caution is needed. All those present are aware of the great and striking variations which occur in so many orders of plants when hybridisation is effected. As everyone knows, it is to those extraordinary "breaks" that we owe perhaps the majority of our modern flowers. Such, for example, are Narcissus, Begonia, Pelargonium, Gladiolus, Streptocarpus, a great number of Orchids, Rhododendron, the Cineraria, and the like. I mention the Cineraria, because I have personal knowledge of these hybrids, and because I notice that the view that our garden Cinerarias are not hybrids is being again repeated, in spite of the clear evidence, both of history and recent experiment, to the contrary.

With such cases in view some may be disposed to say: "Here are the great and striking variations we are seeking. These new forms are like new species—some would even take rank as new genera. May not the natural species have arisen in like manner by hybridisation?" The answer to this question, however, is almost certainly *No.* And herein I believe most, if not all, professed botanists and zoologists will agree. To go into the matter fully here is impossible; but for many reasons, most of which have often been repeated, there is, I think, no good

11-2

evidence for supposing that any natural species, whether of
animal or plant, arose by direct hybridisation. Tempting as it
may at one time have been to hope that we should thus get a
short cut to the origin of species, few, I think, are now sanguine
of such an issue. It is not in this direction that we can look for
that advancement in knowledge which I believe will surely
come from the work of the cross-breeder.

I am far from saying that these striking hybrids are without
scientific interest, or that they have no bearing on the problem
of species. I wish only to say that it is pretty clear that they
have not the direct bearing which they would have if it could
be supposed that natural species arose as similar hybrids.

The interest in the cross-breeder's work lies, as I think, in a
somewhat different field. Whatever view we adopt of the
origin of species—provided that we believe in the doctrine of
Descent at all—we believe that every species has been actually
produced from something like itself in general, though different
in some particular. Wherever these two closely allied varieties
exist, the problem of species is presented in a concrete form:
How did variety A arise from variety B, or B from A, or both from
something else? This question involves two further questions:

1. By what steps—by integral changes of what size—did the
new form come into being?

2. How did the new form persist? How was it perpetuated
when the varying individual or individuals mated with their
fellows? Why did it not regress to the form from which it
sprang, or to an intermediate form?

To those who admit this reasoning it will be clear that the
whole question of the origin of species turns on the relationship
of each species or each variety to its *nearest allies*. We may
not yet have an authentic case of a nascent species that will
satisfy all doubts, but unquestionably we have lots of nascent
varieties. If only we make it our business to observe the way in
which these nascent varieties come into being, and especially
what happens when these varieties are crossed with their nearest
allies, we shall have material from which to answer the main
questions of which the Species problem consists.

It is only quite lately that any systematic study of such
variations has been undertaken from the point of view of the
evolutionist, and already some very clear results have been
perceived.

As the first difficulty in applying the doctrine of Descent
turned on the magnitude of variations, so, as soon as careful
study of Variation is begun it is found that large and distinct

variations are by no means rare, and that in certain classes of characters they are indeed the rule. To this class of variation, in which the variation is found already at its beginning in some degree of perfection, I apply the term *discontinuous*.

We are taught that Evolution is a very slow process, going forward by infinitesimal steps. To the horticulturist it is rarely anything of the kind. In the lifetime of the older men here present it is not Evolution but Revolution that has come about in very many of the best-known Orders of horticultural plants. Even the younger of us have seen vast changes. It may have seemed a slow process to individual men in the case of their own speciality. It may have taken all their lives to obtain and fix a strain; but in Evolution that is nothing. It is going at a gallop!

Whenever, then, it can be shown that a variation comes discontinuously into being, it is no longer necessary to suppose that for its production long generations of selection and gradual accumulation of differences are needed, and the process of Evolution thus becomes much easier to conceive. According to what may be described as the generally received view, this process consists in the *gradual* transition from one normal form to another normal form. This supposition involves the almost impossible hypothesis that every intermediate form has successively been in its turn the normal. Wherever there is discontinuity the need for such a suggestion is wholly obviated.

The first question was: How large are the integral steps by which varieties arise? The second question is: How, when they have arisen, are such variations perpetuated? It is here especially that we appeal to the work of the cross-breeder. He, and he only, can answer this question: Why do not nascent varieties become obliterated by crossing with the type form?

If you study what has been written on these subjects you will find it almost always assumed that such blending and obliteration of characters is the rule in nature. Whole chapters have been compiled with the object of showing how, in a world in which there is such complete blending, evolution might still go on. There has been a word invented to expressly denote this kind of blending; the word is *Panmixia*, a word barbarously and incorrectly formed to denote an idea which is for the most part incorrect likewise. For if instead of abstract ideas the facts of cross-breeding are appealed to, it is found that so far from this blending and gradual obliteration of character being the rule, it is nothing of the kind. In many characters, on the contrary, it is at once found on crossing that the varying character may be transmitted in as perfect a degree as that in which it was found in the parent. It need scarcely be said that there are many structures and conditions which do not thus retain any integrity when crossed, but there are very many that do. Which characters are thus unblending, and which blend, must be determined by careful cross-breeding; and this knowledge can be discovered in no other way.

The recognition of the existence of discontinuity in variation, and of the possibility of complete or integral inheritance when the variety is crossed with the type, is, I believe, destined to simplify to us the phenomenon of evolution, perhaps beyond anything that we can yet foresee. At this time we need no more *general* ideas about evolution. We need *particular* knowledge of the evolution of *particular* forms. What we first require is to know what happens when a variety is crossed with its *nearest allies*. If the result is to have a scientific value, it is almost absolutely necessary that the offspring of such crossing should then be examined *statistically*. It must be recorded how many of the offspring resembled each parent and how many shewed characters intermediate between those of the parents. If the parents differ in several characters, the offspring must be examined statistically, and marshalled, as it is called, in respect of each of those characters separately. Even very rough statistics may be of value. If it can only be noticed that the offspring came, say, half like one parent and half like the other, or that the whole shewed a mixture of parental characters, a few brief notes of this kind may be a most useful guide to the student of evolution. Detailed and full statistics can only be made with great labour, while such rough statistics are easily made. All that is really necessary is that *some* approximate numerical statement of the result should be kept. The horticulturist makes a cross: he is perhaps obliged by want of time and space simply to keep what he wants and throw the rest away; but sometimes surely he might put down a few words as to what that "rest" consisted of. If he would do so he would have the gratitude of many a student hereafter. On looking through the literature of hybridisation one is saddened by the thought that while so much skill and money and effort have been expended, for want of a very little more attention to recording, immeasurable opportunities have been missed.

We have seen that it is likely that those experiments will be found the most fruitful which deal with the relationship subsisting between a given variety or species and its nearest allies. The essential problem of evolution is how any one given step in evolution was accomplished. How did the one form separate from the other? By crossing the two forms together and studying the phenomena of inheritance, as manifested by the cross-bred offspring, we may hope to obtain an important light on the origin of the distinctness of the parents, and the causes which operate to maintain that distinctness.

Useful contributions to the physiology of inheritance may no doubt be made by experimental crossing of forms only remotely connected. Such work, however, will not supply the particular kind of evidence most needed. This can only be got by an exhaustive study of the results of cross-breeding between various forms whose common origin is not very distant. Such experiments must, besides, be repeated sufficiently often to give a fairly extensive series of observations on which to base conclusions. Anyone, therefore, who wishes to work on these lines would do well to restrict himself to an examination of the transmitting properties of a small group of closely allied varieties or species, and to explore these properties thoroughly within that group.

Cross-breeding, then, is a method of investigating *particular* cases of evolution one by one, and determining which variations are discontinuous and which are not, which characters are capable of blending to produce a mean form and which are not. It has sometimes been urged as an objection against this method of investigation that the results are often conflicting. It has been said that such work will only lead to accumulations of contradictory evidence. It is, however, in this very fact of the variety of results that the great promise of the method lies. When varieties and species are tested by this method it is found that their mutual relations are by no means alike, and properties are disclosed which can in no other way be revealed.

In illustration, I will refer to three cases of hairy and smooth varieties. In each case there is a well-marked discontinuity between the two varieties; but, as is shewn by the evidence obtained by cross-breeding, the nature of the relationship[1] of the two forms to each other is different in each case, and the distinctness is maintained by different means.

The plants (produced at the meeting) illustrating the following observations were raised by Miss E. R. Saunders, of

[1] The term "relationship" is somewhat misleading, but I cannot find a better. It is used to denote not simply the blood-relationship of the forms to each other, but those physiological relations subsisting between them which are manifested by experimental crossing. The word is thus used in a sense similar to that which it bears when we speak of the chemical relations of one substance to another.

Newnham College, Cambridge, who is carrying out a large series of experiments on this subject.

The first case is that of *Matthiola incana*, a hoary species, and its smooth variety known in gardens as the wallflower-leaved Stock. Experiments in crossing these two forms were made by Brevor Clarke, and briefly described by him in "Report of gotanical Congress", 1866. Amongst other things his investigations shewed that on crossing these two varieties the offspring consisted entirely of completely hoary and completely glabrous individuals, no intermediate being present. Miss Saunders' work entirely confirms this result. The type-form used by her was procured from seed of presumably wild specimens growing in the Isle of Wight. The glabrous variety was the ordinary garden form the origin of which is not known to us. In this case discontinuity is manifested in its simplest form.

The second example is that of *Lychnis diurna*. There, again, the normal is hairy. A glabrous variety was found by Professor de Vries, and was by him crossed with the type. All the first generation of cross-breds inherited the hairiness in its complete form. When, however, these plants were crossed again with the smooth form, the result was a mixed progeny, of which some were hairy and others smooth. The same result also occurred when the cross-bred plants were bred with each other. Professor de Vries kindly sent seed of his glabrous form to Cambridge, where Miss Saunders repeated the experiments with the same results. In all the cases of mixed progeny there is a sharp discontinuity.

The third case is that of *Biscutella laevigata*. A full account of this important case was published by Miss Saunders in *Proc. Roy. Soc.* LXII, 1897, 11. Briefly the facts are as follows. The species is common as a hairy plant throughout a great part of the Alps. In a few localities a variety occurs having the *surfaces* of the leaves quite devoid of hairs. (There are almost always some hairs on the margins and leaf-teeth.) When present, this smooth form occurs abundantly, mixed with the hairy type. Intermediates are of rare occurrence. If plants of the two kinds breed freely together, as in the natural state we must suppose they do, how is the sharp distinction in their respective characters maintained? The result of artificial cross-breeding went to shew that of the young seedlings of mixed parentage some were hairy, some smooth, and a good many intermediate. But as these seedlings grew, the hairy and the smooth retained their original characters, while the intermediate ones gradually became smooth. The transition was not effected by actual loss of hairs, but after the first few leaves of intermediate character the leaves subsequently produced were smooth.

In all these three cases there is discontinuity, the intermediates between the varieties being absent or relatively scarce. Nevertheless, on examination it is found that the discontinuity is not maintained in the same way in the different cases. The transmitting powers of the one variety in respect of the other are quite different in each case, and it must, I think, be admitted that we have here a fact of great physiological significance. In each of the three cases enumerated the two varieties are seen to stand towards each other in a different relation, and in each the mechanism of inheritance works differently.

From facts like these we perceive how imperfect is the survey of the characteristics of species and varieties which can be obtained by the ordinary methods of anatomy and physiology. There can be no doubt that, tested by the method of breeding and by study of the transmitting powers, the relation of varieties and species would be shewn in an entirely new light. We are accustomed to speak of "variability" as though it were a single phenomenon common to all living things; and just as the older naturalists spoke of *species* in general as all fixed and comparable entities, so many of the present evolutionists speak of "varieties" in general as all comparable. This is a mere slurring of the facts. Not only must variability in respect of *different* characters be a manifestation of distinct physical processes, but, as we have seen, variability, even in what appears to us to be the same character, may be a wholly different matter.

Our business, then, is to test and examine these different kinds of variabilities according to their behaviour when the different varieties are crossed together. By this means we are enabled to investigate the properties of organisms in a way that no other method provides.

If I may be allowed to use a metaphor taken from chemical science, regarding species and varieties as substances, we may investigate their properties and their powers of entering into genetic combinations, just as the chemist investigates the powers of his bodies to enter into chemical combinations.

To lump all the different manifestations of variation together as "varieties", and to rest there, is to give up in despair.

Similarly, it is certain that what we call "species" is a mixture of different phenomena, or rather of different classes of phenomena confounded under one name. I look to the study of cross-breeding to unravel that extraordinary mass of confusion. I look to this method of investigation to deliver us from the eternal debates on the subject of what is specific rank and what is not.

On the one hand we have at the present day many who devote themselves entirely to discussions of this nature, though they know in their hearts that their views correspond to no natural fact whatever. On the other hand, many in disgust and impatience reject the whole thing. "There is no such thing as species", say they. Both sides are surely wrong: there is such a thing as species, and we have to find out what are the properties of species.

It is true that, as to most species and varieties, artificial breeding is impossible, but in numerous cases a beginning can be made. Take merely the phenomenon of local varieties, or local species, or local races, about which such weary discussions have arisen. Each of these offers a particular example of the Evolution problem. In numbers of such cases an investigation of the behaviour on crossing could be practised, and a very few such experiments would, I venture to predict, do more to establish true views of the relation of species and varieties than the labours of systematists will do in ages.

To come much nearer home, we do not know for certain the true relationships—in this special sense—between the varieties of the commonest domestic animals and plants. For example, I have been trying to investigate these relationships between the several kinds of comb in domestic poultry. I have thus far found no one who can tell me for certain what happens when they are crossed. The various forms of comb in our breeds of poultry— simple comb, pea-comb, rose-comb, etc.—are important structural features, which differ from each other very much as many natural species do. The answer generally given is that the result of such crossing is uncertain—that sometimes one result occurs, and sometimes another. This, of course, merely means that the problem must be studied on a scale sufficiently large to give a statistical result. There is here an almost untouched ground on which the properties of specific characters can be investigated. Many similar examples might be given.

True and precise experiments in these fields so ready to our hand have never been made. We appeal to those who have the opportunity to use it for the advancement of this fascinating line of research. It is delightful to form great collections of animals or plants, and to "bring out a novelty" may be an exhilarating sensation; but if anyone will abandon these well-worn pursuits, and devote himself to experimental cross-breeding, he will soon have his reward, for no line of research is likely to prove more fruitful.

PROBLEMS OF HEREDITY AS A SUBJECT FOR HORTICULTURAL INVESTIGATION

Journ. Roy. Hort. Soc. Vol. xxv, Parts 1 and 2, 1900

(Read 8 May, 1900)

An exact determination of the laws of heredity will probably work more change in man's outlook on the world, and in his power over nature, than any other advance in natural knowledge that can be foreseen.

There is no doubt whatever that these laws can be determined. In comparison with the labour that has been needed for other great discoveries it is even likely that the necessary effort will be small. It is rather remarkable that while in other branches of physiology such great progress has of late been made, our knowledge of the phenomena of heredity has increased but little; though that these phenomena constitute the basis of all evolutionary science and the very central problem of natural history is admitted by all. Nor is this due to the special difficulty of such inquiries so much as to general neglect of the subject.

It is the hope of inducing others to pursue these lines of investigation that I take the problems of heredity as the subject of this lecture to the Royal Horticultural Society.

No one has better opportunities of pursuing such work than horticulturists. They are daily witnesses of the phenomena of heredity. Their success depends also largely on a knowledge of its laws, and obviously every increase in that knowledge is of direct and special importance to them.

The want of systematic study of heredity is due chiefly to misapprehension. It is supposed that such work requires a lifetime. But though for adequate study of the complex phenomena of inheritance long periods of time must be necessary, yet in our present state of deep ignorance almost of the outline of the facts, observations carefully planned and faithfully carried out for even a few years may produce results of great value. In fact, by far the most appreciable and definite additions to our knowledge of these matters have been thus obtained.

There is besides some misapprehension as to the kind of knowledge which is especially wanted at this time, and as to the modes by which we may expect to obtain it. The present paper is written in the hope that it may in some degree help to clear the ground of these difficulties by a preliminary consideration of the question, How far have we got towards an exact knowledge of heredity, and how can we get further?

Now this is pre-eminently a subject in which we must distinguish what we *can* do from what we want to do. We *want* to know the whole truth of the matter; we want to know the physical basis, the inward and essential nature, "the causes", as they are sometimes called, of heredity. We want also to know the laws which the outward and visible phenomena obey.

Let us recognise from the outset that as to the essential nature of these phenomena we still know absolutely nothing. We have no glimmering of an idea as to what constitutes the essential process by which the likeness of the parent is transmitted to the offspring. We can study the processes of fertilisation and development in the finest detail which the microscope manifests to us, and we may fairly say that we have now a thorough grasp of the visible phenomena; but of the nature of the physical basis of heredity we have no conception at all. No one has yet any suggestion, working hypothesis, or mental picture that has thus far helped in the slightest degree to penetrate beyond what we see. The process is as utterly mysterious to us as a flash of lightning is to a savage. We do not know what is the essential agent in the transmission of parental characters, not even whether it is a material agent or not. Not only is our ignorance complete, but no one has the remotest idea how to set to work on that part of the problem. We are in the state in which the students of physical science were in the period when it was open to anyone to believe that heat was a material substance or not, as he chose.

But apart from any conception of the essential modes of transmission of characters, we can study the outward facts of the transmission. Here, if our knowledge is still very vague, we are at least beginning to see how we ought to go to work. Formerly naturalists were content with the collection of numbers of isolated instances of transmission—more especially, striking and peculiar cases—the sudden appearance of highly prepotent forms, and the like. We are now passing out of that stage. It is not that the interest of particular cases has in any way diminished—for such records will always have their value —but it has become likely that general expressions will be found capable of sufficiently wide application to be justly called "laws" of heredity. That this is so is due almost entirely to the work of Mr F. Galton, to whom we are indebted for the first systematic attempt to enunciate such a law.

All laws of heredity so far propounded are of a statistical character and have been obtained by statistical methods. If we consider for a moment what is actually meant by a "law of heredity" we shall see at once why these investigations must follow statistical methods. For a "law" of heredity is simply an attempt to declare the course of heredity under given conditions. But if we attempt to predicate the course of heredity we have to deal with conditions and groups of causes wholly unknown to us, whose presence we cannot recognise, and whose magnitude we cannot estimate in any particular case. The course of heredity in particular cases therefore cannot be foreseen.

Of the many factors which determine the degree to which a given character shall be present in a given individual only one is known to us, namely, the degree to which that character is present in the parents. It is common knowledge that there is not that close correspondence between parent and offspring which would result were this factor the only one operating; but that, on the contrary, the resemblance between the two is only a general one.

In dealing with phenomena of this class the study of single instances reveals no regularity. It is only by collection of facts in great numbers, and by statistical treatment of the mass, that any order or law can be perceived. In the case of a chemical reaction, for instance, by suitable means the conditions can be accurately reproduced, so that in every individual case we can predict with certainty that the same result will occur. But with heredity it is somewhat as it is in the case of the rainfall. No one can say how much rain will fall to-morrow in a given place, but we can predict with moderate accuracy how much will fall next year, and for a period of years a prediction can be made which accords very closely with the truth.

Similar predictions can from statistical data be made as to the duration of life and a great variety of events the conditioning causes of which are very imperfectly understood. It is predictions of this kind that the study of heredity is beginning to make possible, and in that sense laws of heredity can be perceived.

We are as far as ever from knowing why some characters are transmitted, while others are not; nor can anyone yet foretell which individual parent will transmit characters to the offspring, and which will not; nevertheless the progress made is distinct.

As yet investigations of this kind have been made in only a few instances, the most notable being those of Galton on human stature, and on the transmission of colours in Basset hounds. In each of these cases he has shewn that the expectation of inheritance is such that a simple arithmetical rule is approximately followed. The rule thus arrived at is that of the whole heritage of the offspring the two parents together on an average contribute one-half, the four grandparents one-quarter, the eight great-grandparents one-eighth, and so on, the remainder being contributed by the remoter ancestors.

Such a law is obviously of practical importance. In any case to which it applies we ought thus to be able to predict the degree with which the purity of a strain may be increased by selection in each successive generation.

To take a perhaps impossibly crude example, if a seedling shew any particular character which it is desired to fix, on the assumption that successive self-fertilisations are possible, according to Galton's law the expectation of purity should be in the first generation of self-fertilisation 1 in 2, in the second generation 3 in 4, in the third 7 in 8, and so on.

But already many cases are known to which the rule in the simple form will not apply. Galton points out that it takes no account of individual prepotencies. There are, besides, numerous cases in which on crossing two varieties the character of one variety is almost always transmitted to the first generation. Examples of these will be familiar to those who have experience in such matters. The offspring of the Polled Angus cow and the Shorthorn bull is almost invariably polled. Seedlings raised by crossing *Atropa belladonna* with the yellow-fruited variety have without exception the blackish purple fruits of the type. In several hairy species when a cross with a glabrous variety is made, the first cross-bred generation is altogether hairy.

Still more numerous are examples in which the characters of one variety very largely, though not exclusively, predominate in the offspring.

These large classes of exceptions—to go no further—indicate that, as we might in any case expect, the principle is not of universal application, and will need various modifications if it is to be extended to more complex cases of inheritance of varietal characters. No more useful work can be imagined than a systematic determination of the precise "law of heredity" in numbers of particular cases.

Until lately the work which Galton accomplished stood almost alone in this field, but quite recently remarkable additions to our knowledge of these questions have been made.

In the present year Professor de Vries published a brief account[1] of experiments which he has for several years been carrying on, giving results of the highest value.

The description is very short, and there are several points as to which more precise information is necessary both as to details of procedure and as to statement of results.[2] Nevertheless it is impossible to doubt that the work as a whole constitutes a marked step forward, and the full publication which is promised will be awaited with great interest.

The work relates to the course of heredity in cases where definite varieties differing from each other in some *one* definite character are crossed together. The cases are all examples of discontinuous variation: that is to say, cases in which actual intermediates between the parent forms are not usually produced on crossing. It is shewn that the subsequent posterity obtained by self-fertilising these cross-breds or hybrids break up into the original parent forms according to fixed numerical rule.

Professor de Vries begins by reference to a remarkable memoir by Gregor Mendel,[3] giving the results of his experiments in crossing varieties of *Pisum sativum*. These experiments of Mendel's were carried out on a large scale, his account of them is excellent and complete, and the principles which he was able to deduce from them will certainly play a conspicuous part in all future discussions of evolutionary problems. It is not a little remarkable that Mendel's work should have escaped notice, and been so long forgotten.

For the purposes of his experiments Mendel selected seven pairs of characters as follows:

1. Shape of ripe seed, whether round, or angular and wrinkled.

2. Colour of "endosperm" (cotyledons), whether some shade of yellow, or a more or less intense green.

3. Colour of the seed-skin, whether various shades of grey and grey-brown, or white.

4. Shape of seed-pod, whether simply inflated, or deeply constricted between the seeds.

5. Colour of unripe pod, whether a shade of green, or bright yellow.

[1] *Comptes Rendus*, 26 March 1900, and *Ber. d. Deutsch. Bot. Ges.* xviii, 1900, p. 83.

[2] For example, I do not understand in what sense de Vries considers that Mendel's law can be supposed to apply even to all "monohybrids", for numerous cases are already known in which no such rule is obeyed.

[3] "Versuche üb. Pflanzenhybriden" in the *Verh. d. Naturf. Ver. Brünn*, iv, 1865.

6. Shape of inflorescence, whether the flowers are arranged along one axis, or are terminal and more or less umbellate.

7. Length of peduncle, whether about 6 or 7 inches long, or about $\frac{3}{4}$ to $1\frac{1}{2}$ inches.

Large numbers of crosses were made between Peas differing in respect of each of these pairs of characters. It was found that in each case the offspring of the cross exhibited the character of one of the parents in almost undiminished intensity, and intermediates which could not be at once referred to one or other of the parental forms were not found.

In the case of each pair of characters there is thus one which in the first cross prevails to the exclusion of the other. This prevailing character Mendel calls the *dominant* character, the other being the *recessive* character.[1]

That the existence of such "dominant" and "recessive" characters is a frequent phenomenon in cross-breeding, is well known to all who have attended to these subjects.

By self-fertilising the cross-breds Mendel next raised another generation. In this generation were individuals which shewed the dominant character, but also individuals which preserved the recessive character. This fact also is known in a good many instances. But Mendel discovered that in this generation the numerical proportion of dominants to recessives is approximately constant, being in fact *as three to one*. With very considerable regularity these numbers were approached in the case of each of his pairs of characters.

There are thus in the first generation raised from the crossbreds 75 per cent. dominants and 25 per cent. recessives.

These plants were again self-fertilised, and the offspring of each plant separately sown. It next appeared that the offspring of the recessives *remained pure recessive*, and in subsequent generations never reverted to the dominant again.

But when the seeds obtained by self-fertilising the dominants were sown it was found that some of the dominants gave rise to pure dominants, while others had a mixed offspring, composed partly of recessives, partly of dominants. Here also it was found that the average numerical proportions were constant, those with pure dominant offspring being to those with mixed offspring as one to two. Hence it is seen that the 75 per cent. dominants really are not all alike, but consist of twenty-five which are pure dominants and fifty which are really cross-

[1] Note that by these useful terms the complications involved by use of the expression "prepotent" are avoided.

breds, though, like the cross-breds raised by crossing the two varieties, they only exhibit the dominant character.

To resume, then, it was found that by self-fertilising the original cross-breds the same proportion was always approached, namely:

25 dominants, 50 cross-breds, 25 recessives, or $1\,D : 2\,DR : 1\,R$.

Like the pure recessives, the pure dominants are thenceforth pure, and only give rise to dominants in all succeeding generations.

On the contrary the fifty cross-breds, as stated above, have mixed offspring. But these, again, in their numerical proportions, follow the same law, namely, that there are three dominants to one recessive. The recessives are pure like those of the last generation, but the dominants can, by further self-fertilisation and cultivation of the seeds produced, be shewn to be made up of pure dominants and cross-breds in the same proportion of one dominant to two cross-breds.

The process of breaking up into the parent forms is thus continued in each successive generation, the same numerical law being followed so far as has yet been observed.

Mendel made further experiments with *Pisum sativum*, crossing pairs of varieties which differed from each other in *two* characters, and the results, though necessarily much more complex, shewed that the law exhibited in the simpler case of pairs differing in respect of one character operated here also.

Professor de Vries has worked at the same problem in some dozen species belonging to several genera, using pairs of varieties characterised by a great number of characters: for instance, colour of flowers, stems, or fruits, hairiness, length of style, and so forth. He states that in all these cases Mendel's law is followed.

The numbers with which Mendel worked, though large, were not large enough to give really smooth results; but with a few rather marked exceptions the observations are remarkably consistent, and the approximation to the numbers demanded by the law is greatest in those cases where the largest numbers were used. When we consider, besides, that Tschermak and Correns announce definite confirmation in the case of *Pisum*, and de Vries adds the evidence of his long series of observations on other species and orders, there can be no doubt that Mendel's law is a substantial reality; though whether some of the cases that depart most widely from it can be brought

B   12

within the terms of the same principle or not, can only be decided by further experiments.

One may naturally ask, How can these results be brought into harmony with the facts of hybridisation as hitherto known; and, if all this is true, how is it that others who have so long studied the phenomena of hybridisation have not long ago perceived this law? The answer to this question is given by Mendel at some length, and it is, I think, satisfactory. He admits from the first that there are undoubtedly cases of hybrids and cross-breds which maintain themselves pure and do not break up. Such examples are plainly outside the scope of his law. Next he points out, what to anyone who has rightly comprehended the nature of discontinuity in variation is well known, that the variations in *each* character must be *separately* regarded. In most experiments in crossing, forms are taken which differ from each other in a multitude of characters —some continuous, others discontinuous, some capable of blending with their contraries, while others are not. The observer on attempting to perceive any regularity is confused by the complications thus introduced. Mendel's law, as he fairly says, could only appear in such cases by the use of over-whelming numbers, which are beyond the possibilities of practical experiment.

Both these answers should be acceptable to those who have studied the facts of variation and have appreciated the nature of species in the light of those facts. That different species should follow different laws, and that the same law should not apply to all characters alike, is exactly what we have every right to expect. It will also be remembered that the principle is only declared to apply to discontinuous characters. As stated also it can only be true where reciprocal crossings lead to the same result. Moreover, it can only be tested when there is no sensible diminution in fertility on crossing.

Upon the appearance of de Vries' papers announcing the "rediscovery" and confirmation of Mendel's law and its extension to a great number of cases two other observers came forward and independently describe series of experiments fully confirming Mendel's work. Of these papers the first is that of Correns,[1] who repeated Mendel's original experiment with Peas having seeds of different colours. The second is a long and very valuable memoir of Tschermak,[2] which gives an account of elaborate researches into the results of crossing a

[1] *Ber. deut. Bot. Ges.* 1900, xviii, 158.
[2] *Zeitschr. f. d. landw. Versuchswesen in Oesterr.* 1900, iii, 465.

number of varieties of *Pisum sativum*. These experiments were in many cases carried out on a large scale, and prove the main fact enunciated by Mendel beyond any possibility of contradiction. Both Correns (in regard to maize) and Tschermak in the case of *P. sativum* have obtained further proof that Mendel's law holds as well in the case of varieties differing from each other in *two* characters, one of each being dominant, though of course a more complicated expression is needed in such cases.[1]

That we are in the presence of a new principle of the highest importance is, I think, manifest. To what further conclusions it may lead us cannot yet be foretold. But both Mendel and the authors who have followed him lay stress on one conclusion, which will at once suggest itself to anyone who reflects on the facts. For it will be seen that the results are such as we might expect if it is imagined that the cross-bred plant produced pollen grains and ovules, each of which bears only *one* of the alternative varietal characters and not both. If this were so, and if on the average the same number of pollen grains and ovules partook of each of the two characters, it is clear that on a random assortment of pollen grain and ovules Mendel's law would be obeyed. For 25 per cent. of "dominant" pollen grains would unite with 25 per cent. "dominant" ovules; 25 per cent. "recessive" pollen grains would similarly unite with 25 per cent. "recessive" ovules; while the remaining 50 per cent. of each kind would unite together. It is this consideration which leads both de Vries and Mendel to assert that these facts of crossing prove that each ovule and each pollen grain is pure in respect of each character to which the law applies. It is highly desirable that varieties differing in the form of their pollen should be made the subject of these experiments, for it is quite possible that in such a case strong confirmation of this deduction might be obtained.

As an objection to this deduction, however, it is to be noted that though true intermediates did not occur, yet the degrees in which the characters appeared did vary, and it is not easy to see how the hypothesis of perfect purity in the reproductive cells can be supported in such cases Be this, however, as it may, there is no doubt we are beginning to get new lights of a most valuable kind on the nature of heredity and the laws which it

[1] Tschermak's investigations were besides directed to a re-examination of the question of the absence of beneficial results on cross-fertilising *P. sativum*, a subject already much investigated by Darwin, and upon this matter also important further evidence is given in great detail.

180 PROBLEMS OF HEREDITY

obeys. It is to be hoped that these indications will be at once followed up by independent workers. Enough has been said to shew how necessary it is that the subjects of experiment should be chosen in such a way as to bring the laws of heredity to a real test. For this purpose the first essential is that the differentiating characters should be few, and that all avoidable complications should be got rid of. Each experiment should be reduced to its simplest possible limits. The results obtained by Galton, and also the new ones especially detailed in this paper, have each been reached by restricting the range of observation to one character or group of characters, and there is every hope that by similar treatment our knowledge of heredity may be rapidly extended.

[NOTE. Since the above was printed further papers on Mendel's law have appeared, namely, de Vries, *Rev. génér. Bot.* 1900, p. 257; Correns, *Bot. Ztg.* 1900, p. 229; and *Bot. Cblt.* LXXXIV, p. 97, containing new matter of importance. Professor de Vries kindly writes to me that in asserting the general applicability of Mendel's law to "monohybrids" (crosses between parents differing in respect of *one* character only), he intends to include cases of discontinuous varieties only, and he does not mean to refer to continuous varieties at all. 31 October 1900.]

AN ADDRESS ON MENDELIAN HEREDITY
AND ITS APPLICATION TO MAN

Delivered before the Neurological Society, London, 1. ii. 1906

Reprinted from *Brain*

MR PRESIDENT AND GENTLEMEN: I need not say that I feel it a very high compliment for me, a layman, to be asked to address a professional meeting of this kind. When I received the invitation I hesitated for a moment as to whether anything I could say would have a sufficiently direct bearing on the subjects in which you are interested to justify me in occupying this place. Though I venture to call my paper "Mendelian Heredity and its Application to Man", I ought to say that that application is rather for the future than for the present. We do know cases in man where the rules of inheritance traced in other animals and plants must certainly apply, but those cases are very few. On the other hand, I feel sure that if members of your profession were to take the matter up and study the phenomena of inheritance with due regard to those points which we now know to be critical, the list will be very soon increased.

The advance in knowledge which I am about to describe is the outcome of the work of Mendel. I have not time to tell you who Mendel was, or how it came about that his paper published in 1865 was completely ignored until 1900, when de Vries and others simultaneously rediscovered it. Nor need I go through the reasoning by which he worked out his discovery. That, I think, will become easily intelligible when you see the practical examples which I shall be able to illustrate by specimens. I will therefore begin at the point we have now reached.

In a very large range of animals and plants we know that bodily characters—points of structure and features of physiological constitution—are treated by the cell divisions in which the germs are formed as *units*. This may be made intelligible by a simple demonstration. Here are draughts, white and black. Suppose the discriminating points whose heredity we are to trace are blackness and whiteness.[1] The draughts are the germ cells. When two are united together they represent

---

[1] Taken as imaginary illustrations. We know, of course, that in man blackness and whiteness do not follow the simple system. In the rabbit, however, they do.

the soma, or "zygote". The body cells, as you know, are double structures, containing twice the number of chromosomes present in the germ cell. A pure black individual is thus represented by two black draughts; a pure white by two white ones. When the two are crossed the hybrid contains both characters—a black draught on a white one. But though the combination of characters in the body of the hybrid may be intimate, yet when that hybrid forms its reproductive cells, the

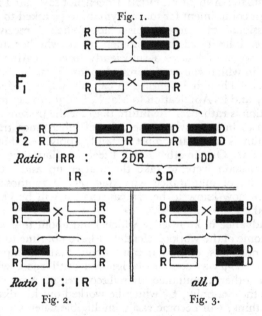

Fig. 1.

F₁

F₂

Ratio 1RR :   2DR   : 1DD

1R :   3D

Ratio 1D : 1R   all D

Fig. 2.   Fig. 3.

processes of cell division are such that the blackness is segregated from the whiteness. Each germ cell then carries either blackness or whiteness, not both; and an equal number of each kind are formed. Such segregating characters or *allelomorphs* as they are called, are, so far as we know, always constituted in pairs. In respect of each pair, therefore, there are only three kinds of individuals, the pure, or *homozygous*, of each kind, and the impure, or *heterozygous*, containing both.

It is often found that one member of an allelomorphic pair is so conspicuous in the heterozygotes that these impure individuals may pass for pure. The character which thus prevails is then called the dominant (D), the other being the recessive (R). The gametes, or germ cells, may be represented by the single letters D, D, D, etc., and R, R, R, etc.; but to represent the zygotes, namely, the individuals resulting from the union of two germ cells in fertilisation, *two* letters, DD, or DR, or RR, are required. In the diagram black is supposed to be dominant, the black draughts lying on the white ones. The results of the various possible unions are then shewn. $RR \times DD$ gives all DR, which if dominance be complete may pass for pure D. Consequently $DR \times DR$ *seems* to give on an average $3D : 1R$ (Fig. 1).

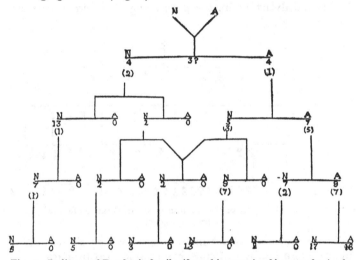

Fig. 4. Pedigree of Farabee's family (from his paper). N, normal; A, abnormal, with only two phalanges in fingers and toes.

In reality, of the $3D$ two are impure on an average, and one is pure. Similarly $DR \times RR$ gives an equal number of DR and RR (Fig. 2), while $DR \times DD$ appears to give all D (Fig. 3).

The most striking consequence of allelomorphism is that pure individuals may be bred from impure ones. For example, whenever an R germ meets an R germ in fertilisation, the individual is pure R, and the same is true for the D germs. Consequently, when an R individual appears in the family, the D element is thrown out of that individual for good and all.

The simplest illustration is that of tall and dwarf "Cupid" sweet peas crossed together. The first filial generation—F_1, as we call it—are all tall, for tallness is dominant. In the germ cells the characters, tallness and dwarfness, segregate.

By self-fertilisation the second generation—F_2—is produced, consisting on an average of three talls to one dwarf. The dwarfs are then pure dwarfs, giving no more talls in their posterity; but of the talls, one on an average will be pure tall, and two will be heterozygous, giving the same mixture in the next generation. In the photograph (Plate I, fig. 1), members of the F_2 generation are shewn. All came from one pod of seed. The dwarfs are now pure dwarfs, though they are the offspring of a tall plant, and though the grandparent was also tall. But of the talls some are pure tall, and some are hybrids of tall and dwarf again, the pure being to the impure on an average 1 : 2.

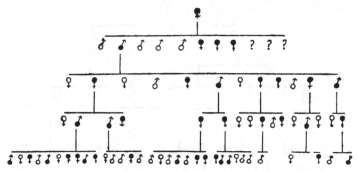

Fig. 5. Offspring of abnormal members of Farabee's family[1]
(from his paper).

The next, which is a human case, is exactly similar. It is the case of a family in Pennsylvania described by Farabee(1), many of whose members had a deformity of the digits—the fingers and toes all having two phalanges like the hallux and pollex (Plate III, fig. 4). This character was like tallness in the peas, dominant. Every individual that bore the character exhibited it. Those that had normal digits did not transmit the peculiarity, because they were formed from two germ cells not bearing it. Fig. 4 shews the pedigree of the whole family. Though there is even a first-cousin marriage between normals, no abnormal appears in their posterity. Fig. 5 shews the descent from abnormal members, the other parent being in each case a normal person. The 14 families bred in this way contained 33 normals, 36 abnormals, a close approach to the expected equality. It is therefore an even chance that the child of one of the abnormals will be normal, or abnormal;

[1] In this and subsequent pedigrees the *black* symbols represent abnormal individuals.

whereas if one of the normals were to consult a medical man and ask what was his chance of having a child with the deformity, he might be confidently assured that the chance was infinitesimal, and no greater for a normal member of this curious family than for any normal person in the general population.

Here is one of a collection of cases made by Mr Nettleship (2), dealing with inheritance in congenital cataract (Fig. 6). I have to thank Mr Nettleship for calling my attention to these cases, and for other help in connection with the inquiry. The

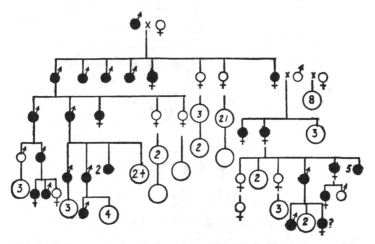

Fig. 6. Pedigree of congenital cataract, after Nettleship, *l.c.* p. 40. The figures in the circles shew the total number of unaffected persons.

same account applies to it as to the last. Here again a dominant character is passed on; roughly speaking, the affected have children of both kinds in equal numbers. The actual numbers in this series are 29 (? − 1), and 26 + ?. But the normals have exclusively normal offspring.

A chart of another family, also in Mr Nettleship's collection (2) (Fig. 7), shews a similar distribution. It differs from the last in one respect, because at the top of the series there is a normal with a number of affected persons in his family. That, of course, is against our rule, and constitutes a definite exception to it. Though this family and that shewn in Fig. 6 are in good agreement with the Mendelian scheme, several other cases in Nettleship's series present more or less difficulty. (See especially Gjersing's family (2), which shews such an excess of affected persons, that unless it is to be supposed that normals have been partially omitted—which is improbable— the case cannot possibly be brought within our rules.)

Such cases as that shewn in Fig. 7 suggest a further question which we cannot answer, namely, What is the origin of these variations? As to that we know nothing at all. Whether a variation be dominant or recessive, we cannot form any clear

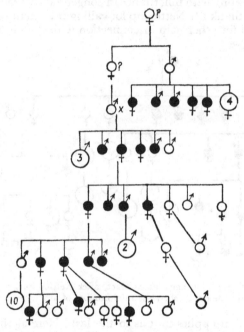

Fig. 7. Congenital cataract, from Nettleship after Berry.
The female at the top of the pedigree came of an affected family, but she was probably normal; and her son, from whom the rest descend, is said to have been certainly without cataract. Such descent of the condition through the unaffected is exceptional.

idea as to what is the cause which leads it to begin. But having begun, it usually follows the ordinary course. The dominants, viz. the affected, have affected and unaffected offspring in equal numbers, and the unaffected have unaffected offspring. This is characteristic of the distribution of dominant characters.

I will now pass rapidly through cases that we know more about, in plants and animals which are amenable to experiment. The first is a case in the Chinese primrose (*Primula sinensis*). The original species has a palmate leaf (Plate II, fig. 2). About 1860 a variety appeared as a sport from seed, having a pinnate or "fern" leaf (Fig. 2 (*R*)). The palmate form is dominant, and the fern leaf is recessive, breeding true whenever it appears. The deformed sweet pea known as the "Snapdragon" variety is a similar recessive.

In wheat, a long series of characters following these rules has been determined by Biffen [3] and Spillman. Beardless, for instance, is dominant to bearded, and consequently the bearded offspring extracted from a cross-bred are pure to the bearded type.

Another of Biffen's cases has considerable theoretical importance, and is especially interesting as bearing on conceptions of evolution. It occurs in a cross between 2-row and 6-row barley. The difference between them is, briefly, that in the 2-row form certain lateral flowers develop no seed because they are *male* only, having no female organs. In the 6-row all the flowers are both male and female, and all set seed.

When these two are crossed together what should we expect? Certainly I should expect that the positive characteristic, the presence of both male and female parts, would dominate over the negative characteristic, the absence of female organs. But it is not so. It is the negative characteristic which dominates. Though in the "blood" of the cross-bred there is the positive characteristic, yet the negative form dominates and excludes it from appearing. It is as though an inhibitory process had occurred; that there was some substance or arrangement, some peculiarity—we know not what—present in that organism which prevents development of the female organs. Consequently it is 2-rowed, like the 2-rowed parent.

One of the simplest cases which first attracted attention is that of the cross between the wrinkled sugar-maize and the common or starchy maize with round seeds. In this case the round form dominates. Every "wrinkled" egg cell fertilised by a pollen grain bearing the "round" character becomes a round seed, developing a starchy endosperm instead of a sugary endosperm. Such a seed, when sown, becomes a plant which, on self-fertilisation, will have a cob containing a mixture of starchy and sugary seeds in the ratio 3 : 1. Every one of the wrinkled seeds would give a plant of pure sugary or "wrinkled" type; but of the round seeds two would be impure "rounds" for one which was pure (Plate II, fig. 3).

The next is the classical experiment of Mendel, which consisted in crossing a wrinkled pea with a round pea, shewing that the same series of results follows as in the case of the wrinkled and round maize.

So far all is very simple. From this point various extensions have been made. The first was to find other animals and plants, and other characteristics which followed these rules. Some very interesting ones have been found, for example, a case which will probably have an interest for the medical profession. Biffen has found in wheat that the condition of greater resistance to the rust disease is a recessive character. I have seen myself his wheats growing and can attest that that is so. He crossed a breed of wheat which has comparative immunity, or a greater resistance to rust, with one which has less resistance, whose leaves are yellow with rust. The first generation is affected heavily with rust, like the bad parent. In the next generation one sees the yellowed plants, with green or recessive ones standing among them comparatively immune; and when the seeds of the resisting plants are saved—they are self-fertilised before the flower opens—the offspring are comparatively disease-resisting. That is the only case of this kind which has been properly attested, but I have no doubt plenty of others will be found when they are looked for. Another case which is rather interesting occurs in the primrose. Most people know from Darwin's books that primroses exist in two forms. In one form the anthers are in the top of the tube, and the style is short. In the other form the anthers are at the point of contraction of the tube, and the style is long. Primroses in nature exist in about equal numbers in the two forms. And we find that the thrum, or short style, is dominant over the long style, and these characters are segregated in the ordinary way. Long-styled plants being recessive never produce short-styled offspring unless they are pollinated from a short-styled plant. That is a complete answer to the objection which was made when the Mendelian facts were first discovered, that wild animals and plants might not follow the rules applying to domesticated breeds.

An illustration of interest to neurologists is that of the Japanese "waltzing" mice. The peculiar vertiginous movements exhibited by these creatures are probably connected with malformations of the semi-circular canals. The "waltzing" character is recessive to the normal, for the offspring of normal × waltzer is always normal, and in F_2 waltzers reappear [4].

Another interesting case, also of a recessive character, is that of sterile anthers in the sweet pea. This curious pathological condition is partially coupled with the absence of colour in the axils of the leaves.

The application of Mendelian rules to mankind has not made the progress that was to have been expected. In particular, we have not clear evidence that these rules apply to the transmission of colour characteristics in man. We know that they do in the rabbit, the mouse, the horse, and many other creatures; but in man there is no clear evidence that the colour characteristics follow the Mendelian rules. There is evidently a complication, probably an important one, in that case. Those who are familiar with the subject will know that though we can work out the simple rules for the blacks, chocolates, and albinos in the mouse and rabbit, we have great difficulty in tracing the rules for yellows and browns. Yellows and browns do not segregate clean when crossed with black. I think that perhaps some difficulty of this kind exists in regard to human pigments, though we should probably be wrong in imagining that there was no segregation.

Even in regard to albinism—usually the most obvious Mendelian recessive in animals and plants—the human evidence is not yet clear. Albinos have been recorded with some frequency among the offspring of consanguineous matings— a circumstance characteristic of recessive features; but the existence of complications is indicated both by the many degrees in which human albinism may present itself, and by the frequent association of the peculiarity with various forms of disease—an association not usual among domesticated animals.

I now pass to another subject. So far we have only been concerned with the distribution, in their posterity, of characters obviously existing in the original parents from whom the descent is traced. We knew long ago that such original parental types might reappear in F_2, together with the cross-bred form produced in the first generation. But we knew also that in other cases F_2 might contain a number of *new* forms appearing for the first time. This phenomenon is the "variability" caused by crossing with which all breeders are familiar. Within the last two years this problem has been greatly elucidated. In all the cases which have been properly examined we now know that these new forms are created by simple recombinations of characters brought in by the original parents.

We cross, for example, a red variety of some plant, say a stock, with a cream-coloured variety. The red variety is characterised by red sap, the cream variety is characterised by yellow corpuscles, surrounded by colourless sap. In the red those corpuscles are represented by colourless corpuscles floating in the red sap. What will happen when these two are crossed together? As a matter of fact, we find that the red is dominant. But in the second generation, the F_2 generation, we have nine red, three red with cream, three *white*, and one cream as before. What is the white? White was not put in. Apparently we have produced it *de novo*—an albino by crossing two coloured forms: cream was a corpuscle colour, red was a sap colour, but the white have *colourless* corpuscles floating in *colourless* sap. It is evident what has happened. Which are the factors which segregate in the formation of the germ cells? They are (*a*) red sap from colourless sap, and (*b*) white corpuscles from yellow corpuscles; so that when the possible combinations of those two pairs of characters are made, colourless corpuscles may coincide with colourless sap, and a white flower is the result. I think the answer is quite clear.

The operations may be shewn in a tabular form, thus:

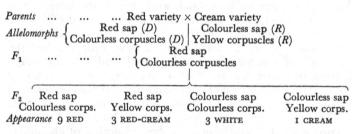

Parents ... ... Red variety × Cream variety			
Allelomorphs { Red sap (D)	Colourless sap (R)		
Colourless corpuscles (D)	Yellow corpuscles (R)		
F_1 ... ... { Red sap			
Colourless corpuscles			
F_2 Red sap	Red sap	Colourless sap	Colourless sap
Colourless corps.	Yellow corps.	Colourless corps.	Yellow corps.
Appearance 9 RED	3 RED-CREAM	3 WHITE	I CREAM

Plate III, fig. 5, shews another example of a novelty arising in F_2 by recombination. In the Chinese Primula we have a modern variety with an extraordinarily big yellow eye or flush spreading up the petals, quite different from the small yellow pentagon characteristic of normal types. This variety has the peculiarity that though its pollen and anthers are those of the ordinary pin-eyed type, yet in it the style is short and the stigma is down the tube at the anther level.

Mr Gregory and I crossed this "homostyle" plant, as it was called by Darwin, with the small-eyed short-styled form. F_2 is short-styled with the small eye. In the F_2 generation we got out four types: (*a*) Short style with small eye; (*b*) short style with big eye; (*c*) *long style* with small eye; (*d*) homostyle with big eye, in the ratio 9*a* : 3*b* : 3*c* : 1*d*. Now, the long-styled or pin type, which apparently was not put in at all, is due to a recombination of the characters. The two pairs of characters are:

      Thrum type (Dom.)   Pin type (Rec.)

      Small eye (Dom.)   Large eye (Rec.)

The homostyle is the form which the pin type assumes when the large eye is developed; but when in F_2 the pin type meets the small eye, the ordinary pin, or long-styled, form is produced.

The combinations of two such pairs of characters Cc and Rr can be represented diagrammatically as shown in Fig. 8. The figure contains sixteen squares, presenting every combination of the two pairs of characters.

There are therefore four groups of individuals produced in F_2, viz. nine of the kind that contains C and R; three that contain C only; three that contain R only; and one with neither.

A complication we often meet in the application of the rules of heredity lies in the fact that characters belonging to distinct allelomorphic pairs react on each other. A particular appearance, for instance, may depend on the co-existence of both C and R, and either of these factors alone may be unable to manifest any influence on the individual in the absence of the other. In that case there will be nine shewing the appearance in question, say a colour, for seven which are without it. Or, again, the factor C may produce an effect alone; while R, though imperceptible in the absence of C, may modify the effect of C when C and R co-exist. There will then be nine of the $C \times R$ class; three of the C class; and four all alike because C is absent, though their gametic composition is really diverse. This is a result well seen in the colour heredity of mice and rabbits. For example, grey × albino gives grey F_1, with, in F_2, three grey: one albino. But F_2 may be instead nine grey: three black: four albino. The latter result indicates that the factor which determines the colour to be grey was absent in the albino. The meaning of these occurrences was first pointed out by Cuénot.[1]

In illustration of these descents I exhibit skins of some of the remarkable series of rabbits bred by Mr C. C. Hurst (5), who has kindly lent them for this meeting. Some of his F_2 families were

[1] *Arch. zool. exp. et gén.* 1904; (4) II, *Notes et Rev.* p. xlv. Cuénot, in his argument, speaks of the determiner for grey as present in the grey mouse, and of the black determiner as present in the albino, regarding the two determiners as allelomorphic to each other. For various reasons, however, it is probably more correct to regard the grey determiner in the grey mouse as allelomorphic to its absence, and to imagine the black determiner as present in both the grey and the albino.

three grey : one albino, while others were nine grey : three black : four albino; in either case, of course, three coloured : one albino. The albinos can give no more coloured. The blacks may be pure blacks, *or* they may give blacks and albinos.

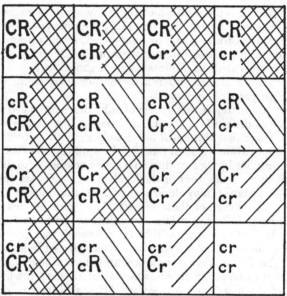

Fig. 8. Scheme of distribution of two pairs of factors in F_2, shewing how the F_2 ratio 9 : 3 : 3 : 1 is produced. From the same distribution the F_2 ratio 9 : 3 : 4 results in cases where one factor produces no visible appearance unless the other is present; while, if neither factor can be perceived unless both are present, the F_2 generation is 9 : 7.

The big letter in each case denotes the presence, the small letter the absence, of the given factor.

The greys may be pure greys; or they may give greys and albinos; or greys and blacks; or greys, blacks and albinos as before. If in the diagram (Fig. 8) C stands for colour, c for albino, R for grey and r for black, the nature of this series will be understood.[1] Inspection shews that of the four albinos, for

---

[1] As implied in the previous note, the account in the text is too simple, though it may suffice as a rough presentation of the phenomena. The case should be expressed thus: The allelomorphs are C, colour; c, albino; G, grey determiner; g, absence of ditto; B, black determiner; b, absence of ditto. The grey rabbit is CGB; the albino is cgB. B is common to both, so that in F_2 the arithmetical result, nine grey : three black : four albino occurs as it would do if G were allelomorphic to B so that the original parents could be expressed as CG and cB.

instance, which appear on an average in 16 F_2, one will be carrying the grey determiner, one the black determiner, and two will have both grey and black determiners. By suitable cross-matings the condition of each albino can be exactly determined. For example, when bred to a pure black, a G albino will give greys only; a GB albino will give equal numbers of greys and blacks; while a B albino will give blacks only. Similarly the exact composition of each coloured rabbit may be determined by experimental breeding.

The particular colour of a rabbit or a mouse is therefore not a simple character depending on the presence of a single factor, but a double one, depending on interaction between one distinct factor and another, each factor being transmitted independently in heredity.

In certain plants we can go beyond this. Two white-flowered sweet peas, for instance, may when crossed give a coloured F_1, which by self-fertilisation will produce nine coloured : seven white. This result proves that the colour depends for its appearance on the co-existence of two complementary elements or factors. The nine are the nine squares in Fig. 8 containing both C and R; the seven are the three with C, the three with R, and the one with neither. Either factor alone is insufficient to cause colour. C and R are not allelomorphic to each other, but each is allelomorphic to its own absence. This conclusion has been tested and confirmed by an elaborate series of experiments made by Miss Saunders, Mr Punnett, and myself. A further complication is due to the fact that colour, once formed by the meeting of the two complementary factors, is modified by the operations of distinct determining elements, just as is that of a rabbit. Thus, for instance, the nine coloured : seven white commonly forms the series twenty-seven purple : nine red : twenty-eight white.

In the stock,[1] which has formed the subject of a long series of experiments by Miss Saunders, a still further complication is met with. Colour in the stock, as in the sweet pea, requires the co-existence of two independent factors, each of which in the germ formation is allelomorphic to its own absence. In addition, the development of the hoariness or felting of hairs upon the leaves is also produced by another similar pair of factors, either of which alone may be present without a single hair being formed. But in the stocks employed ("ten-week stocks") both these factors for hoariness may be present, but no hairs are developed *unless the flowers are coloured*, that is to say,

[1] *Matthiola.*

B   13

unless the complementary pair which form pigment are also present. To go through the elaborate steps by which this conclusion has been experimentally established is impossible on this occasion, but I mention it to illustrate the kind of complexity which is to be met with in these investigations, and at the same time the far-reaching precision that may be attained by Mendelian analysis.

Evolutionists will note that in the stock and the sweet pea we have examples of Darwin's famous "reversion on crossing". Two white-flowered plants crossed may give the wild, purple flower. Two glabrous stocks may reproduce the original hoary, ancestral type. Such reversion is nothing but the meeting of

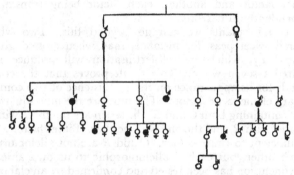

Fig. 9. Pedigree of a family containing colour-blind members observed by Dr Rivers, among the Todas, an Indian hill-tribe, illustrating the "knight's move" descent of that condition.

two parted complementary elements, which have somehow been separated by variation.

The various factors, then, which interact to produce what we formerly regarded as a single character, e.g. colour, are *separately* transmitted in heredity. Such interaction constitutes a remarkable phenomenon from the standpoint of general physiology. We cannot doubt that this inter-relationship between these distinct units plays a great part in the determination of bodily features. As these complex inter-relationships are gradually resolved, we shall acquire a new light on the problems of evolution, and especially on the genesis of variation. What the nature of such elements may be we must leave as yet to conjecture. It is natural, however, that we should think of their mutual actions as comparable with that of a ferment acting upon a definite material.

To the student of disease, illustrations of a comparable nature will be familiar in the case of the sex-limited inheritances. In these families the peculiarity is manifested generally, if not always, by members of one sex, say the male; but the females, though unaffected themselves, may transmit it to their male offspring. Haemophilia and colour-blindness are the most famous examples of this "knight's move" descent, as it may be called. The possible bearing of the new ideas upon

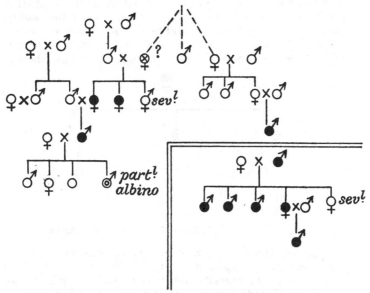

Fig. 10. Pedigrees of two families containing colour-blind females, from Mr Nettleship's collection.

this curious phenomenon may be shewn by an illustration seen in crossing sheep.

Mr T. B. Wood[6] made a cross between the Dorset Horn and the Suffolks. The Dorsets are horned in both sexes, the Suffolks are hornless in both sexes. In F_1 the males are horned, the females hornless. In other words, horns are dominant in males, recessive in females. But in F_2 there are both horned and hornless males, and also horned and hornless females; because two "hornless" germs have had the opportunity of meeting in presence of "maleness", and two "horned" germs in presence of "femaleness". Undoubtedly when these animals have been tested, the hornless F_2 males and the horned

F_2 females will be proved to be each pure to their respective characters. (Plate III, fig. 6.)

Though the evidence as to sex-limited descents is imperfect, and in some respects obscure, it is obvious that there is a general resemblance to what we have seen in this case of horns in the sheep. If the affected sex is the male, the unaffected males do not transmit the condition, though the unaffected females[1] may do so. We may suppose that the unaffected males in fact are, by segregation, freed from the peculiarity, just as the hornless rams in F_2 are freed from the horned character. To test the applicability of these comparisons much evidence is required which is not yet obtained, but the tests which must be applied are clear. Supposing the members of

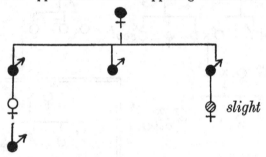

Fig. 11. Pedigree of colour-blindness, from Burckhardt.

the affected family mate with persons not bearing the peculiarity, we should expect that: (1) half the sons of (a) *affected males*, and of (b) *unaffected females who had any affected sons*, would be affected; that (2) half the daughters of (a) and (b) would transmit the peculiarity to half their sons. Moreover (3), affected females should be pure in respect of the character, and consequently *all* their sons should exhibit it.

We are far from possessing adequate evidence that this simple system holds in any sex-limited descent. The most that can be asserted is that the general course of the inheritance is in rough agreement with some such scheme.

As regards colour-blindness the pedigrees are few, but it is noticeable that the only examples—three—of colour-blind women with sons that are known to me are, so far as they go, in agreement with our hypothesis. Two (Fig. 10) are cases

---

[1] The popular belief that such a condition is transmitted *only* by females is, of course, a mistake. In haemophilia, for example, there are many cases of transmission by affected males.

observed by Mr Nettleship. In each a colour-blind woman had one son, who was colour-blind. The other case is that of Burckhardt (7)—a colour-blind woman with three sons, all colour-blind (Fig. 11).

As to haemophilia the evidence is more abundant, and in respect of criteria (1) and (2) is usually not very discordant with expectation. With rarest exceptions unaffected males do not transmit the disease. Nevertheless, many more males are affected than our rule expects;[1] and unless it is supposed that

Males.			Females.	Sex not given.
Bleeders.	Non-bleeders.	Died young.	Non-bleeders.	Non-bleeders.
188	65	20	177	44

normal families have been freely disregarded by the recorders, too many of the females transmit. Moreover, of some thirty-five groups of families, at least twelve cannot be brought into line at all. The subject is complicated by the circumstance that various pathological conditions have been included under the name haemophilia, as has been suggested by Legge in reference to female haemophilia, and a proper analysis of the evidence could only be made by an expert.

Some of these families, especially those in mountain villages, were doubtless much in-bred, a circumstance which may have introduced complications. The excess both of bleeding males and of transmitting females is, however, far too large to admit of any simple explanation. (For extreme cases, see Stahel, *Inaug. Diss.* Zurich, 1880, and Sadler, *Birm. Med. Rev.* 1898 and 1903.) The phenomena could only be brought under the Mendelian scheme by the supposition that a complex system of coupling exists, for which we have as yet no justification. I hope at some time to discuss this subject in more detail.

The same features are presented by another sex-limited disease, peroneal atrophy (8), with the added difficulty that too few of the sons of affected males were affected; while, as before, too many of the women transmit and too many of their sons are affected.[2]

[1] I got the following very rough figures by adding all the published families (containing "bleeder" males) which can be at all supposed to be comparable. Parents non-bleeders in each case.

[2] I have since examined many published pedigrees of nervous diseases to which my attention was called by Dr Batten and others. In several of these transmission is exclusively, or almost exclusively, through the affected, so the characters may well be dominants, but the ratio of affected to unaffected is nearly always too high. Normals may doubtless have sometimes been omitted, but the deficiency of normals is so persistent as almost to forbid a recourse to

A source of complication in understanding these statistics may arise through complete or partial *coupling* between distinct characters. We have proof that in certain cases a character, say of shape, may be so linked or coupled with another character, say of colour, that all or a majority of the germs which carry the one carry the other also. To this fact I can now only refer, but the existence of such coupling constitutes one of the most interesting, as well as one of the most difficult, branches of these inquiries.

Finally, I would say something as to the way in which evidence must be collected if it is to be used in the study of heredity. First, the facts must be so reported as to be capable of analysis. It is for want of such analysis that all examination of the facts by pre-Mendelian methods failed. The tabulations must present each family separately. Miscellaneous statistics are of little use. Secondly, *it is absolutely necessary that the normal or unaffected members should be recorded*, together, if possible, with information as to their offspring. In the records hitherto published these essentials have too often been omitted, the doctor's attention having been more or less exclusively directed to the individuals manifesting the disease. Next, if similar families are to be added together, it is scarcely necessary to insist that the cases added must be in reality similar. For instance, there are abundant genealogies of deaf mutism, but the various families present such inconsistencies in the heredity rules which they follow that there can be no doubt that not one, but many, pathological states are concerned. Accurate diagnosis is the first preliminary in dealing with these phenomena.

When complete information has been obtained, it is to be expected that definite rules of transmission will not rarely be traceable. Suppose the disease dependent on a simple dominant character, the fact will manifest itself in exclusive transmission through the affected; and—if the other parent be unaffected—in the frequent production of affected and normal offspring in equal numbers. The case of dominance limited by sex has already been considered.

Recessive characters may be distinguished by the frequency

this hypothesis. The possibility suggests itself that some of these diseases may be due to actual transmission of a disease germ through the reproductive cells (like *pébrine* of silkworms studied by Pasteur). I am told that this suggestion is highly improbable; but in view of the statistics for some of these family diseases I feel considerable doubt whether they should be regarded as illustrations of heredity, pure and simple, in the sense in which a naturalist employs the term.

with which they appear in the offspring of normal parents, and especially as the result of consanguineous marriages. The identification of recessive characters will be less easy, direct experiment being impossible.

One condition probably of this nature may be mentioned—that of alkaptonuria, studied by Dr Garrod (9). The high proportion of cases of this rare condition resulting from first-cousin marriages makes it very likely that this peculiar "chemical sport", as Dr Garrod has called it, is recessive. First-cousin marriages give an opportunity for the meeting of recessive germs. The same element or character, as we may say, meets itself; and thus the hidden characteristic, being brought in by both male and female germs, may appear, though it would still have remained hidden if one of the two germs had not carried it. Think of the tall peas bearing dwarfness as a recessive character. Bred together they give one dwarf on an average in four, though bred with a pure tall individual no dwarf would be given. So when first cousins marry, it gives a chance for the hidden character to come out, because each parent may be carrying that character derived from the common grandparent. The chance even so, however, is a very small one. I do not suggest it should be considered in practical life. Certainly there is not enough to base advice upon respecting consanguineous marriages. The chance is remote that the hidden character will be present on both sides, and perhaps, too, some recessive characters may be very desirable ones!

In conclusion, I should warn you that the detection of these systems of heredity is, except in a few cases, not a simple matter, even when experimental breeding is applicable. The rules perceived in the sweet pea and the stock, for instance, have only appeared after five years' careful breeding, and, as these cases shew, results which at first seem hopelessly conflicting may ultimately be proved to be entirely consistent. We cannot suppose that the complexities there encountered can be absent in the case of man. We have seen how the presence of one character may either excite or prevent the manifestation of another independent character. Other features are undoubtedly dependent on excitation by external influence—in a wide sense. There are many indications that this is true of some nervous diseases. Lastly, it sometimes happens that two characters, one dominant, the other recessive, to a third character may be scarcely distinguishable from each other. For instance, pure white in fowls (white Leghorns) is dominant to colours; but a white associated with a few minute specks of colour (white rose-comb Bantams) is recessive to colours. Distinctions of this kind are not perceived on superficial examination, and can only be unravelled by most exact and specific investigation.

My object has been to put before you the results of work on lines somewhat remote from yòur own, though, as you have seen, presenting points of possible contact. If this discourse may lead to a closer study of the hereditary transmission of disease, I am confident that positive results may be expected, and that I shall feel that my opportunity has not been wasted.

DESCRIPTION OF PLATES

PLATE I

Fig. 1. F_2 family from the cross tall sweet pea × a dwarf procumben "Cupid" variety.
The plants, five talls and three dwarf (on the left of the photograph), came from one pod of seed.

PLATE II

Fig. 2. Palm-leaved (dominant) and fern-leaved (recessive) *Primula sinensis* in the F_2 generation from a cross between these two varieties. In F_1 all are like the upper, palm-leaved, plant.

Fig. 3. A cob of maize produced by Mr R. H. Lock, in F_2, from the cross round-seed × wrinkled seed.

PLATE III

Fig. 4. Skiagraph of hand in Farabee's family copied from his paper.

Fig. 5. The four types of *Primula* flower in F_2 from the cross large-eyed, homostyle × small-eyed short style. A, small-eyed, long style; B, large-eyed, homostyle; C, small-eyed, short style; D, large-eyed, short style.

Fig. 6. The four types of sheep in F_2 from Dorset Horn × Suffolk (hornless) from lambs bred by Mr T. B. Wood. (For parents and F_1 types, see his paper.) A, horned male; B, hornless male; C, hornless female; D, horned female. In F_1 all males are horned, and all females are hornless.

LIST OF REFERENCES

(1) FARABEE, W. C. *Papers Peabody Mus. Amer. Archaeol.* 1905, 69.
(2) NETTLESHIP, E. *Rep. Roy. Lond. Ophth. Hosp.* XVI, pt. iii, 23, 40 and 63.
(3) BIFFEN, R. H. *Journ. Agric. Sci.* 1905, I, 40.
(4) DARBISHIRE, A. D. *Biometrika*, III, 20. The fact that waltzers are only about 1 in 6 instead of 1 in 4 in F_2 is without doubt to be ascribed to their great delicacy.
(5) HURST, C. C. *Journ. Linn. Soc.* XXIX, 283.
(6) WOOD, T. B. *Journ. Agric. Sci.* 1905, I, 364, Pl. IV.
(7) BURCKHARDT. *Verh. Naturf. Ges. Basel*, 1873, 566.
(8) HERRINGHAM. *Brain*, XI, 230.
(9) GARROD, A. E. *Lancet*, 13 December 1902, and other papers.

PLATE I

Dwarf (R)   Fig. 1   Tall (D)

PLATE II

Palm (*D*)

Fern (*R*)

Fig. 2

Fig. 3

PLATE III

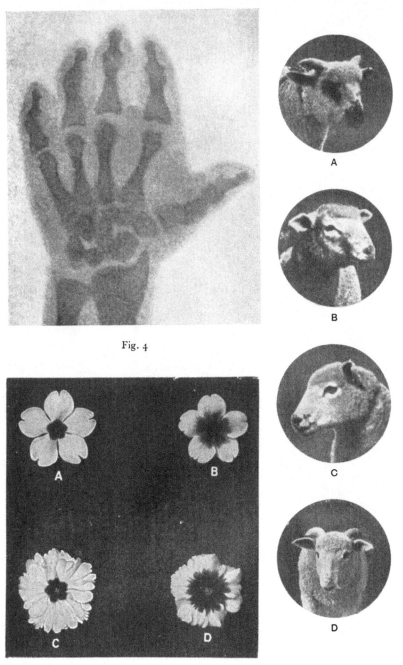

Fig. 4

Fig. 5

Fig. 6

GAMETE AND ZYGOTE. A LAY DISCOURSE

The Henry Sidgwick Memorial Lecture, 1917
(Unpublished)

The two words which I have chosen as the title of this discourse are technical terms. Round them whole regions of modern biological discovery are grouped. I am warned that terms unfamiliar to the laity may repel an audience; but since the common speech as yet contains no good equivalents by which the concepts gamete and zygote may be rendered, and my purpose this afternoon is to set forth the leading ideas which Science has gained from those concepts, I therefore make so bold as to begin by naming them. The ideas we derive from a knowledge of the inter-relations of gamete and zygote are of so wide an application that they cannot be ignored by any person who aspires to form just views of the nature of man and of life. No one having learnt these ideas can look on the world with quite the same eyes as before. They have something of that fundamental quality which we associate with chemistry, and just as it is impossible to talk easily or freely of any natural phenomena with those who are wholly ignorant of chemistry, so are we conscious of difficulty when we discuss not merely formal problems of evolution or of heredity, but any broad question of social organisation with those wholly ignorant of genetic fact.

The words *gamete* and *zygote* are not fortunate expressions. They were introduced, according to the haphazard fashion which scientific men commonly follow in naming their creatures, to denote on the one hand the single, unpaired germs or germ cells, and on the other the double or paired bodies which result from the union of two germ cells in fertilisation. The bodies of men and of most of the other animals and plants with which unprofessional people are acquainted are all *zygotes*, the product, that is to say, of two *gametes*, "marrying", or germ cells, "yoked" together in a common system, the body. In the higher animals and plants it is easy to perceive that the two gametes which unite in fertilisation are apparently of different natures, the egg or female cell being comparatively large and commonly not capable of free movement, while the male cell, or sperm, is generally exceedingly minute and often free-moving. But in many lower organisms, both plant and animal, no such distinctions can be found, and there is as yet no critical evidence to shew that as regards their intrinsic powers of transmission and their essential contributions to the zygotic body, the male and female cells are in any way different from each other. The distinctions between the two kinds of gametes are secondary, the egg being a gamete to which food material is added, the male cell being without such additions.

The attributes and powers, whether bodily or mental, which the animal or plant possesses are all, as we can no longer doubt, conveyed to it by one, or the other, or both of the gametes combining to form the double cell from which the body grows. Of life unattached to matter we have no evidence at all, and, however minute they may be, it is by the matter which the two germ cells contribute and by nothing else, that the powers of the animal and plant are fixed. Many competent biologists have satisfied themselves that these powers are conferred alone by the nuclei of the germ cells. Others going still further declare that each property of the organism is determined by a specific particle of nuclear material, and believe that as the result of certain very remarkable experiments they are even able to decide the order in which these particles are grouped. I mention this interesting line of inquiry to illustrate the scope of modern genetic analysis, though for the present I am unconvinced of the cogency of the arguments employed. But whatever part or parts of the germ cells are responsible we are sure that the properties of the zygote are decided by the gametes. To take simple illustrations we know that two red-haired parents have exclusively red-haired children, whatever the ancestry of those parents may have been. Two feeble-minded parents have none but feeble-minded children. A horse whose running gait is "pacing" mated with a pacing mare produces pacing progeny only, and no trotters. Hens raised from parents belonging to non-sitting strains will never brood their eggs, being destitute of the maternal instinct. I take these examples to shew that as regards the finality of physiological determination there is no distinction between physical and mental attributes.

So clearly can all this be perceived in the simpler cases of zygotes "pure", as we call it, for definite attributes, that we may speculatively but not unreasonably extend the same method of interpretation to some forms of instability in opinion, and other mental characteristics.

In birds the female deriving femaleness from one gamete only, may in later life develop male characteristics. Conversely many frogs ultimately male were in later larval life of indifferent sex. So may a good deal of mental fluctuation be suspected of being merely an indication of composite gametic origin. I have been told that such materialistic doctrine is scientific Calvinism, and the phrase is well applied.

Gametes are formed afresh by a resolution of the zygotic material. In becoming a full-grown creature the zygote cell takes much from the outer world both in material and experience, but save in certain exceptional cases which I cannot consider here, the most careful study reveals not only no positive indication of transmission of these experiences but no deflexion attributable to them. We may concede at once that, if such evidence were forthcoming, the course of evolution would be more easy to apprehend. The old difficulty of the origin of adapted structures would be largely removed. But upon that easy course we have no warrant to embark.

It has been held by some that although there is no real *proof* that evolutionary change proceeds by direct response to accumulated experience, yet a general survey of natural phenomena compels the adoption of some such view. Their contention in effect is that though we can neither perceive evidence of direct transmission of experience, nor form any sort of conjecture as to the mode in which such transmission might be effected, we must nevertheless hold that faith. An eminent predecessor in this Lectureship defended the Lamarckian creed by instituting a comparison with gravitation, a phenomenon the mode of action of which is also still obscure. But the parallel surely fails: for whereas the action of gravitation can be easily witnessed and accurately measured, forming the basis of absolutely reliable prediction, all that we know of heredity shews that its course is not disturbed by environmental influence, that there is consequently no effect which can be measured, and that if you predict an effect in any special case your prediction will be falsified. The evidence by which heredity is represented as unconscious memory is purely circumstantial. The evidence by which we perceive the still mysterious force of gravitation is direct. The living body is indeed a wonderful machine but we doubt if it can transform effort into structure, or produce an adapted instrument out of the raw material, necessity. It is, I understand, mainly from psychological considerations that Dr James Ward has been led to believe in unconscious memory, as Mr Edmond Holmes has been converted to metempsychosis. Offered these views as an alternative I frankly prefer Genesis, which at least is simple, and makes no pretence of appeal to observation. The doctrine of transmitted experience as the origin of modification asks us to turn away from the whole course of genetic experiment, nor can it be maintained in face of that determinism which is so impressively forced upon the minds of those engaged in recording the natural facts.

Concerned as we are solely with things that we can touch and see, we cheerfully postpone the attempt to solve the mystery of adaptation, dismissing it from our minds as a subject of barren speculation. We find whenever we can test the matter, that the properties of the parents are distributed among the offspring according to rules which are commonly definite; and whatever methods the future may provide, for the present we must be content to follow this concrete analysis. The zygote is endowed with a vast number of properties, and the problem of heredity, as a matter of concrete biology, is the manner of distribution of those properties among the gametes formed from the zygote. Gametes are cells formed by a process of division from the zygote, and therefore their distribution is in some way effected by cell divisions. Every new light on cell division is a new light on heredity. In many cell divisions that occur in animals and plants the products of division are in all respects alike. These are *divisions of equality or repetition*. In others the products are dissimilar, and we may speak of them as *divisions of differentiation or diversity*.

The permanence of type and the multiformity of living creatures are the expression and the consequence of these phenomena of cell division. That is an indisputable conclusion, and the aim of the science of genetics is to trace the working of repetitive divisions and differentiating divisions in the concrete manifestations of physiological uniformity and diversity of type in the living world.

Given the fact that cell divisions occur, we can figure to ourselves how it is that two light-eyed parents have all light-eyed children, for all the gametes of such individuals are alike in that ingredient. They result from repetitive divisions. On the other hand a dark-eyed person born from the union of a dark-eyed parent with a light-eyed parent has gametes of two kinds, bearing respectively the dark and the light characters. He is the product of two dissimilar gametes, and consequently the divisions by which his gametic cells are formed will be differentiating, sorting out the ingredients which combined to make him.

The germ cells of animals—as opposed to plants—are formed

from material definitely reserved from a very early stage in individual development. This fragment of material undergoes unnumbered divisions, of which the free eggs or sperm are the final products. It is in some of these divisions that the ingredients of the germ cells are determined.

In the history of all germinal tissues there is one critical division, known technically as the *reduction-division*, in which the doubleness of the zygotic cells is resolved. The elements of the nucleus or chromosomes then are visibly reduced to one set instead of two sets as they had remained ever since the moment of fertilisation. Each daughter cell then contains one set of chromosomes, not two as the zygotic cells did, and there is good evidence that the set which each such reduced cell receives is made up of elements taken at random from the sets which the father and mother had respectively provided. Naturally therefore as the visible processes strongly suggest this conclusion, so far at least as nuclear material is concerned, most authorities incline to the belief that it is in this critical division that the ingredients of the gametes are decided.

The sterility common among hybrids is in part a consequence of their inability to sort out into gametes the ingredients which are united in them. Such hybrid zygotes cannot then make gametes, and consistently with this view it is often found in these cases that the germinal processes have a normal course until the reduction-division should occur, when the nuclear materials fail to divide properly and deformed cells result.

Whether, taking the whole range of facts into consideration, we are right in supposing the reduction-division the only possible moment at which this differentiation, or *segregation*, as we call it, can be effected, I doubt. My hesitation is due chiefly to phenomena seen in the genetics of plants, but into that vexed question I do not propose to enter now.

Segregation is somehow accomplished in cell division, and attention therefore centres primarily on the nature of that process, and especially on the mode by which the determining elements are distributed among the daughter cells from which the next generation will be constituted. How is that distribution effected? Though we are able thus to limit the inquiry, and in so doing clear away countless obscurities among which our predecessors were lost, we are very far from foreseeing an answer. The spontaneous separation of the living cell into two halves has no clear parallel in the physical world. All efforts to resolve this process into simpler terms have failed. Note that the segmentation of the fertilised ovum is only a particular case of this power. It is possessed at some time in their existence by most living cells. The old idea that the division of the ovum is caused or provoked by the entrance of the sperm was a mistake. The work of many experimenters, especially Loeb, has shewn that the unfertilised egg may divide and undergo all the processes of development in response to various mechanical and chemical stimuli. The prick of a clean needle is enough to start the development of a frog's egg without any spermatozoon, and complete frogs raised from such eggs have lived more than a year. The properties of cells in these respects are far greater than we formerly supposed, but hitherto with extension of knowledge interpretation has become if anything more difficult than it was. Eggs begin their development by dividing into two, four, eight segments, and in several animals any one of these segments if detached from the rest has the power of regenerating the whole. True twins, whose resemblance to each other may amount almost to identity, arise from such divisions of fertilised eggs. In one of the Armadillos this division is a normal process, and four young are regularly born together all from one original egg and all, of course, of the same sex. The power of division to form similar halves, though most simply seen in single cells, may also be manifested by groups of cells. In some lower animals the embryos, already many-celled, can divide. Budding both in animals and in the higher plants is accomplished in the same way. From a group of young cells—young, in the sense that they retain their divisibility—a smaller group separates. This daughter-group may have the power of developing axes of symmetry forming a complete set of organs. It is astonishing that these very exactly ordained properties can be allotted among a collection of undifferentiated cells, but that pheno-menon is one of the limiting facts of biological dynamics. The cells take their places in an ordered pattern almost like iron filings in a magnetic field, with in addition, the vast compli-cation that each assumes special chemical properties. Un-differentiated before, they then assume the elaborate differen-tiation which we associate with the organs of a living creature. I may remind those who are biologists of the curious instance of this power exemplified in Winkler's graft hybrids, in which indifferent cells contributed by two species of plants arrange themselves in the appropriate layers, though the whole has the form and organs of a normal plant.

H. V. Wilson's regenerated sponges are wonderful enough.

The live tissues of a sponge were pulped through the meshes of a fine cloth, falling as a layer of cells to the bottom of a vessel. These cells then joined up again into small masses, and those derived from the different layers migrated into their appropriate positions, thus forming new sponges from the *débris*. This coherence may perhaps be regarded as realising the suggestion made by the late Richard Assheton,[1] that the shapes of young cells are such as to imply the existence of forces of attraction acting on each other. But of course in this case the cells were already differentiated before the dissolution, whereas when groups of embryonic cells separate to form a bud they carry with them also the power of differentiation.

In the budding of embryos or of growing points we see results comparable with those which led Driesch to formulate his idea of an ordering principle—an "entelechy" as he calls it, using an Aristotelian term—peculiar to living things. Segmenting embryos, up to certain stages, can be so treated that the cells composing them are rearranged, and nevertheless a normal larva results. If they are divided artificially, a number of larvae would be formed, as many as the parts into which the whole was divided—within limits.

Driesch, abandoning all hope of a natural account of these curious phenomena, appeals at once to metaphysic, or if we may use the Latin equivalent, the supernatural. Still, I suppose we may go on trying, in a humble, mundane way. If these processes are supernatural it is to be remarked that we reach the limits of the supernatural surprisingly soon. The distribution of these extraordinary powers is strictly limited by specific rules. Entelechy comes to the rescue of the mutilated newt and regenerates its limbs, but the mutilated frog can no more grow a new arm or leg than we can. Cut off the tail of an earthworm and it grows again and you make the cut further and further forwards for some distance and still a new tail grows. But ultimately a point is reached at which instead of a new *tail* which it needs, "Entelechy" provides it with a new backwardly-directed *head*, and of course this two-headed piece soon perishes. The inference most fairly drawn from these and other equally surprising facts is not that the mechanisms of regulation are outside physics and chemistry, but rather that with more knowledge it will be possible to map the peculiar mechanics of development and growth with some accuracy, and till then we may leave the decision of the question whether

[1] *Growth in Length*, Cambridge, 1916.

these occurrences are amenable to ordinary physics and chemistry or not.

There is one remarkable distinction between reproductions, whether of the whole body or of its parts, effected by divisions of the zygote and those which result from gametes: that in gametic reproduction geometrical relation to the axes of the parent is lost, whereas in *zygotic*, or as we call it *somatic* repetition, whether normal or abnormal, relations of form are, I believe, always maintained. Without attempting fully to illustrate this proposition I may shew what I mean by reference to the case of plant spirals. The parts of all higher plants are arranged in spirals. Each seedling is from the start either right or left-handed, and the spiral of each branch is definitely related to that of the main stem. But the spirals of the seedlings are independent. Even in the ears of barley where the seeds stand necessarily in a formal relation to the plant axis, any seed irrespective of its position on the mother plant may be a right or a left.

In the attempt to find out what is really happening in cell division and thence to reach an understanding of heredity and variation the visible geometrical phenomena offer the most promising clue.

A close connection between heredity and symmetry is suggested at once by the genetic behaviour of plants which are in any respect mosaics or patchworks. Varieties with colours arranged in definite patterns, for instance picotee carnations, can breed true or be "fixed" as it is called. In them the colour is treated as part of the geometrical system of the plant, and is regularly represented by the gametes. But if the colour is *not* definitely arranged, appearing only in irregular segments and stripes, as in Bizarre carnations, no amount of selection can make it breed true. Many instances of the same phenomenon may be seen in the genetics of variegated plants. The genetic behaviour depends on the pattern of distribution. To those who have not thought about such things it may seem paradoxical that many kinds of *animals* with piebald coats can at once be fixed as breeds. Not only is this true of horses, cows, dogs and mice, etc., with large irregular masses of colour, but of some in which the colour is reduced to a mere speck. White Bantams have, I believe always, a very few coloured feathers sometimes in one place, sometimes in another. According to the conventional view of variation such a trifling feature might be supposed to be especially inconstant, to be on the point of disappearing, and so on.

Of those illusions about variation nothing now remains. In the animal such a feature can be fixed, but in the plant it is an expression of a composite or mosaic nature.

This striking distinction between animals and plants is a consequence of a difference in geometrical plan. The animal is a closed system growing by intercalation, the plant is an open system growing by division at the apices. Since nothing in the world grows at all like either animals or plants no illustration is really in point, but in some ways an animal zygote is like a hollow ball with gametes inside it, able to become similar balls when liberated, while a plant is like a stocking, knitting itself forwards at both ends by means of growing points. The growing point has perpetually the power of making gametes. When stripes on the stocking are geometrically regular the whole pattern can be represented in the gametes, but when the stripes are irregular some gametes carry on one colour and some another, according to the stripe in which they arise. Perhaps I ought to say that this is a mere post-impressionist picture of the facts, but like that pictorial method, though bearing no inspection and having no permanent value it may serve to attract attention.

The formal rhythms by which repeated organs of both plants and animals are related to each other have points of superficial resemblance, though different in essential respects. Adapting a passage of Cuvier's, Michael Foster used to say that a living creature is a vortex of chemical and molecular change, and I know no other description which contains so much of the truth. We commonly think of animals and plants as matter, but they are really systems through which matter is continually passing. The orderly relations of their parts are as much under geometrical control as the concentric waves spreading from a splash in a pool. If we could in any real way identify or analyse the causation of growth, biology would become a branch of physics. Till then we are merely collecting diagrams which some day the physicist will interpret. He will I think work on the geometrical clue.

When we try to understand how gametes come to unite in fertilisation we speak of the process as a consequence of "attraction" or "affinity", or in some similar terms. The attraction evidently enough exists, and a visible coalescence of nuclei occurs almost always. (In a few remarkable exceptions the two nuclei lie throughout zygotic life in contact without amalgamating.) But at the reduction-division there is obviously a repulsion as strong as the attraction or affinity had

B 14

been in fertilisation, for the two halves come apart again. In the separation the ingredients contributed by the two original parents are redistributed according to the systems which Mendelian analysis has disclosed. The repelling influences may be represented, perhaps, as dealing with these ingredients as eddies deal with suspended material derived from two confluents, and I suspect that our "unit factors" have the varying degrees of coherence or divisibility which miscellaneous materials so amalgamated would, according to their specific nature, possess.

I have said that physics offers no parallel to the spontaneous *division* of the cell, and the *union* of gametes is almost equally unique. Attempts to represent union and separation as cyclical events have had no success.

In this connection a reference to the geometrical forms of gametes may be in place, though the facts do not advance us much.

Gametes are nearly always markedly asymmetrical, whereas on fertilisation symmetry is often restored. Why should the nucleus of the matured—i.e. "reduced"—ovum, awaiting fertilisation, always, even in the most yolk-free egg, be eccentrically placed? On fertilisation the nucleus immediately migrates to the centre of the egg and there undergoes the first division. More than this: in artificial parthenogenesis, whether caused by fatty acids or otherwise, the nucleus, though unfertilised, similarly moves to the centre before division. It cannot be drawn there by the sperm, for no sperm is present. This is shewn in all the figures I have found, though the fact seems not to have been specially noticed by the observers. Such a change in the position of the nucleus seems to me a more significant event than the deposition of a membrane of which so much has been made. Is it not strongly suggestive of the release of some strain?

The asymmetry of spermatozoa is generally even more evident than that of ova, and a more or less obvious corkscrew spiral is one of the commonest forms in most dissimilar animal groups.

In passing I may remark that I can find no comprehensive statement as to the direction of these spirals, nor even whether they vary in the same species. From the published drawings it is evident that some forms, e.g. Passerine birds, have all left-handed spirals whereas the duck and many mammals have right-handed spirals. Mixture of both forms in any one species appears to be exceedingly rare, the rat being the only animal in which such mixture has been recorded, so far as I know.

It would be tempting to suppose that the male and female gametes were essentially counterparts in this respect, but there are indications that even so we should be no further advanced. For whatever be the rationale of fertilisation, it must take into account the conjugation of ciliated infusoria and the unicellular algae. Among ciliates screw-forms are common, but the individuals of each species are alike. Consequently since organisms spiral in the same direction can conjugate, it follows that if male and female gametes are in any way counterparts of each other their complementary relation must be much more profound than outward form, and probably independent of it.

Another fact that must ultimately be reckoned with is the existence of what are called "Giant" forms. I am not alluding to gigantism as we know it in human beings, which is a manifestation of disease of the pituitary body, but to a kind of multiplicity, in which the material or ingredients of four gametes combine to produce a single zygote instead of two. Such bodies can be described as a pair of twins arranged as one person. We are compelled to believe that through some accident, probably in the maturation divisions, giant or double gametes have appeared, and by their union have been formed giant zygotes. Among both animals and plants many such cases have now been seen. As the best known example I may mention the giant Primulas studied by Mr R. P. Gregory. But more unexpected still is the statement of Tahara that several species of Chrysanthemum are related to each other as multiples of 9, the series 9, 18, 27, 36, 45 having been found as normal numbers. It is strange enough that *even* multiples of the base number should occur so freely, but how the *odd* multiples can settle down to form a normal plant is utterly beyond our comprehension. There are nevertheless parallel instances, as for example the races of banana in which Tischler found respectively not only 8 and 16 chromosomes but also 24. Were it not for the existence of these[1] *odd* multiples we might look on the giant form as a peculiar and special manifestation of duplicity. We might say, indeed, that a wave of division having reached a certain development had faded out again, somewhat as Driesch has seen in his twinning echinoderm larvae.

[1] The odd multiples in the gametic series have now also been shewn in certain cases to be due to a doubling in the zygote resulting from a cross between diploid and tetraploid parents. Cp. for example Goodspeed, *Genetics*, x, 1925. [C.B.B.]

The observations of Geoffrey Smith on hybrid pigeons strongly suggest that giant *male* gametes are actually the consequence of the suppression of the last division. Just as the two halves of an embryo may each on occasion become a whole, and so produce a pair of twins, so in these "giants" may two sets of material, which ought to become two individuals, group themselves as one. As to the production of giant female gametes we have no evidence.

The giant forms, representing as they do, not two gametes only, but higher multiples, are genetically of much interest. We know now that a zygote is *pure*-bred in any given respect only when the contributions received from each parent were alike in that respect. In all ordinary cases, therefore, as regards a given character only three classes of zygotes can be made, two pure-bred and one cross-bred. Among the giant forms there are many more possible combinations. Even the double giants have, as it were, four parents, and the numerical schemes which are followed in these cases have not yet been determined with any certainty.

If the picture of gametes and zygotes which I have put before you is true, how can anything new be formed? On other occasions I have argued that we have no perfectly clear warrant for asserting that change can come except by loss. Naturally this view has been the subject of some criticism but as yet I have seen nothing which positively shews it to be wrong. The difficulties which the modern evolutionist has to meet are in reality two. He has to shew how a form, once true-breeding, made by the union of similar gametes, can vary *at all*, since at each resolution it must be expected to make only gametes like those from which it grew; and secondly, if he is to avoid the paradox of which I have spoken, he must produce unimpeachable proof that in a hitherto true-breeding strain a new variation may arise which cannot be attributable to loss. The first difficulty can, I think, be dealt with on the evidence. Original variation is not abundant as we used to suppose, but there are examples whose authenticity has not been successfully impugned. Original variation may be large and striking, or it may be seemingly trivial, and in segregation the integrity of the element to which that variation was due may be maintained or may be impaired. Some of you have doubtless seen the attempt lately made by Jennings to revive the belief that the existence of a continuous series of transitional forms justifies the older theories of evolution by gradual selection. Everyone familiar with genetics knows that the unity of a character may for long be maintained, though on occasion the same character may be disintegrated. Of such disintegration we have abundant illustrations. The best is perhaps the sweet pea, of which there were three or four very distinct forms for centuries. The inter-grades were a modern production. It is noticeable that Jennings founds his argument on facts almost equally signifi-cant, teaching as they do the same lesson in terms hardly mistakable. The fly, *Drosophila*, normally red-eyed, produced a white-eyed variety. In generations of intercrossing between the red-eyed and the white-eyed, pink-eyed flies occasionally appeared. By more intercrossing cherry-coloured, eosin-coloured, etc., etc., were produced, until there is now a series which may reasonably be described as continuous. The ingredient on which the red eye depends is *commonly* treated in division as an integral unit, but sometimes its integrity is impaired. Offered such a series the evolutionist of 25 years ago would of course have inverted the truth, seeing clearly that the white was derived from the red by the occur-rence of a gradual change towards white, whereas the truth was that the white came directly from the red and the transitional forms afterwards. Examples of such disintegra-tion abound. It is the proof of *integration* that is doubtful and elusive.

I am not without hope that in comprehensive experiments on the mechanical separation of various materials parallels would be found for the phenomena of genetic segregation, and the attempt may be commended to those who are expert in laboratory technique. Any model of this kind, which works, may not improbably assist analysis. In critical cell divisions, especially those immediately antecedent to the formation of gametes, segregation happens, and we must express that process as a breaking up, a tearing apart, a sorting out, or by the use of some similar figure. Nevertheless we should keep our minds open to the possibility that the thing broken up is not actual material, whether of the nucleus or the cell plasm. Many of our American colleagues interpret the visible features of cell division as literally conforming with this material scheme, and it is not to be denied that following that guiding principle they have been led to far-reaching positive discoveries. Complex, however, as the living cell must be I am unable to see in it a material heterogeneity so vast; and I incline to the expectation that the heterogeneity of the determining elements as factors lies rather in forces, of which the cell materials are the vehicle, than in the nature of the material itself.

In examining the relations of the gamete to the zygote we are observing the concrete phenomena of life at the point where they are confined to their narrowest channel. Even there their course is still of infinite subtlety and significance. I can only trust that the greatness of the theme may atone somewhat for the imperfection of the survey, and render this discourse a not wholly unworthy offering to the memory of the acute and catholic student of life in whose honour we are assembled to-day.

HEREDITY AND VARIATION IN MODERN LIGHTS

From *Darwin and Modern Science*
(Cambridge University Press, 1909)

Darwin's work has the property of greatness in that it may be admired from more aspects than one. For some the perception of the principle of Natural Selection stands out as his most wonderful achievement to which all the rest is subordinate. Others, among whom I would range myself, look up to him rather as the first who plainly distinguished, collected, and comprehensively studied that new class of evidence from which hereafter a true understanding of the process of Evolution may be developed. We each prefer our own standpoint of admiration; but I think that it will be in their wider aspect that his labours will most command the veneration of posterity.

A treatise written to advance knowledge may be read in two moods. The reader may keep his mind passive, willing merely to receive the impress of the writer's thought; or he may read with his attention strained and alert, asking at every instant how the new knowledge can be used in a further advance, watching continually for fresh footholds by which to climb higher still. Of Shelley it has been said that he was a poet for poets: so Darwin was a naturalist for naturalists. It is when his writings are used in the critical and more exacting spirit with which we test the outfit for our own enterprise that we learn their full value and strength. Whether we glance back and compare his performance with the efforts of his predecessors, or look forward along the course which modern research is disclosing, we shall honour most in him not the rounded merit of finite accomplishment, but the creative power by which he inaugurated a line of discovery endless in variety and extension. Let us attempt thus to see his work in true perspective between the past from which it grew, and the present which is its consequence. Darwin attacked the problem of Evolution by reference to facts of three classes: Variation; Heredity; Natural Selection. His work was not as the laity suppose, a sudden and unheralded revelation, but the first fruit of a long and hitherto barren controversy. The occurrence of variation from type, and the hereditary transmission of such variation had of course been long familiar to practical men, and inferences as to the possible bearing of those phenomena on the nature of specific difference had been from time to time drawn by naturalists.

Maupertuis, for example, wrote: *Ce qui nous reste à examiner, c'est comment d'un seul individu, il a pu naître tant d'espèces si différentes.* And again: *La Nature contient le fonds de toutes ces variétés: mais le hasard ou l'art les mettent en œuvre. C'est ainsi que ceux dont l'industrie s'applique à satisfaire le goût des curieux, sont, pour ainsi dire, créateurs d'espèces nouvelles.*[1]

Such passages, of which many (though few so emphatic) can be found in eighteenth-century writers, indicate a true perception of the mode of Evolution. The speculations hinted at by Buffon,[2] developed by Erasmus Darwin, and independently proclaimed above all by Lamarck, gave to the doctrine of descent a wide renown. The uniformitarian teaching which Lyell deduced from geological observation had gained acceptance. The facts of geographical distribution[3] had been shewn to be obviously inconsistent with the Mosaic legend. Prichard, and Lawrence, following the example of Blumenbach, had successfully demonstrated that the races of man could be regarded as different forms of one species, contrary to the opinion up till then received. These treatises all begin, it is true, with a profound obeisance to the sons of Noah, but that performed, they continue on strictly modern lines. The question of the mutability of species was thus prominently raised.

Those who rate Lamarck no higher than did Huxley in his contemptuous phrase "buccinator tantum", will scarcely deny that the sound of the trumpet had carried far, or that its note was clear. If then there were few who had already turned to evolution with positive conviction, all scientific men must at least have known that such views had been promulgated; and many must, as Huxley says, have taken up his own position of "critical expectancy".[4]

[1] *Venus Physique, contenant deux Dissertations, l'une sur l'origine des Hommes et des Animaux: Et l'autre sur l'origine des Noirs,* La Haye, 1746, pp. 124 and 129. For an introduction to the writings of Maupertuis I am indebted to an article by Professor Lovejoy in *Popular Sci. Monthly,* 1902.
[2] For the fullest account of the views of these pioneers of Evolution, see the works of Samuel Butler, especially *Evolution, Old and New* (2nd ed.), 1882. Butler's claims on behalf of Buffon have met with some acceptance; but after reading what Butler has said, and a considerable part of Buffon's own works, the word "hinted" seems to me a sufficiently correct description of the part he played. It is interesting to note that in the chapter on the ass, which contains some of his evolutionary passages, there is a reference to "plusieurs idées très-élevées sur la génération" contained in the Letters of Maupertuis.
[3] See especially W. Lawrence, *Lectures on Physiology,* London, 1823, pp. 213f.
[4] See the chapter contributed to the *Life and Letters of Charles Darwin,* II, p. 195. I do not clearly understand the sense in which Darwin wrote (Autobiography, *ibid.* I, p. 87): "It has sometimes been said that the success of the

Why, then, was it, that Darwin succeeded where the rest had failed? The cause of that success was two-fold. First, and obviously, in the principle of Natural Selection he had a suggestion which would work. It might not go the whole way, but it was true as far as it went. Evolution could thus in great measure be fairly represented as a consequence of demonstrable processes. Darwin seldom endangers the mechanism he devised by putting on it strains much greater than it can bear. He at least was under no illusion as to the omnipotence of Selection; and he introduces none of the forced pleading which in recent years has threatened to discredit that principle.

For example, in the latest text of the *Origin*[1] we find him saying:

But as my conclusions have lately been much misrepresented, and it has been stated that I attribute the modification of species exclusively to natural selection, I may be permitted to remark that in the first edition of this work, and subsequently, I placed in a most con-

---

Origin proved 'that the subject was in the air', or 'that men's minds were prepared for it'. I do not think that this is strictly true, for I occasionally sounded not a few naturalists, and never happened to come across a single one who seemed to doubt about the permanence of species". This experience may perhaps have been an accident due to Darwin's isolation. The literature of the period abounds with indications of "critical expectancy". A most interesting expression of that feeling is given in the charming account of the "Early Days of Darwinism" by Alfred Newton, *Macmillan's Magazine*, LVII, 1888, p. 241. He tells how in 1858 when spending a dreary summer in Iceland, he and his friend, the ornithologist John Wolley, in default of active occupation, spent their days in discussion. "Both of us taking a keen interest in Natural History, it was but reasonable that a question, which in those days was always coming up wherever two or more naturalists were gathered together, should be continually recurring. That question was, 'What is a species?' and connected therewith was the other question, 'How did a species begin?'....Now we were of course fairly well acquainted with what had been published on these subjects". He then enumerates some of these publications, mentioning among others T. Vernon Wollaston's *Variation of Species*—a work which has in my opinion never been adequately appreciated. He proceeds: "Of course we never arrived at anything like a solution of these problems, general or special, but we felt very strongly that a solution ought to be found, and that quickly, if the study of Botany and Zoology was to make any great advance". He then describes how on his return home he received the famous number of the *Linnean Journal* on a certain evening. "I sat up late that night to read it; and never shall I forget the impression it made upon me. Herein was contained a perfectly simple solution of all the difficulties which had been troubling me for months past....I went to bed satisfied that a solution had been found".

But he knew Herbert. *Trans. Hort. Soc.* IV, 1819, pp. 16–17. Also see *Samuel Butler, A Memoir*, I, p. 165.

[1] *Origin*, 6th ed. (1882), p. 421.

spicuous position—namely, at the close of the Introduction—the following words: "I am convinced that natural selection has been the main but not the exclusive means of modification".

But apart from the invention of this reasonable hypothesis, which may well, as Huxley estimated, "be the guide of biological and psychological speculation for the next three or four generations", Darwin made a more significant and imperishable contribution. Not for a few generations, but through all ages he should be remembered as the first who shewed clearly that the problems of Heredity and Variation are soluble by observation, and laid down the course by which we must proceed to their solution.[1] The moment of inspiration did not come with the reading of Malthus, but with the opening of the "first note-book on Transmutation of Species".[2] Evolution is a process of Variation and Heredity. The older writers, though they had some vague idea that it must be so, did not study Variation and Heredity. Darwin did, and so begat not a theory, but a science.

The extent to which this is true, the scientific world is only beginning to realise. So little was the fact appreciated in Darwin's own time that the success of his writings was followed by an almost total cessation of work in that special field. Of the causes which led to this remarkable consequence I have spoken elsewhere. They proceeded from circumstances peculiar to the time; but whatever the causes there is no doubt that this statement of the result is historically exact, and those who make it their business to collect facts elucidating the physiology of Heredity and Variation are well aware that they will find little to reward their quest in the leading scientific journals of the Darwinian epoch.

In those thirty years the original stock of evidence current and in circulation even underwent a process of attrition. As in the story of the Eastern sage who first wrote the collected learning of the universe for his sons in a thousand volumes, and by successive compression and burning reduced them to one,

---

[1] Whatever be our estimate of the importance of Natural Selection, in this we all agree. Samuel Butler, the most brilliant, and by far the most interesting of Darwin's opponents—whose works are at length emerging from oblivion—in his Preface (1882) to the 2nd edition of *Evolution, Old and New*, repeats his earlier expression of homage to one whom he had come to regard as an enemy: "To the end of time, if the question be asked, 'Who taught people to believe in Evolution?' the answer must be that it was Mr Darwin. This is true, and it is hard to see what palm of higher praise can be awarded to any philosopher".

[2] *Life and Letters*, I, pp. 83 and 276.

and from this by further burning distilled the single ejaculation of the Faith, "There is no god but God and Mohamed is the Prophet of God", which was all his maturer wisdom deemed essential—so in the books of that period do we find the *corpus* of genetic knowledge dwindle to a few prerogative instances, and these at last to the brief formula of an unquestioned creed.

And yet in all else that concerns biological science this period was, in very truth, our Golden Age, when the natural history of the earth was explored as never before; morphology and embryology were exhaustively ransacked; the physiology of plants and animals began to rival chemistry and physics in precision of method and in the rapidity of its advances; and the foundations of pathology were laid.

In contrast with this immense activity elsewhere the neglect which befell the special physiology of Descent, or Genetics as we now call it, is astonishing. This may of course be interpreted as meaning that the favoured studies seemed to promise a quicker return for effort, but it would be more true to say that those who chose these other pursuits did so without making any such comparison; for the idea that the physiology of Heredity and Variation was a coherent science, offering possibilities of extraordinary discovery, was not present to their minds at all. In a word, the existence of such a science was well-nigh forgotten. It is true that in ancillary periodicals, as for example those that treat of entomology or horticulture, or in the writings of the already isolated systematists,[1] observations with this special bearing were from time to time related, but the class of fact on which Darwin built his conceptions of Heredity and Variation was not seen in the highways of biology. It formed no part of the official curriculum of biological students, and found no place among the subjects which their teachers were investigating.

[1] This isolation of the systematists is the one most melancholy *sequela* of Darwinism. It seems an irony that we should read in the peroration to the *Origin* that when the Darwinian view is accepted "Systematists will be able to pursue their labours as at present; but they will not be incessantly haunted by the shadowy doubt whether this or that form be a true species. This, I feel sure, and I speak after experience, will be no slight relief. The endless disputes whether or not some fifty species of British brambles are good species will cease" (*Origin*, 6th ed. (1882), p. 425). True they have ceased to attract the attention of those who lead opinion, but anyone who will turn to the literature of systematics will find that they have not ceased in any other sense. Should there not be something disquieting in the fact that among the workers who come most into contact with specific differences, are to be found the only men who have failed to be persuaded of the unreality of those differences?

During this period nevertheless one distinct advance was made, that with which Weismann's name is prominently connected. In Darwin's genetic scheme the hereditary transmission of parental experience and its consequences played a considerable rôle. Exactly how great that rôle was supposed to be, he with his habitual caution refrained from specifying, for the sufficient reason that he did not know. Nevertheless much of the process of Evolution, especially that by which organs have become degenerate and rudimentary, was certainly attributed by Darwin to such inheritance, though since belief in the inheritance of acquired characters fell into disrepute, the fact has been a good deal overlooked. The *Origin* without "use and disuse" would be a materially different book. A certain vacillation is discernible in Darwin's utterances on this question, and the fact gave to the astute Butler an opportunity for his most telling attack. The discussion which best illustrates the genetic views of the period arose in regard to the production of the rudimentary condition of the wings of many beetles in the Madeira group of islands, and by comparing passages from the *Origin*[1] Butler convicts Darwin of saying first that this condition was in the main the result of Selection, with disuse aiding, and in another place that the main cause of degeneration was disuse, but that Selection had aided. To Darwin however I think the point would have seemed one of dialectics merely. To him the one paramount purpose was to shew that somehow an Evolution by means of Variation and Heredity might have brought about the facts observed, and whether they had come to pass in the one way or the other was a matter of subordinate concern.

To us moderns the question at issue has a diminished significance. For over all such debates a change has been brought by Weismann's challenge for evidence that use and disuse have any transmitted effects at all. Hitherto the transmission of many acquired characteristics had seemed to most naturalists so obvious as not to call for demonstration.[2] Weismann's demand for facts in support of the main proposition revealed at once

[1] 6th ed. pp. 109 and 401. See Butler, *Essays on Life, Art, and Science*, p. 265, reprinted 1908, and *Evolution, Old and New*, chap. xxii (2nd ed.), 1882.
[2] W. Lawrence was one of the few who consistently maintained the contrary opinion. Prichard, who previously had expressed himself in the same sense, does not, I believe, repeat these views in his later writings, and there are signs that he came to believe in the transmission of acquired habits. See Lawrence, *Lect. Physiol.* 1823, pp. 436–437, 447; Prichard, *Edin. Inaug. Disp.* 1808 (not seen by me), quoted *ibid.* and *Nat. Hist. Man*, 1843, pp. 34 f. See also Godron *passim.*

that none having real cogency could be produced. The time-honoured examples were easily shewn to be capable of different explanations. A few certainly remain which cannot be so summarily dismissed, but—though it is manifestly impossible here to do justice to such a subject—I think no one will dispute that these residual and doubtful phenomena, whatever be their true nature, are not of a kind to help us much in the interpret-ation of any of those complex cases of adaptation which on the hypothesis of unguided Natural Selection are especially difficult to understand. Use and disuse were invoked expressly to help us over these hard places; but whatever changes can be induced in offspring by direct treatment of the parents, they are not of a kind to encourage hope of real assistance from that quarter. It is not to be denied that through the collapse of this second line of argument the Selection hypothesis has had to take an increased and perilous burden. Various ways of meeting the difficulty have been proposed, but these mostly resolve them-selves into improbable attempts to expand or magnify the powers of Natural Selection.

Weismann's interpellation, though negative in purpose, has had a lasting and beneficial effect, for through his thorough demolition of the old loose and distracting notions of inherited experience, the ground has been cleared for the construction of a true knowledge of heredity based on experimental fact.

In another way he made a contribution of a more positive character, for his elaborate speculations as to the genetic meaning of cytological appearances have led to a minute in-vestigation of the visible phenomena occurring in those cell divisions by which germ cells arise. Though the particular views he advocated have very largely proved incompatible with the observed facts of heredity, yet we must acknowledge that it was chiefly through the stimulus of Weismann's ideas that those advances in cytology were made; and though the doctrine of the continuity of germ plasm cannot be maintained in the form originally propounded, it is in the main true and illuminating.[1]

[1] It is interesting to see how nearly Butler was led by natural penetration, and from absolutely opposite conclusions, back to this underlying truth: "So that each ovum when impregnate should be considered not as descended from its ancestors, but as being a continuation of the personality of every ovum in the chain of its ancestry, which every ovum *it actually is* quite as truly as the octogenarian *is* the same identity with the ovum from which he has been developed. This process cannot stop short of the primordial cell, which again will probably turn out to be but a brief resting-place. We therefore prove each one of us to *be actually* the primordial cell which never died nor dies, but has differentiated itself into the life of the world, all living beings whatever, being one with it and members one of another" (*Life and Habit*, 1878, p. 86).

Nevertheless, in the present state of knowledge we are still as a rule quite unable to connect cytological appearances with any genetic consequence and save in one respect (obviously of extreme importance—to be spoken of later) the two sets of phenomena might, for all we can see, be entirely distinct.

I cannot avoid attaching importance to this want of connection between the nuclear phenomena and the features of bodily organisation. All attempts to investigate Heredity by cytological means lie under the disadvantage that it is the nuclear changes which can alone be effectively observed. Important as they must surely be, I have never been persuaded that the rest of the cell counts for nothing. What we know of the behaviour and variability of chromosomes seems in my opinion quite incompatible with the belief that they alone govern form, and are the sole agents responsible in heredity.[1]

If, then, progress was to be made in Genetics, work of a different kind was required. To learn the laws of Heredity and Variation there is no other way than that which Darwin himself followed, the direct examination of the phenomena. A beginning could be made by collecting fortuitous observations of this class, which have often thrown a suggestive light, but such evidence can be at best but superficial and some more penetrating instrument of research is required. This can only be provided by actual experiments in breeding.

The truth of these general considerations was becoming gradually clear to many of us when in 1900 Mendel's work was rediscovered. Segregation, a phenomenon of the utmost novelty, was thus revealed. From that moment not only in the problem of the origin of species, but in all the great problems of biology a new era began. So unexpected was the discovery that many naturalists were convinced it was untrue, and at once proclaimed Mendel's conclusions as either altogether mistaken, or, if true, of very limited application. Many fantastic notions about the workings of Heredity had been asserted as general principles before: this was probably only another fancy of the same class.

Nevertheless those who had a preliminary acquaintance with the facts of Variation were not wholly unprepared for some such revelation. The essential deduction from the discovery of segregation was that the characters of living things are de-

---

[1] This view is no doubt contrary to the received opinion. I am, however, interested to see it lately maintained by Driesch (*Science and Philosophy of the Organism*, London, 1907, p. 233), and from the recent observations of Godlewski it has received distinct experimental support.

pendent on the presence of definite elements or factors, which
are treated as units in the processes of Heredity. These factors
can thus be recombined in various ways. They act sometimes
separately, and sometimes they interact in conjunction with
each other, producing their various effects. All this indicates
a definiteness and specific order in heredity, and therefore in
variation. This order cannot by the nature of the case be
dependent on Natural Selection for its existence, but must be
a consequence of the fundamental chemical and physical
nature of living things. The study of Variation had from the
first shewn that an orderliness of this kind was present. The
bodies and the properties of living things are cosmic, not
chaotic. No matter how low in the scale we go, never do we
find the slightest hint of a diminution in that all-pervading
orderliness, nor can we conceive an organism existing for a
moment in any other state. Moreover not only does this order
prevail in normal forms, but again and again it is to be seen in
newly sprung varieties, which by general consent cannot have
been subjected to a prolonged Selection. The discovery of
Mendelian elements admirably coincided with and at once
gave a rationale of these facts. Genetic Variation is then
primarily the consequence of additions to, or omissions
from, the stock of elements which the species contains.
The further investigation of the species problem must thus
proceed by the analytical method which breeding experiments
provide.

In the nine years which have elapsed since Mendel's clue
became generally known, progress has been rapid. We now
understand the process by which a polymorphic race maintains
its polymorphism. When a family consists of dissimilar
members, given the numerical proportions in which these
members are occurring, we can represent their composition
symbolically and state what types can be transmitted by the
various members. The difficulty of the "swamping effects of
intercrossing" is practically at an end. Even the famous puzzle
of sex-limited inheritance is solved, at all events in its more
regular manifestations, and we know now how it is brought about
that the normal sisters of a colour-blind man can transmit the
colour-blindness while his normal brothers cannot transmit it.

We are still only on the fringe of the inquiry. It can be seen
extending and ramifying in many directions. To enumerate
these here would be impossible. A whole new range of possibilities is being brought into view by study of the inter-relations
between the simple factors. By following up the evidence as to

segregation, indications have been obtained which can only be interpreted as meaning that when many factors are being simultaneously redistributed among the germ cells, certain of them exert what must be described as a repulsion upon other factors. We cannot surmise whither this discovery may lead.

In the new light all the old problems wear a fresh aspect. Upon the question of the nature of sex, for example, the bearing of Mendelian evidence is close. Elsewhere I have shewn that from several sets of parallel experiments the conclusion is almost forced upon us that, in the types investigated, of the two sexes the female is to be regarded as heterozygous in sex, containing one unpaired dominant element, while the male is similarly homozygous in the absence of that element.[1] It is not a little remarkable that on this point—which is the only one where observations of the nuclear processes of gametogenesis have yet been brought into relation with the visible characteristics of the organisms themselves—there should be diametrical opposition between the results of breeding experiments and those derived from cytology.

Those who have followed the researches of the American school will be aware that, after it had been found in certain insects that the spermatozoa were of two kinds according as they contained or did not contain the accessory chromosome, E. B. Wilson succeeded in proving that the sperms possessing this accessory body were destined to form *females* on fertilisation, while sperms without it form males, the eggs being apparently indifferent. Perhaps the most striking of all this series of observations is that lately made by T. H. Morgan,[2] since confirmed by von Baehr, that in a Phylloxeran two kinds of spermatids are formed, respectively with and without an accessory (in this case, *double*) chromosome. Of these, only those possessing the accessory body become functional spermatozoa, the others degenerating. We have thus an elucidation of the puzzling fact that in these forms fertilisation results in the formation of *females* only. How the males are formed—for of course males are eventually produced by the parthenogenetic females—we do not know.

If the accessory body is really to be regarded as bearing the factor for femaleness, then in Mendelian terms female is DD and male is DR. The eggs are indifferent and the spermatozoa

---

[1] In other words, the ova are each *either* female, *or* male (i.e. non-female), but the sperms are all non-female.

[2] Morgan, *Proc. Soc. Exp. Biol. Med.* v, 1908, and von Baehr, *Zool. Anz.* XXXII, 1908, p. 507.

are each male, *or* female. But according to the evidence derived from a study of the sex-limited descent of certain features in other animals the conclusion seems equally clear that in them female must be regarded as *DR* and male as *RR*. The eggs are thus each either male or female and the spermatozoa are indifferent. How this contradictory evidence is to be reconciled we do not yet know. The breeding work concerns fowls, canaries, and the currant moth (*Abraxas grossulariata*). The accessory chromosome has been now observed in most of the great divisions of insects,[1] except, as it happens, Lepidoptera. At first sight it seems difficult to suppose that a feature apparently so fundamental as sex should be differently constituted in different animals, but that seems at present the least improbable inference. I mention these two groups of facts as illustrating the nature and methods of modern genetic work. We must proceed by minute and specific analytical investigation. Wherever we look we find traces of the operation of precise and specific rules.

In the light of present knowledge it is evident that before we can attack the species problem with any hope of success there are vast arrears to be made up. He would be a bold man who would now assert that there was no sense in which the term Species might not have a strict and concrete meaning in contradistinction to the term Variety. We have been taught to regard the difference between species and variety as one of degree. I think it unlikely that this conclusion will bear the test of further research. To Darwin the question, What is a variation? presented no difficulties. Any difference between parent and offspring was a variation. Now we have to be more precise. First we must, as de Vries has shewn, distinguish real, genetic, variation from *fluctuational* variations, due to environmental and other accidents, which cannot be transmitted. Having excluded these sources of error the variations observed must be expressed in terms of the factors to which they are due before their significance can be understood. For example, numbers of the variations seen under domestication, and not a few witnessed in nature, are simply the consequence of some ingredient being in an unknown way omitted from the composition of the varying individual. The variation may on the contrary be due to the

[1] As Wilson has proved, the unpaired body is not a universal feature even in those orders in which it has been observed. Nearly allied types may differ. In some it is altogether unpaired. In others it is paired with a body of much smaller size, and by selection of various types all gradations can be demonstrated ranging to the condition in which the members of the pair are indistinguishable from each other.

addition of some new element, but to prove that it is so is by no means an easy matter. Casual observation is useless, for though these latter variations will always be dominants, yet many dominant characteristics may arise from another cause, namely the meeting of complementary factors, and special study of each case in two generations at least is needed before these two phenomena can be distinguished.

When such considerations are fully appreciated it will be realised that medleys of most dissimilar occurrences are all confused together under the term Variation. One of the first objects of genetic analysis is to disentangle this mass of confusion.

To those who have made no study of heredity it sometimes appears that the question of the effect of conditions in causing variation is one which we should immediately investigate, but a little thought will shew that before any critical inquiry into such possibilities can be attempted, a knowledge of the working of heredity under conditions as far as possible uniform must be obtained. At the time when Darwin was writing, if a plant brought into cultivation gave off an albino variety, such an event was without hesitation ascribed to the change of life. Now we see that albino *gametes*, germs, that is to say, which are destitute of the pigment-forming factor, may have been originally produced by individuals standing an indefinite number of generations back in the ancestry of the actual albino, and it is indeed almost certain that the variation to which the appearance of the albino is due cannot have taken place in a generation later than that of the grandparents. It is true that when a new *dominant* appears we should feel greater confidence that we were witnessing the original variation, but such events are of extreme rarity, and no such case has come under the notice of an experimenter in modern times, as far as I am aware. That they must have appeared is clear enough. Nothing corresponding to the Brown-breasted Game fowl is known wild, yet that colour is a most definite dominant, and at some moment since *Gallus bankiva* was domesticated, the element on which that special colour depends must have at least once been formed in the germ cell of a fowl; but we need harder evidence than any which has yet been produced before we can declare that this novelty came through over-feeding, or change of climate, or any other disturbance consequent on domestication. When we reflect on the intricacies of genetic problems as we must now conceive them there come moments when we feel almost thankful that the Mendelian principles were unknown to Darwin.

The time called for a bold pronouncement, and he made it, to our lasting profit and delight. With fuller knowledge we pass once more into a period of cautious expectation and reserve.

In every arduous enterprise it is pleasanter to look back at difficulties overcome than forward to those which still seem insurmountable, but in the next stage there is nothing to be gained by disguising the fact that the attributes of living things are not what we used to suppose. If they are more complex in the sense that the properties they display are throughout so regular[1] that the selection of minute random variations is an unacceptable account of the origin of their diversity, yet by virtue of that very regularity the problem is limited in scope and thus simplified.

To begin with, we must relegate Selection to its proper place. Selection permits the viable to continue and decides that the non-viable shall perish; just as the temperature of our atmosphere decides that no liquid carbon shall be found on the face of the earth: but we do not suppose that the form of the diamond has been gradually achieved by a process of Selection. So again, as the course of descent branches in the successive generations, Selection determines along which branch Evolution shall proceed, but it does not decide what novelties that branch shall bring forth. *La Nature contient le fonds de toutes ces variétés, mais le hasard ou l'art les mettent en œuvre*, as Maupertuis truly said.

Not till knowledge of the genetic properties of organisms has attained to far greater completeness can evolutionary speculations have more than a suggestive value. By genetic experiment, cytology and physiological chemistry aiding, we may hope to acquire such knowledge. In 1872 Nathusius wrote:[2] *Das Gesetz der Vererbung ist noch nicht erkannt; der Apfel ist noch nicht vom Baum der Erkenntniss gefallen, welcher, der Sage nach, Newton auf den rechten Weg zur Ergründung der Gravitationsgesetze führte.* We cannot pretend that the words are not still true, but in Mendelian analysis the seeds of that apple-tree at last are sown.

If we were asked what discovery would do most to forward our inquiry, what one bit of knowledge would more than any

---

[1] I have in view, for example, the marvellous and specific phenomena of regeneration, and those discovered by the students of *Entwicklungsmechanik*. The circumstances of its occurrence here preclude any suggestion that this regularity has been brought about by the workings of Selection. The attempts thus to represent the phenomena have resulted in mere parodies of scientific reasoning.

[2] *Vorträge über Viehzucht und Rassenerkenntniss*, Berlin, 1872, 120.

other illuminate the problem, I think we may give the answer without hesitation. The greatest advance that we can foresee will be made when it is found possible to connect the geometrical phenomena of development with the chemical. The geometrical symmetry of living things is the key to a knowledge of their regularity, and the forces which cause it. In the symmetry of the dividing cell the basis of that resemblance we call Heredity is contained. To imitate the morphological phenomena of life we have to devise a system which can divide. It must be able to divide, and to segment as—grossly— a vibrating plate or rod does, or as an icicle can do as it becomes ribbed in a continuous stream of water; but with this distinction, that the distribution of chemical differences and properties must simultaneously be decided and disposed in orderly relation to the pattern of the segmentation. Even if a model which would do this could be constructed it might prove to be a useful beginning.

This may be looking too far ahead. If we had to choose some one piece of more proximate knowledge which we would more especially like to acquire, I suppose we should ask for the secret of inter-racial sterility. Nothing has yet been discovered to remove the grave difficulty, by which Huxley in particular was so much oppressed, that among the many varieties produced under domestication—which we all regard as analogous to the species seen in nature—no clear case of inter-racial sterility has been demonstrated. The phenomenon is probably the only one to which the domesticated products seem to afford no parallel. No solution of the difficulty can be offered which has positive value, but it is perhaps worth considering the facts in the light of modern ideas. It should be observed that we are not discussing incompatibility of two species to produce offspring (a totally distinct phenomenon), but the sterility of the offspring which many of them do produce.

When two species, both perfectly fertile severally, produce on crossing a sterile progeny, there is a presumption that the sterility is due to the development in the hybrid of some substance which can only be formed by the meeting of two complementary factors. That some such account is correct in essence may be inferred from the well-known observation that if the hybrid is not totally sterile but only partially so, and thus is able to form some good germ cells which develop into new individuals, the sterility of these daughter-individuals is sensibly reduced or may be entirely absent. The fertility once re-established, the sterility does not return in the later progeny, a fact strongly suggestive of segregation. Now if the sterility of the cross-bred be really the consequence of the meeting of two complementary factors, we see that the phenomenon could only be produced among the divergent offspring of one species by the acquisition of at least *two* new factors; for if the acquisition of a single factor caused sterility the line would then end. Moreover each factor must be separately acquired by distinct individuals, for if both were present together, the possessors would by hypothesis be sterile. And in order to imitate the case of species each of these factors must be acquired by distinct breeds. The factors need not, and probably would not, produce any other perceptible effects; they might, like the colour-factors present in white flowers, make no difference in the form or other characters. Not till the cross was actually made between the two complementary individuals would either factor come into play, and the effects even then might be unobserved until an attempt was made to breed from the cross-bred.

Next, if the factors responsible for sterility were acquired, they would in all probability be peculiar to certain individuals and would not readily be distributed to the whole breed. Any member of the breed also into which *both* the factors were introduced would drop out of the pedigree by virtue of its sterility. Hence the evidence that the various domesticated breeds say of dogs or fowls can when mated together produce fertile offspring, is beside the mark. The real question is, Do they ever produce sterile offspring? I think the evidence is clearly that sometimes they do, oftener perhaps than is commonly supposed. These suggestions are quite amenable to experimental tests. The most obvious way to begin is to get a pair of parents which are known to have had any sterile offspring, and to find the proportions in which these steriles were produced. If, as I anticipate, these proportions are found to be definite, the rest is simple.

In passing, certain other considerations may be referred to. First, that there are observations favouring the view that the production of totally sterile cross-breds is seldom a universal property of two species, and that it may be a matter of individuals, which is just what on the view here proposed would be expected. Moreover, as we all know now, though incompatibility may be dependent to some extent on the degree to which the species are dissimilar, no such principle can be demonstrated to determine sterility or fertility in general. For example, though all our finches can breed together, the hybrids are all sterile. Of ducks some species can breed together

230 HEREDITY AND VARIATION

without producing the slightest sterility; others have totally sterile offspring, and so on. The hybrids between several *genera* of orchids are perfectly fertile on the female side, and some on the male side also, but the hybrids produced between the turnip (*Brassica napus*) and the swede (*Brassica campestris*), which, according to our estimates of affinity, should be nearly allied forms, are totally sterile.[1] Lastly, it may be recalled that in sterility we are almost certainly considering a meristic phenomenon. *Failure to divide* is, we may feel fairly sure, the immediate "cause" of the sterility. Now, though we know very little about the heredity of meristic differences, all that we do know points to the conclusion that the less-divided is dominant to the more-divided, and we are thus justified in supposing that . there are factors which can arrest or prevent cell division. My conjecture therefore is that in the case of sterility of cross-breds we see the effect produced by a complementary pair of such factors. This and many similar problems are now open to our analysis.

The question is sometimes asked, Do the new lights on Variation and Heredity make the process of Evolution easier to understand? On the whole the answer may be given that they do. There is some appearance of loss of simplicity, but the gain is real. As was said above, the time is not ripe for the discussion of the origin of species. With faith in Evolution unshaken—if indeed the word faith can be used in application to that which is certain—we look on the manner and causation of adapted differentiation as still wholly mysterious. As Samuel Butler so truly said: "To me it seems that the 'Origin of Variation', whatever it is, is the only true 'Origin of Species'",[2] and of that Origin not one of us knows anything. But given Variation—and it is given: assuming further that the variations are not guided into paths of adaptation—and both to the Darwinian and to the modern school this hypothesis appears to be sound if unproven—an evolution of species proceeding by definite steps is more, rather than less, easy to imagine than an evolution proceeding by the accumulation of indefinite and insensible steps. Those who have lost themselves in contemplating the miracles of Adaptation (whether real or spurious) have not unnaturally fixed their hopes rather on the indefinite than on the definite changes. The reasons are obvious. By suggesting that the steps through which an adaptative mechanism arose were indefinite and insensible, all further

[1] See Sutton, A. W., *Journ. Linn. Soc.* xxxviii, 1908, 341.
[2] *Life and Habit*, London, 1878, 263.

trouble is spared. While it could be said that species arise by an insensible and imperceptible process of variation, there was clearly no use in tiring ourselves by trying to perceive that process. This labour-saving counsel found great favour. All that had to be done to develop evolution theory was to discover the good in everything, a task which, in the complete absence of any control or test whereby to check the truth of the discovery, is not very onerous. The doctrine *que tout est au mieux* was therefore preached with fresh vigour, and examples of that illuminating principle were discovered with a facility that Pangloss himself might have envied, till at last even the spectators wearied of such dazzling performances.

But in all seriousness, why should indefinite and unlimited variation have been regarded as a more probable account of the origin of Adaptation? Only, I think, because the obstacle was shifted one plane back, and so looked rather less prominent. The abundance of Adaptation, we all grant, is an immense, almost an unsurpassable difficulty in all non-Lamarckian views of Evolution; but if the steps by which that adaptation arose were fortuitous, to imagine them insensible is assuredly no help. In one most important respect indeed, as has often been observed, it is a multiplication of troubles. For the smaller the steps, the less could Natural Selection act upon them. Definite variations—and of the occurrence of definite variations in abundance we have now the most convincing proof—have at least the obvious merit that they can make and often do make a real difference in the chances of life.

There is another aspect of the Adaptation problem to which I can only allude very briefly. May not our present ideas of the universality and precision of Adaptation be greatly exaggerated? The fit of organism to its environment is not after all so very close—a proposition unwelcome perhaps, but one which could be illustrated by very copious evidence. Natural Selection is stern, but she has her tolerant moods.

We have now most certain and irrefragable proof that much definiteness exists in living things apart from Selection, and also much that may very well have been preserved and so in a sense constituted by Selection. Here the matter is likely to rest. There is a passage in the sixth edition of the *Origin* which has I think been overlooked. On page 70 Darwin says, "The tuft of hair on the breast of the wild turkey-cock cannot be of any use, and it is doubtful whether it can be ornamental in the eyes of the female bird". This tuft of hair is a most definite and unusual structure, and I am afraid that the remark that it "cannot be of any use" may have been made inadvertently; but it may have been intended, for in the first edition the usual qualification was given and must therefore have been deliberately excised.[1] Anyhow I should like to think that Darwin did throw over that tuft of hair, and that he felt relief when he had done so. Whether however we have his great authority for such a course or not, I feel quite sure that we shall be rightly interpreting the facts of Nature if we cease to expect to find purposefulness wherever we meet with definite structures or patterns. Such things are, as often as not, I suspect rather of the nature of tool-marks, mere incidents of manufacture, benefiting their possessor not more than the wire-marks in a sheet of paper, or the ribbing on the bottom of an oriental plate renders those objects more attractive in our eyes.

If Variation may be in any way definite, the question once more arises, may it not be definite in direction? The belief that it is has had many supporters, from Lamarck onwards, who held that it was guided by need, and others who, like Nägeli, while laying no emphasis on need, yet were convinced that there was guidance of some kind. The latter view under the name of "Orthogenesis", devised I believe by Eimer, at the present day commends itself to some naturalists. The objection to such a suggestion is of course that no fragment of real evidence can be produced in its support. On the other hand, with the experimental proof that variation consists largely in the unpacking and repacking of an original complexity, it is not so certain as we might like to think that the order of these events is not pre-determined. For instance the original "pack" may have been made in such a way that at the nth division of the germ cells of a sweet pea a colour factor might be dropped, and that at the $n + n'$ division the hooded variety be given off, and so on. I see no ground whatever for holding such a view, but in fairness the possibility should not be forgotten, and in the light of modern research it scarcely looks so absurdly improbable as before.

No one can survey the work of recent years without perceiving that evolutionary orthodoxy developed too fast, and that a great deal has got to come down; but this satisfaction at least remains, that in the experimental methods which Mendel inaugurated, we have means of reaching certainty in regard to the physiology of Heredity and Variation upon which a more lasting structure may be built.

[1] *Origin*, 1st edition, p. 90, "which can hardly be either useful or ornamental to this bird".

PRESIDENTIAL ADDRESS TO THE ZOOLOGICAL SECTION, BRITISH ASSOCIATION

Cambridge Meeting. 1904

In choosing a subject for this Address I have availed myself of the kindly usage which permits a sectional president to divert the attention of his hearers into those lines of inquiry which he himself is accustomed to pursue. Nevertheless, in taking the facts of breeding for my theme, I am sensible that this privilege is subjected to a certain strain.

Heredity—and Variation too—are matters of which no naturalist likes to admit himself entirely careless. Everyone knows that, somewhere hidden among the phenomena denoted by these terms, there must be principles which, in ways untraced, are ordering the destinies of living things. Experiments in Heredity have thus, as I am told, a universal fascination. All are willing to offer an outward deference to these studies. The limits of that homage, however, are soon reached, and, though all profess interest, few are impelled to make even the moderate mental effort needed to apprehend what has been already done. It is understood that Heredity is an important mystery, and Variation another mystery. The naturalist, the breeder, the horticulturist, the sociologist, man of science and man of practice alike, has daily occasion to make and to act on assumptions as to Heredity and Variation, but many seem well content that such phenomena should remain for ever mysterious.

The position of these studies is unique. At once fashionable and neglected, nominally the central common ground of botany and zoology, of morphology and physiology, belonging specially to neither, this area is thinly tenanted. Now, since few have leisure for topics with which they cannot suppose themselves concerned, I am aware that, when I ask you in your familiar habitations to listen to tales of a no man's land, I must forgo many of those supports by which a speaker may maintain his hold on the intellectual sympathy of an audience.

Those whose pursuits have led them far from their companions cannot be exempt from that differentiation which is the fate of isolated groups. The stock of common knowledge and common ideas grows smaller till the difficulty of intercommunication becomes extreme. Not only has our point of view changed, but our materials are unfamiliar, our methods of inquiry new, and even the results attained accord little with the common expectations of the day. In the progress of sciences we are used to be led from the known to the unknown, from the half-perceived to the proven, the expectation of one year becoming the certainty of the next. It will aid appreciation of the change coming over evolutionary science if it be realised that the new knowledge of Heredity and Variation rather replaces than extends current ideas on those subjects.

Convention requires that a president should declare all well in his science; but I cannot think it a symptom indicative of much health in our body that the task of assimilating the new knowledge has proved so difficult. An eminent foreign professor lately told me that he believed there were not half a dozen in his country conversant with what may be called Mendelism, though he added hopefully, "I find these things interest my students more than my colleagues". A professed biologist cannot afford to ignore a new life-history, the okapi, or the other last new version of the old story; but phenomena which put new interpretations on the whole, facts witnessed continually by all who are working in these fields, he may conveniently disregard as matters of opinion. Had a discovery comparable in magnitude with that of Mendel been announced in physics or in chemistry, it would at once have been repeated and extended in every great scientific school throughout the world. We could come to a British Association audience to discuss the details of our subject—the polymorphism of extracted types, the physiological meaning of segregation, its applicability to the case of sex, the nature of non-segregable characters, and like problems with which we are now dealing—sure of finding sound and helpful criticism, nor would it be necessary on each occasion to begin with a popular presentation of the rudiments. This state of things in a progressive science has arisen, as I think, from a loss of touch with the main line of inquiry. The successes of descriptive zoology are so palpable and so attractive, that, not unnaturally, these which are the means of progress have been mistaken for the end. But now that the survey of terrestrial types by existing methods is happily approaching completion, we may hope that our science will return to its proper task, the detection of the fundamental nature of living things. I say *return*, because, in spite of that perfecting of the instruments of research characteristic of our time, and an extension of the area of scrutiny, the last generation was nearer the main quest. No one can study the history of biology without perceiving that in some essential respects the spirit of the naturalists of fifty years ago was truer in aim, and that their methods of inquiry were more direct and more fertile—so far, at least, as the problem of Evolution is concerned—than those which have replaced them.

If we study the researches begun by Kölreuter and continued with great vigour till the middle of the 'sixties, we cannot fail to see that had the experiments he and his successors undertook been continued on the same lines, we should by now have advanced far into the unknown. More than this: if a knowledge of what those men actually accomplished had not passed away from the memory of our generation, we should now be able to appeal to an informed public mind, having some practical acquaintance with the phenomena, and possessing sufficient experience of these matters to recognise absurdity in statement and deduction, ready to provide that healthy atmosphere of instructed criticism most friendly to the growth of truth.

Elsewhere I have noted the paradox that the appearance of the work of Darwin, which crowns the great period in the study of the phenomena of species, was the signal for a general halt. The *Origin of Species*, the treatise which for the first time brought the problem of species fairly within the range of human intelligence, so influenced the course of scientific thought that the study of this particular phenomenon—specific difference— almost entirely ceased. That this was largely due to the simultaneous opening up of lines of research in many other directions may be granted; but in greater measure, I believe, it is to be ascribed to the substitution of a conception of species which, with all the elements of truth it contains, is yet barren and unnatural. It is not wonderful that those who held that specific difference must be a phenomenon of slowest accumulation, proceeding by steps needing generations for their perception, should turn their attention to subjects deemed more amenable to human enterprise.

The indiscriminate confounding of all divergences from type into one heterogeneous heap under the name "Variation" effectually concealed those features of order which the phenomena severally present, creating an enduring obstacle to the progress of evolutionary science. Specific normality and distinctness being regarded as an accidental product of exigency, it was thought safe to treat departures from such normality as comparable differences: all were "variations" alike. Let us illustrate the consequences. "Princess of Wales" is a large modern violet, single, with stalks a foot long or more.

"Marie Louise" is another, with large double flowers, pale colour, short stalks, peculiar scent, leaf, etc. We call these "varieties", and we speak of the various fixed differences between these two, and between them and wild *odorata*, as due to Variation; and, again, the transient differences between the same *odorata* in poor, dry soil, or in a rich hedge-bank, we call Variation, using but the one term for differences, quantitative or qualitative, permanent or transitory, in size, number of parts, chemistry, and the rest. We might as well use one term to denote the differences between a bar of silver, a stick of lunar caustic, a shilling, or a teaspoon. No wonder that the ignorant tell us they can find no order in Variation.

This prodigious confusion, which has spread obscurity over every part of these inquiries, is traceable to the original misconception of the nature of specific difference, as a thing imposed and not inherent. From this, at least, the earlier experimenters were free; and the undertakings of Gärtner and his contemporaries were informed by the true conception that the properties and behaviour of species were themselves specific. Free from the later fancy that but for Selection the forms of animals and plants would be continuous and indeterminate, they recognised the definiteness of species and variety, and boldly set themselves to work out case by case the manifestations and consequences of that definiteness.

Over this work of minute and largely experimental analysis, rapidly growing, the new doctrine that organisms are mere conglomerates of adaptative devices descended like a numbing spell. By an easy confusion of thought, faith in the physiological definiteness of species and variety passed under the common ban which had at last exorcised the demon Immutability. Henceforth no naturalist must hold communion with either, on pain of condemnation as an apostate, a danger to the dynasty of Selection. From this oppression we in England, at least, are scarcely beginning to emerge. Bentham's *Flora*, teaching very positively that the primrose, the cowslip, and the oxlip are impermanent varieties of one species, is in the hand of every beginner, while the British Museum Reading Room finds it unnecessary to procure Gärtner's *Bastarderzeugung*.

And so this mass of specific learning has passed out of account. The evidence of the collector, the horticulturist, the breeder, the fancier, has been treated with neglect, and sometimes, I fear, with contempt. That wide field whence Darwin drew his wonderful store of facts has been some forty years untouched. Speak to professional zoologists of any breeder's matter, and how many will not intimate to you politely that fanciers are unscientific persons, and their concerns beneath notice? For the concrete in Evolution we are offered the abstract. Our philosophers debate with great fluency whether between imaginary races sterility could grow up by an imaginary Selection; whether Selection working upon hypothetical materials could produce sexual differentiation; how under a system of Natural Selection bodily symmetry may have been impressed on formless protoplasm—that monstrous figment of the mind, fit starting-point for such discussions. But by a physiological irony enthusiasm for these topics is sometimes fully correlated with indifference even to the classical illustrations; and for many whose minds are attracted by the abstract problem of inter-racial sterility there are few who can name for certain ten cases in which it has been already observed.

And yet in the natural world, in the collecting box, the seed bed, the poultry yard, the places where Variation, Heredity, Selection may be seen in operation and their properties tested, answers to these questions meet us at every turn—fragmentary answers, it is true, but each direct to the point. For if any one will stoop to examine Nature in those humble places, will do a few days' weeding, prick out some rows of cabbages, feed up a few score of any variable larva, he will not wait long before he learns the truth about Variation. If he go further and breed two or three generations of almost any controllable form, he will obtain immediately facts as to the course of Heredity which obviate the need for much laborious imagining. If strictly trained, with faith in the omnipotence of Selection, he will not proceed far before he encounters disquieting facts. Upon whatever character the attention be fixed, whether size, number, form of the whole or of the parts, proportion, distribution of differentiation, sexual characters, fertility, precocity or lateness, colour, susceptibility to cold or to disease—in short, all the kinds of characters which we think of as best exemplifying specific difference, we are certain to find illustrations of the occurrence of departures from normality, presenting exactly the same definiteness elsewhere characteristic of normality itself. Again and again the circumstances of their occurrence render it impossible to suppose that these striking differences are the product of continued Selection, or, indeed, that they represent the results of a gradual transformation of any kind. Whenever by any collocation of favouring circumstances such definite novelties possess a superior viability, supplanting their "normal" relatives, it is obvious that new types will be created.

The earliest statement of this simple inference is, I believe, that of Marchant,[1] who in 1719, commenting on certain plants of *Mercurialis* with laciniated and hair-like leaves, which for a time established themselves in his garden, suggested that species may arise in like manner. Though the same conclusion has appeared inevitable to many, including authorities of very diverse experience, such as Huxley, Virchow, F. Galton, it has been strenuously resisted by the bulk of scientific opinion, especially in England. Lately, however, the belief in Mutation, as de Vries has taught us to call it, has made notable progress,[2] owing to the publication of his splendid collection of observations and experiments, which must surely carry conviction of the reality and abundance of Mutation to the minds of all whose judgments can be affected by evidence.

That the dread test of Natural Selection must be passed by every aspirant to existence, however brief, is a truism which needs no special proof. Those who find satisfaction in demonstrations of the obvious may amply indulge themselves by starting various sorts of some annual, say French poppy, in a garden, letting them run to seed, and noticing in a few years how many of the finer sorts are represented; or by sowing an equal number of seeds taken from several varieties of carnation, lettuce, or auricula, and seeing in what proportions the fine kinds survive in competition with the common.

Selection is a true phenomenon; but its function is to *select*, not to create. Many a white-edged poppy may have germinated and perished before Mr Wilks saved the individual which in a few generations gave rise to the "Shirleys". Many a black *Amphidasys betularia* may have emerged before, some sixty years ago, in the urban conditions of Manchester the black var. *doubledayaria* found its chance, soon practically superseding the type in its place of origin, extending itself over England, and reappearing even in Belgium and Germany.

[1] Marchant, *Mém. Ac. roy. des Sci.* for 1719, 1721, p. 59, Pls. 6–7. I owe this reference to Coutagne, "L'hérédité chez les vers à soie" (*Bull. sci. Fr. Belg.* 1902).
[2] This progress threatens to be rapid indeed. Since these lines were written Professor Hubrecht, in an admirable exposition (*Pop. Sci. Monthly*, July 1904) of de Vries' *Mutations-theorie*, has even blamed me for having ten years ago attached *any* importance to continuous Variation. Nevertheless, when the unit of segregation is small, something mistakably like continuous Evolution must surely exist. (Cp. Johannsen, *Ueb. Erblichkeit in Populationen und in reinen Linien*, 1903.)

Darwin gave us sound teaching when he compared man's selective operations with those of Nature. Yet how many who are ready to expound Nature's methods have been at the pains to see how man really proceeds? To the domesticated form our fashions are what environmental exigency is to the wild. For years the conventional Chinese primrose threw sporadic plants of the loose-growing *stellata* variety, promptly extirpated because repugnant to mid-Victorian primness. But when taste, as we say, revived, the graceful "Star" primula was saved by Messrs Sutton, and a stock raised which is now of the highest fashion. I dare assert that few botanists meeting *P. stellata* in nature would hesitate to declare it a good species. This and the "Shirleys" precisely illustrate the procedure of the raiser of novelties. His operations start from a definite beginning. As in the case of *P. stellata*, he may notice a mutational form thrown off perfect from the start, or, as in the Shirleys, what catches his attention may be the first indication of that flaw which if allowed to extend will split the type into a host of new varieties each with its own peculiarities and physiological constitution.

Let anyone who doubts this try what he can do by selection without such a definite beginning. Let him try from a pure strain of black and white rats to raise a white one by breeding from the whitest, or a black one by choosing the blackest. Let him try to raise a dwarf "Cupid" sweet pea from a tall race by choosing the shortest, or a crested fowl by choosing the birds with most feather on their heads. To formulate such suggestions is to expose their foolishness.

The creature is beheld to be very good after, not before its creation. Our domesticated races are sometimes represented as so many incarnations of the breeder's prophetic fancy. But except in recombinations of pre-existing characters—now a comprehensible process—and in such intensifications and such finishing touches as involve variations which analogy makes probable, the part played by prophecy is small. Variation leads; the breeder follows. The breeder's method is to notice a desirable novelty, and to work up a stock of it, picking up other novelties in his course—for these genetic disturbances often spread—and we may rest assured the method of Nature is not very different.

The popular belief that Evolution, whether natural or artificial, is effected by mass-selection of impalpable differences arises from many errors which are all phases of one—imperfect analysis—though the source of the error differs with the circumstances of its exponent. When the scientific advocate

professes that he has statistical proofs of the continuity of
Variation, he is usually availing himself of that comprehensive
use of the term Variation to which I have referred. Statistical
indications of such continuity are commonly derived from the
study, not of nascent varieties, but of the fluctuations to which
all normal populations are subject. Truly varying material
needs care in its collection, and if found is often sporadic or in
some other way unsuitable for statistical treatment. Sometimes
it happens that the two phenomena are studied together in in-
extricable entanglement, and the resulting impression is a blur.

But when a practical man, describing his own experience,
declares that the creation of his new breed has been a very long
affair, the scientist, feeling that he has found a favourable
witness, puts forward this testimony as conclusive. But on
cross-examination it appears that the immense period deposed
to seldom goes back beyond the time of the witness's grandfather, covering, say, seventy years; more often ten, or eight,
or even five years will be found to have accomplished most of
the business. Next, in this period—which, if we take it at
seventy years, is a mere point of time compared with the epochs
of which the selectionist discourses—a momentous transformation has often been effected, not in one character but many.
Good characters have been added, it may be, of form, fertility,
precocity, colour, and other physiological attributes, un-
desirable qualities have been eliminated, and all sorts of defects
"rogued" out. On analysis these operations can be proved to
depend on a dozen discontinuities. Be it, moreover, remembered that within this period, besides *producing* his mutational
character and combining it with other characters (or it may be
groups of characters), the breeder has been working up a *stock*,
reproducing in quantity that quality which first caught his
attention, thus converting, if you will, a phenomenon of individuals into a phenomenon of a mass, to the future mystifi-
cation of the careless.

Operating among such phenomena the gross statistical
method is a misleading instrument; and, applied to these
intricate discriminations, the imposing Correlation Table into
which the biometrical Procrustes fits his arrays of unanalysed
data is still no substitute for the common sieve of a trained
judgment. For nothing but minute analysis of the facts by an
observer thoroughly conversant with the particular plant or
animal, its habits and properties, checked by the test of crucial
experiment, can disentangle the truth.

To prove the reality of Selection as a factor in Evolution is, as I have said, a work of supererogation. With more profit may experiments be employed in defining the *limits* of what Selection can accomplish. For whenever we can advance no further by Selection, we strike that hard outline fixed by the natural properties of organisms. We come upon these limits in various unexpected places, and to the naturalist ignorant of breeding nothing can be more surprising or instructive.

Whatever be the mode of origin of new types, no theoretical evolutionist doubts that Selection will enable him to fix his character when obtained. Let him put his faith into practice. Let him set about breeding canaries to win in the class for "Clear Yellow Norwich" at the Crystal Palace Show. Being a selectionist, his plan will be to pick up winning yellow cocks and hens at shows and breed them together. The results will be disappointing. Not getting what he wants, he may buy still better clear yellows and work them in, and so on till his funds are exhausted, but he will pretty certainly breed no winner, be he never so skilful. For no selection of winning yellows will make them into a breed. They must be formed afresh by various combinations of colours appropriately crossed and worked up. Though breeders differ as to the system of combinations to be followed, all would agree that selection of birds representing the winning type was a sure way to fail. The same is true for nearly all canary colours except in "Lizards", and, I believe, for some pigeon and poultry colours also.

Let this scientific fancier now go to the Palace Poultry Show and buy the winning Brown Leghorn cock and hen, breed from them, and send up the result of such a mating year after year. His chance of a winner is not quite, but almost *nil*. For in its wisdom the fancy has chosen one type for the cock and another for the hen. They belong to distinct strains. The hen corresponding to the winning cock is too bright, and the cock corresponding to the winning hen is too dull for the judge's taste. The same is the case in nearly every breed where the sex colours differ markedly. Rarely winners of both sexes have come in one strain—a phenomenon I cannot now discuss—but the contrary is the rule. Does anyone suppose that this system of "double mating" would be followed, with all the cost and trouble it involves, if Selection could compress the two strains into one? Yet current theory makes demands on Selection to which this is nothing.

The tyro has confidence in the power of Selection to fix type, but he never stops to consider what fixation precisely means. Yet a simple experiment will tell him. He may go to a great show and claim the best pair of Andalusian fowls for any number of guineas. When he breeds from them he finds, to his disgust, that only about half their chickens, or slightly more, come blue at all, the rest being blacks or splashed whites. Indignantly, perhaps, he will complain to the vendor that he has been supplied with no selected breed, but worthless mongrels. In reply he may learn that beyond a doubt his birds come from blues only in the direct line for an indefinite number of generations, and that to throw blacks and splashed whites is the inalienable property of blue Andalusians. But now let him breed from his "wasters", and he will find that the extracted blacks are *pure* and give blacks only, that the splashed whites similarly give only whites or splashed whites—but if the two sorts of "wasters" are crossed together *blues only* will result. *Selection* will never make the blues breed true; nor can this ever come to pass unless a blue be found whose germ cells are bearers of the blue character—which may or may not be possible. If the selectionist reflect on this experience he will be led straight to the centre of our problem. There will fall, as it were, scales from his eyes, and in a flash he will see the true meaning of fixation of type, variability, and mutation: vaporous mysteries no more.

Owing to the unhappy subdivisions of our studies, such phenomena as these—constant companions of the breeder— come seldom within the purview of modern science, which, forced for a moment to contemplate them, expresses astonishment and relapses into indolent scepticism. It is in the hope that a little may be done to draw research back into these forgotten paths that I avail myself of this great opportunity of speaking to my colleagues with somewhat wider range of topic than is possible within the limits of a scientific paper. For I am convinced that the investigation of Heredity by experimental methods offers the sole chance of progress with the fundamental problems of Evolution.

In saying this I mean no disrespect to that study of the physiology of reproduction by histological means, which, largely through the stimulus of Weismann's speculations, has of late made such extraordinary advances. It needs no penetration to see that, by an exact knowledge of the processes of maturation and fertilisation, a vigorous stock is being reared, upon which some day the experience of the breeder will be firmly grafted, to our mutual profit. We, who are engaged in experimental breeding, are watching with keenest interest the researches of Strasburger, Boveri, Wilson, Farmer, and their many fellow-workers and associates in this difficult field, sure that in the near future we shall be operating in common. We know already that the experience of the breeder is in no way opposed to the facts of the histologist; but the point at which we shall unite will be found when it is possible to trace in the maturing germ an indication of some character afterwards recognisable in the resulting organism. Till then, in order to pursue directly the course of Heredity and Variation, it is evident that we must fall back on those tangible manifestations which are to be studied only by field observation and experimental breeding.

The breeding-pen is to us what the test-tube is to the chemist —an instrument whereby we examine the nature of our organisms and determine empirically what for brevity I may call their genetic properties. As unorganised substances have their definite properties, so have the several species and varieties which form the materials of our experiments. Every attempt to determine these definite properties contributes immediately to the solution of that problem of problems, the physical constitution of a living organism. In those morphological studies which I suppose most of us have in our time pursued, we sought inspiration from the belief that in the examination of present normalities we were tracing the past, the phylogenetic order of our types, the history—as we conceived—of Evolution. In the work which I am now pressing upon your notice we may claim to be dealing not only with the present and the past, but with the future also.

On such an occasion as this it is impossible to present to you in detail the experiments—some exceedingly complex—already made in response to this newer inspiration. I must speak of results, not of methods. At a later meeting, moreover, there will be opportunities of exhibiting practically to those interested some of the more palpable illustrations. It is also impossible to-day to make use of the symbolic demonstrations by which the lines of analysis must be represented. The time cannot be far distant when ordinary Mendelian formulae will be mere *as in praesenti* to a biological audience. Nearly five years have passed since this extraordinary re-discovery was made known to the scientific world by the practically simultaneous papers of de Vries, Correns, and Tschermak, not to speak of thirty-five years of neglect endured before. Yet a phenomenon comparable in significance with any that biological science has revealed remains the intellectual possession

16-2

of specialists. We still speak sometimes of Mendel's hypothesis or theory, but in truth the terms have no strict application. It is no theory that water is made up of hydrogen and oxygen, though we cannot watch the atoms unite, and it is no theory that the blue Andalusian fowl I produce was made by the meeting of germ cells bearing respectively black and a peculiar white. Both are incontrovertible facts deduced from observation. The two facts have this in common also, that their perception gives us a glimpse into that hidden order out of which the seeming disorder of our world is built. If I refer to Mendelian "theory" therefore, in the words with which Bacon introduced his Great Instauration, "I entreat men to believe that it is not an opinion to be held, but a work to be done; and to be well assured that I am labouring to lay the foundation, not of any sect or doctrine, but of human utility and power".

In the Mendelian method of experiment the one essential is that the posterity of each *individual* should be traced separately. If individuals from necessity are treated collectively, it must be proved that their composition is identical. In direct contradiction to the methods of current statistics, Mendel saw by sure penetration that masses must be avoided. Obvious as this necessity seems when one is told, no previous observer had thought of it, whereby the discovery was missed. As Mendel immediately proved in the case of peas, and as we have now seen in many other plants and animals, it is often impossible to distinguish by inspection individuals whose genetic properties are totally distinct. Breeding gives the only test.

Segregation.

Where the proper precautions have been taken, the following phenomena have been proved to occur in a great range of cases, affecting many characters in some thirty plants and animals. The qualities or characters whose transmission in heredity is examined are found to be distributed among the germ cells, or gametes, as they are called, according to a definite system. This system is such that these characters are treated by the cell divisions (from which the gametes result) as existing in pairs, each member of a pair being alternative or *allelomorphic* to the other in the composition of the germ. Now, as every zygote—that is, any ordinary animal or plant—is formed by the union of two gametes, it may either be made by the union of two gametes bearing similar members of any pair, say two blacks or two whites, in which case we call it *homo-*

zygous in respect of that pair, or the gametes from which it originates may be bearers of the dissimilar characters, say a black and a white, when we call the resulting zygote *heterozygous* in respect of that pair. If the zygote is homozygous, no matter what its parents or their pedigree may have been, it breeds true indefinitely unless some fresh variation occurs.

If, however, the zygote be heterozygous, or gametically crossbred, its gametes in their formation separate the allelomorphs again, so that each gamete contains only one allelomorphic character of each pair. At least one cell division in the process of gametogenesis is therefore a differentiating or *segregating* division, out of which each gamete comes sensibly pure in respect of the allelomorph it carries, exactly as if it had not been formed by a heterozygous body at all. That, translated into modern language, is the essential discovery that Mendel made. It has now been repeated and verified for numerous characters of numerous species, and, in face of heroic efforts to shake the evidence or to explain it away, the discovery of gametic segregation is, and will remain, one of the lasting triumphs of the human mind.

In extending our acquaintance of these phenomena of segregation we encounter several principal types of complication.

Segregation Absent or Incomplete.—From our general knowledge of breeding we feel fairly well satisfied that true absence of segregation is the rule in certain cases. It is difficult, for instance, to imagine any other account of the facts respecting the American mulattos, though even here sporadic occurrence of segregation seems to be authenticated. Very few instances of genuine absence of segregation have been critically studied. The only one I can cite from my own experience is that of *Pararge egeria* and *egeriades*, "climatic" races of a butterfly. When crossed together, they give the common intermediate type of North-western France, which, though artificially formed, breeds in great measure true. This crossed back with either type has given, as a rule, simple blends between intermediate and type. My evidence is not, however, complete enough to warrant a positive statement as to the total absence of segregation, for in the few families raised from pairs of artificial intermediates some dubious indications of segregation have been seen.

The rarity of true failure of segregation when pure strains are crossed may be judged by the fact that since the revival of interest in such work hardly any thoroughly satisfactory cases have been witnessed. The largest body of evidence on this subject is that provided by de Vries. These cases, however, present so many complexities that it is impossible to deal with them now. While so little is definitely known regarding non-segregating characters, it appears to me premature to attempt any generalisation as to what does or does not segregate.

Most of the cases of failure of segregation formerly alleged are evidently spurious, depending on the appearance of homo-zygotes in the second generation (F_2).

One very important group of cases exists, in which the appearance of a *partial* failure of segregation after the second generation (F_2) is really due to another phenomenon. The visible character of a zygote may, for instance, depend on the co-existence in it of two characters belonging to distinct allelo-morphic pairs, each capable of being independently segregated from its fellow, and forming independent combinations. For the demonstration of this important fact we are especially in-debted to Cuénot.[1] We have indications of the existence of such a phenomenon in a considerable range of instances (mice, rabbits (Hurst), probably stocks and sweet peas).

Nevertheless, there are other cases, not always easy to dis-tinguish from these, where *some* of the gametes of F_1 certainly carry on heterozygous characters unsegregated. As an example, which seems to me indisputable, I may mention the so-called "walnut" comb, normal to Malay fowls. This can be made artificially by crossing rose comb with pea comb, and the cross-bred then forms gametes, of which one in four bears the com-pound unsegregated.[2] We may speak of this as a true *synthesis*.

In another type of cases segregation occurs, but is not sharp. The gametes may then represent a full series ranging from the one pure form to the other. Such cases occur in regard to some colours of *Primula sinensis*, and the leg-feathering of fowls (Hurst). In the second generation a nearly complete series of intermediate

[1] When $abc...\times a\beta\gamma...$ gives in F_1 or F_2 a character (not seen in the original parents), which from F_2 or later may breed true: not because aa, $b\beta$, $c\gamma$ do not severally segregate, but through simultaneous homozygosis of, say, aa and $\beta\beta$, giving a zygote $aa\beta\beta c\gamma...$ which will breed true to the character $a\beta$.

[2] Owing to this behaviour, and to the simultaneous production of single comb (? by resolution), there are, even in pure Malays, five types of in-dividuals, all with "walnut" combs—as yet indistinguishable—formed by gametic unions $r \times p$, $rp \times rp$, $rp \times r$, $rp \times p$, $rp \times s$. Of these kinds three can at once be distinguished by crossing with single; but whether $r \times p$ can be distinguished from $rp \times s$ we do not yet know. [r, rose; p, pea; s, single; rp, walnut.] In this example four allelomorphs are simultaneously segregated, one being compound. Neglecting sexual differentiation, there are therefore *ten* gametically distinct types theoretically possible; but of these only *four* are distinguishable by inspection.

zygotes may result, though the two pure extremes (if the case be one of blending characters) may still be found to be pure.

Resolution and Disintegration.—Besides these cases, the features of which we now in great measure comprehend, we encounter frequently a more complex segregation, imperfectly understood, by which gametes of new types, sometimes very numerous, are produced by the cross-bred. Each of these new types has its own peculiarities. We shall, I think, be compelled to regard these phenomena as produced either by a *resolution* of compound characters introduced by one or both parents, or by some process of *disintegration*, effected by a breaking-up of the integral characters followed by recombinations. It seems impossible to imagine simple recombinations of pre-existing characters as adequate to produce many of these phenomena. Such a view would involve the supposition that the number of characters pre-existing as units was practically infinite—a difficulty that as yet we are not obliged to face. However that may be, we have the fact that resolutions and disintegrations of this kind—or recombinations, if that conception be preferred—are among the common phenomena following crossing, and are the sources of most of the breeder's novelties. As bearing on the theoretical question to which I have alluded, we may notice that it is among examples of this complex breaking-up that a great proportion of the cases of partial sterility have been seen.

No quite satisfactory proof as to the actual moment of segregation yet exists, nor have we any evidence that all characters are segregated at the same cell division. Correns has shewn that in maize the segregation of the starch character from the sugar character must happen before the division forming the two generative nuclei, for both bear the same character. The reduction-division has naturally been suggested as the critical moment. The most serious difficulty in accepting this view, as it seems to me, is the fact that somatic divisions appear sometimes to segregate allelomorphs, as in the case of Datura fruits, and some colour cases.

In concluding this brief notice of the complexities of segregation I may call attention to the fact that we are here engaged in no idle speculation. For it is now possible by experimental means to distinguish almost always with which phenomenon we are dealing, and each kind of complication may be separately dealt with by a determination of the properties of the extracted forms. Illustrations of a practical kind will be placed before you at a subsequent meeting.

The consequence of segregation is that in cases where it occurs we are rid of the interminable difficulties which beset all previous attempts to unravel Heredity. On the older view, the individuals of any group were supposed to belong to an indefinite number of classes, according to the various numerical proportions in which various types had entered into their pedigree. We now recognise that when segregation is allelomorphic, as it constantly is, the individuals are of three classes only in respect of each allelomorphic pair—two homozygous and one heterozygous. In all such cases, therefore, fixity of type, instead of increasing gradually generation by generation, comes suddenly, and is a phenomenon of individuals. Only by the separate analysis of individuals can this fact be proved. The supposition that progress towards fixity of type was gradual arose from the study of masses of individuals, and the gradual purification witnessed was due in the main to the gradual elimination of impure individuals, whose individual properties were wrongly regarded as distributed throughout the mass.

We have at last the means of demonstrating the presence of integral characters. In affirming the integrity of segregable characters we do not declare that the size of the integer is fixed eternally, as we suppose the size of a chemical unit to be. The integrity of our characters depends on the fact that they *can* be habitually treated as units by gametogenesis. But even where such unity is manifested in its most definite form, we may, by sufficient searching, generally find a case where the integrity of the character has evidently been impaired in gametogenesis, and where one such individual is found the disintegration can generally be propagated. That the size of the unit may be changed by unknown causes, though a fact of the highest significance in the attempt to determine the physical nature of Heredity, does not in the least diminish the value of the recognition of such units, or lessen their part in governing the course of Evolution.

The existence of unit-characters had, indeed, long been scarcely doubtful to those practically familiar with the facts of Variation,[1] but it is to the genius of Mendel that we owe the proof. We knew that characters could behave as units, but we did not know that this unity was a phenomenon of gametogenesis. He has revealed to us the underworld of gametes. Henceforth, whenever we see a preparation of germ cells we shall remember that, though all may look alike, they may in

1 Cp. de Vries, *Intracellulare Pangenesis*, 1889.

reality be of many and definite kinds, differentiated from each other according to regular systems.

Numerical Relations of Gametes and their Significance.

In addition to the fact of segregation, Mendel's experiments proved another fact nearly as significant; namely, that when characters are allelomorphic, the gametes bearing each member of a pair generally are formed in equal numbers by the heterozygote, if an average of cases be taken. This fact can only be regarded as a consequence of some numerical symmetry in the cell divisions of gametogenesis. We already know cases where individual families shew such departure from normal expectation that either the numbers produced must have been unequal, or subsequent disturbance must have occurred. But so far no case is known *for certain* where the average of families does not point to equality.

The fact that equality is so usual has a direct bearing on conceptions of the physical nature of Heredity. I have compared our segregation with chemical separation, but the phenomenon of numerically symmetrical disjunction as a feature of so many and such different characters seems scarcely favourable to any close analogy with chemical processes. If each special character owed its appearance to the handing on of some complex molecule as a part of one chemical system, we should expect, among such a diversity of characters and forms of life, to encounter some phenomenon of valency, manifested as numerical inequality between members of allelomorphic pairs. So far, equivalence is certainly the rule, and where the characters are simply paired and no resolution has taken place, this rule appears to be universal as regards averages. On the other hand, there are features in the distribution of characters after resolution, when the second generation (F_2) is polymorphic in a high degree, which are not readily accounted for on any hypothesis of simple equivalence; but none of these cases are as yet satisfactorily investigated.

It is doubtful whether segregation is rightly represented as the separation of *two* characters, and whether we may not more simply imagine that the distinction between the allelomorphic gametes is one of presence or absence of some distinguishing element. De Vries has devoted much attention to this question in its bearings on his theory of Pangenesis, holding that cases of both kinds occur, and attempting to distinguish them. Indications may certainly be enumerated pointing in either direction, but for the present I incline to defer a definite opinion.

If we may profitably seek in the physical world for some parallel to our gametic segregations, we shall, I think, find it more close in mechanical separations, such as those which may be effected between fluids which do not freely mix, than in any strictly chemical phenomenon. In this way we might roughly imitate both the ordinary segregation, which is sensibly perfect, and the curious impurity occasionally perceptible even in the most pronounced discontinuities, such as those which divide male from female, petal from sepal, albino from coloured, horn from hair, and so on.

Gametic Unions and their Consequences.

Characters being then distributable among gametes according to regular systems, the next question concerns the properties and features presented by the zygotes formed by the union of gametes bearing different characters.

As to this no rule can as yet be formulated. Such a heterozygote may exhibit one of the allelomorphic characters in its full intensity (even exceeding it in special cases, perhaps in connection with increased vigour), or it may be intermediate between the two, or it may present some character not recognisable in either parent. In the latter case it is often, though not always, reversionary. When one character appears in such intensity as to conceal or exclude the other it is called *dominant*, the other being *recessive*. It may be remarked that frequently, but certainly not universally (as has been stated), the phylogenetically older character is dominant. A curious instance to the contrary is that of the peculiar arrangement of colours seen in a breed of game fowls called Brown-breasted, which in combination with the purple face, though certainly a modern variation, dominates (most markedly in females) over the Black-breasted type of *Gallus bankiva*.

In a few cases irregularity of dominance has been observed as an exception. The clearest illustration I can offer is that of the extra toe in fowls. Generally this is a dominant character, but sometimes, as an exceptional phenomenon, it may be recessive, making subsequent analysis very difficult. The nature of this irregularity is unknown. A remarkable instance is that of the blue colour in maize seeds (Correns; R. H. Lock). Here the dominance of blue is frequently imperfect, or absent, and the figures suggest that some regularity in the phenomenon may be discovered.

Mendel is often represented as having enunciated dominance as a general proposition. That this statement should still be repeated, even by those who realise the importance of his discoveries, is an extraordinary illustration of the oblivion that has overwhelmed the work of the experimental breeders. Mendel makes the specific statement in regard to certain characters in peas which do behave thus, but his proposition is not general. To convict him of such a delusion it would be necessary to prove that he was exceptionally ignorant of breeding, though on the face of the evidence he seems sufficiently expert.

A generalisation respecting the consequences of heterozygosis possessing greater value is this. When a pair of gametes unites in fertilisation the characters of the zygote depend directly on the constitution of these gametes, and not on that of the parents from which they came. To this generalisation we know as yet only two clear exceptions. These very curious cases are exactly alike in that, though segregation obviously occurs in a seed character, the seeds borne by the hybrid (F_1) all exhibit the hybrid character, and the consequences of segregation in the particular seed character are not evident till the seeds (F_3) of the second (F_2) generation are determinable. Of these the first is the case of indent peas investigated especially by Tschermak. Crossed with wrinkled peas I have found the phenomena normal, but when the cross is made with a round type the exceptional phenomenon occurs. The second case is that discovered by Biffen in the cross between the long-grained wheat called Polish and short-grained Rivet wheat, demonstrations of which will be laid before you. No satisfactory account of these peculiarities has been yet suggested, but it is evident that in some unexplained way the maternal plant characters control the seed characters for each generation. It is, of course, likely that other comparable cases will be found.

Appearances have been seen in at least four cases (rats, mice, stocks, sweet peas) suggesting at first sight that a heterozygosis between two gametes, *both* extracted, may give, e.g., dominance; while if one, or both, were pure, they would give a reversionary heterozygote. If this occurrence is authenticated on a sufficient scale, we shall of course recognise that the fact proves the presence in these cases of some pervading and non-segregating quality, distributed among the extracted gametes formed by the parent heterozygote. As yet, however, I do not think the evidence enough to warrant the conclusion that such a pervading quality is really present, and I incline to attribute the appearances to redistribution of characters belonging to independent pairs in the manner elucidated by Cuénot. The point will be easily determined, and meanwhile we must note the two possibilities.

Following, therefore, our first proposition that the gametes belong to definite classes, comes the second proposition, that the unions of members of the various classes have specific consequences. Nor is this proposition simply the truistical statement that different causes have different effects; for by its aid we are led at once to the place where the different cause is to be sought—Gametogenesis. While formerly we hoped to determine the offspring by examining the *ancestry* of the parents, we now proceed by investigating the *gametic composition* of the parents. Individuals may have identical ancestry (and sometimes, to all appearances, identical characters), but yet be quite different in gametic composition; and, conversely, individuals may be identical in gametic composition and have very different ancestry. Nevertheless, those that are identical in gametic composition are the same, whatever their ancestry. Therefore, where such cases are concerned, in any considerations of the physiology of heredity, ancestry is misleading and passes out of account. To take the crudest illustration, if a hybrid is made between two races, *A*, *B*, and another hybrid between two other races, *C*, *D*, it might be thought that when the two hybrids *AB* and *CD* are bred together, four races, *A*, *B*, *C*, and *D*, will be united in their offspring. This expectation may be entirely falsified, for the cell divisions of gametogenesis may have split *A* from *B* and *C* from *D*, so that the final product may contain characters of only two races after all, being either *AC*, *BC*, *AD*, or *BD*. In practice, however, we are generally dealing with *groups* of characters, and the union of all the *A* group, for instance, with all the *C* group will be a rare coincidence.

It is the object of Mendelian analysis to state each case of heredity in terms of gametic composition, and thence to determine the laws governing the distribution of characters in the cell divisions of gametogenesis.

There are, of course, many cases which still baffle our attempts at such analysis, but some of the most paradoxical exceptions have been reduced to order by the accumulation of facts. The consequences of heterozygosis are curiously specific, and each needs separate investigation. A remarkable case occurred in stocks, shewing the need for caution in dealing with contradictory results. Hoary leaves and glabrous leaves are a pair of allelomorphic characters. When glabrous races were crossed with cross-breds, sometimes the results agreed with simple expectation, while in other cases the offspring were all hoary when, in accordance with similar expectation, this should be impossible. By further experiment, however, Miss Saunders has found that *certain* glabrous races crossed together give nothing but hoary heterozygotes, which completely elucidates such exceptions. There is every likelihood that wherever segregation occurs similar analysis will be successful.

Speaking generally, in every case the first point to be worked out is the magnitude of the character units recognised by the critical cell divisions of gametogenesis, and the second is the specific consequence of all the possible combinations between them. When this has been done for a comprehensive series of types and characters, it will be time to attempt further generalisation, and perhaps to look for light on that fundamental physiological property, the power of cell division.

Segregation and Sex.—Acquaintance with Mendelian phenomena irresistibly suggests the question whether in *all* cases of families composed of distinct types the distinctness may not be primarily due to gametic segregation. Of all such distinctions none is so universal or so widespread as that of sex: may it not be possible that sex is due to a segregation occurring between gametes, either male, female, or both? It will be known to you that several naturalists have been led by various roads to incline to this view. We still await the proof of crucial experiments; but without taking you over more familiar ground, it may be useful to shew how the matter looks from our standpoint. As regards actual experiment, all results thus far are complicated by the occurrence of some sterility in the hybrid generation. Correns, fertilising ♀ *Bryonia dioica* with pollen from ☿ *B. alba* obtained offspring (F_1) *either* ♂ or ♀, with only one doubtful exception. Gärtner found a similar result in *Lychnis diurna* ♀ × ☿ *L. Flos-cuculi* as ♂, but only raised six plants (4 ♂, 2 ♀). From *L. diurna* ♀ × ☿ *Silene noctiflora* as ♂ he got only two plants, spoken of as females which developed occasional anthers. These results give a distinct suggestion that sex may be determined by differentiation among the male gametes, but satisfactory and direct proofs can only be obtained from some case where sterility does not ensue.

Apart, however, from such decisive evidence—which, indeed, would be more satisfactory if relating to *animals*—several circumstances suggest that sex is a segregation phenomenon.

Professor Castle in a valuable essay has called attention to distinct evidence of disturbance in the heredity of certain moths (*Aglia tau* and *lugens*, Standfuss's experiments; *Tephrosia*, experiments of Bacot and others, summarised by Tutt),[1] where the disturbance is pretty certainly connected with sexual differentiation. Mr Punnett and I are finding suggestions of the same thing in certain poultry cases. Mr Doncaster has pointed out that the evidence of Mr Raynor clearly indicates that a certain variety of *Abraxas grossulariata*, usually peculiar to the female, is a Mendelian recessive. It is scarcely doubtful that this will be shewn to hold also for some other female varieties, e.g. *Colias edusa*, var. *helice*, etc. We can therefore feel no doubt that there is some entanglement between sex and gametically segregable characters. A curious instance of a comparable nature is that of the "cinnamon" canary (Norduijn, etc.), and similar complications are alleged as regards the descent of colour-blindness and haemophilia.

In one remarkable group of facts we come very near to the phenomenon of sex. Experiments made in conjunction with Mr R. P. Gregory have shewn that the familiar heterostylism of *Primula* is a phenomenon of Mendelian segregation. Short style, or "thrum", is a dominant—with a complication;[2] long style, or "pin", is recessive; while equal, or "homostyle", is recessive to both.

Even nearer we come in a certain sweet-pea example, where abortion of anthers behaves as an ordinary Mendelian recessive character.[3] By a slight exaggeration we might even speak of a hermaphrodite with barren anthers as a "female".

Consider also how like the two kinds of differentiation are. The occasional mosaicism in Lepidoptera, called "gynandromorphism", may be exactly paralleled by specimens where the two halves are two colour varieties, instead of the two sexes.

[1] *Trans. Ent. Soc. Lond.*, 1898.
[2] It is doubtful if "thrum" ever breeds true, as both the other types can do. Perhaps "thrum" is a *Halbrasse* of de Vries.
[3] Neglecting minor complications, the descent is as follows: Lady Penzance ♀ × Emily Henderson (long pollen) ♂ gave purple F_1. In one F_2 family, with rare exceptions, *coloured* plants with *dark* axils were fertile, those with *light* axils having ♂ sterile, whites being either fertile or sterile. The ratios indicated are 9 coloured, dk. ax., fertile ♂: 3 coloured, lt. ax., sterile ♂: 3 white, fertile ♂: 1 white, sterile ♂. The fertile whites, therefore, though light-axilled (as whites almost always are) presumably bear the dark-axil character, which generally cannot appear except in association with coloured flowers. This can be proved next year. Some at least of the plants with sterile ♂ are fertile on the ♀ side, and when crossed with a coloured light-axilled type will presumably give only light-axilled plants.

Patches of *Silene inflata* in this neighbourhood commonly consist of hairy and glabrous individuals,[1] a phenomenon proved in *Lychnis* to be dependent on Mendelian segregation. The same patch consists also of female plants and hermaphrodite plants. Is it not likely that both phenomena are similar in nature? How otherwise would the differentiation be maintained? The sweet-pea case I have spoken of is scarcely distinguishable from this. I therefore look forward with confidence to the elucidation of the real nature of sex—that redoubtable mystery.

We now move among the facts with an altogether different bearing. *Animals and Plants under Domestication*, from being largely a narration of inscrutable prodigies, begins to take shape as a body of coherent evidence. Of the old difficulties many disappear finally. Others are inverted. Darwin says he would have expected "from the law of reversion" that nectarines being the newer form would more often produce peaches than peaches nectarines, which is the commoner occurrence. Now, on the contrary, the unique instance of the Carclew nectarine tree bearing peaches is more astonishing than all the other evidence together!

Though the progress which Mendelian facts make possible is so great, it must never be forgotten that as regards new characters involving the addition of some new factor to the pre-existing stock we are almost where we were. When they have been added by Mutation, we can now study their transmission; but we know not whence or why they come. Nor have we any definite light on the problem of adaptation; though here there is at least no increase of difficulties.

Besides these outstanding problems, there remain many special points of difficulty which on this occasion I cannot treat —curiosities of segregation, obscure aberrations of fertilisation[2] (occasionally met with), coupling of characters, and the very serious possibility of disturbance through gametic selection. Let us employ the space that remains in returning to the problem of variation, already spoken of above, and considering how it looks in the light of the new facts as to Heredity.

[1] This excellent illustration was shewn me by Mr A. W. Hill and Mr A. Wallis. A third form, glabrous, with hairy edges to the leaves, also occurs.

[2] In view of Ostenfeld's discovery of parthenogenesis in *Hieracium*, the possibility that this phenomenon plays a part in some non-segregating cases needs careful examination.

The problem of Heredity is the problem of the manner of distribution of characters among germ cells. So soon as this problem is truly formulated, the nature of Variation at once appears. For the first time in the history of evolutionary thought, Mendel's discovery enables us to form some picture of the process which results in genetic variation. It is simply the segregation of a new kind of gamete, bearing one or more characters distinct from those of the type. We can answer one of the oldest questions in philosophy. In terms of the ancient riddle, we may reply that the Fowl's egg existed before the Fowl; and if we hesitate about the Fowl, we may be sure about the Bantam. The parent zygote, whose offspring display variation, is giving off new gametes, and in its gametogenesis a segregation of their new character, more or less pure, is taking place. The significance and origin of the Discontinuity of Variation is therefore in great measure evident. So far as pre-existing elements are concerned, it is an expression of the power of cell division to distribute character units among gametes. The initial purity of so many nascent mutations is thus no longer surprising, and, indeed, that such initial purity has not been more generally observed we may safely ascribe to imperfections of method.

It is evident that the resemblance between the parent originating a variety and a heterozygote is close, and the cases need the utmost care in discrimination. If, for instance, we knew nothing more of the Andalusian fowl than that it throws blacks, blues, and whites, how should we decide whether the case was one of heterozygosis or of nascent mutation? The second (F_2) generation from Brown leghorn × White leghorn contains an occasional Silver-grey or Duckwing female. Is this a *mutation* induced by crossing, or is it simply due to a recombination of pre-existing characters? We cannot yet point to a criterion which will certainly separate the one from the other; but perhaps the statistical irregularity usually accompanying mutation, contrasted with the numerical symmetry of the gametes after normal heterozygosis, may give indications in simple cases—though scarcely reliable even there. These difficulties reach their maximum in the case of types which are *continually* giving off a second form with greater or less frequency as a concomitant of their ordinary existence. This extraordinarily interesting phenomenon, pointed out first by de Vries, and described by him under the head of *Halb-* and *Mittel-Rassen*, is too imperfectly understood for me to do more than refer to it, but in the attempt to discover what is actually taking place in variation it must play a considerable part.

Just as that normal truth to type, which we call Heredity, is in its simplest elements only an expression of that qualitative symmetry characteristic of all non-differentiating cell divisions, so is genetic variation the expression of a qualitative asymmetry beginning in gametogenesis. Variation is a novel cell division.[1] So soon as this fact is grasped we shall hear no more of Heredity and Variation as opposing "factors" or "forces"—a metaphor which has too long plagued us.

We cease, then, to wonder at the suddenness with which striking variations arise. Those familiar with the older literature relating to domesticated animals and plants will recall abundant instances of the great varieties appearing early in the history of a race, while the finer shades had long to be waited for. In the sweet pea the old purple, the red bi-colour, and the white have existed for generations, appearing soon after the cultivation of the species; but the finer splitting which gave us the blues, pinks, etc., is a much rarer event, and for the most part only came when crossing was systematically undertaken. If any of these had been seen before by horticulturists, we can feel no doubt whatever they would have been saved. An observer contemplating a full collection of modern sweet peas, and ignorant of their history, might suppose that the extreme types had resulted from selective and more or less continuous intensification of these intermediates, exactly inverting the truth.

We shall recognise among the character groups lines of cleavage, along which they easily divide, and other finer subdivisions harder to effect. Rightly considered, the sudden appearance of a total albino or a bi-colour should surprise us less than the fact that the finer shades can appear at all.

At this point comes the inevitable question, What makes the character group split? Crossing, we know, may do this; but if there be no crossing, what is the *cause* of Variation? With this question we come sharply on the edge of human knowledge. But certain it is that if causes of Variation are to be found by penetration, they must be specific causes. A mad dog is not "caused" by July heat, nor a moss rose by progressive culture.

[1] The parallel between the differentiating divisions by which the parts of the normal body are segregated from each other, and the segregating processes of gametogenesis, must be very close. Occasionally we even see the segregation of Mendelian characters among zygotic cells.

We await our Pasteur; founding our hope of progress on the aphorism of Virchow, that every variation from type is due to a pathological accident, the true corollary of *Omnis cellula e cellula.*

In imperfect fashion I have now sketched the lines by which the investigation of Heredity is proceeding, and some of the definite results achieved. We are asked sometimes, Is this new knowledge any use? That is a question with which we, here, have fortunately no direct concern. Our business in life is to find things out, and we do not look beyond. But as regards Heredity, the answer to this question of use is so plain that we may give it without turning from the way.

We may truly say, for example, that even our present knowledge of Heredity, limited as it is, will be found of extraordinary use. Though only a beginning has been made, the powers of the breeder of plants and animals are vastly increased. Breeding is the greatest industry to which science has never yet been applied. This strange anomaly is over; and, so far at least as fixation or purification of types is concerned, the breeder of plants and animals may henceforth guide his operations with a great measure of certainty.

There are others who look to the science of Heredity with a loftier aspiration; who ask, Can any of this be used to help those who come after to be better than we are—healthier, wiser, or more worthy? The answer depends on the meaning of the question. On the one hand it is certain that a competent breeder, endowed with full powers, by the aid even of our present knowledge, could in a few generations breed out several of the morbid diatheses. As we have got rid of rabies and pleuro-pneumonia so we could exterminate the simpler vices. Voltaire's cry *Écraser l'infâme* might well replace Archbishop Parker's Table of Forbidden Degrees, which is all the instruction Parliament has so far provided. Similarly, a race may conceivably be bred true to some physical and intellectual characters considered good. The positive side of the problem is less hopeful, but the various species of mankind offer ample material. In this sense science already suggests the way. No one, however, proposes to take it; and so long as, in our actual laws of breeding, superstition remains the guide of nations, rising ever fresh and unhurt from the assaults of knowledge, there is nothing to hope or to fear from these sciences.

But if, as is usual, the philanthropist is seeking for some external application by which to ameliorate the course of descent, knowledge of Heredity cannot help him. The answer to his question is *No*, almost without qualification. We have no experience of any means by which transmission may be made to deviate from its course; nor from the moment of fertilisation can teaching, or hygiene, or exhortation pick out the particles of evil in that zygote, or put in one particle of good. From seeds in the same pod may come sweet peas climbing five feet high, while their own brothers lie prone upon the ground. The stick will not make the dwarf peas climb, though without it the tall can never rise. Education, sanitation, and the rest, are but the giving or withholding of opportunity. Though in the matter of Heredity every other conclusion has been questioned, I rejoice that in this we are all agreed.

PRESIDENTIAL ADDRESS TO THE AGRICULTURAL SUB-SECTION, BRITISH ASSOCIATION

Portsmouth Meeting. 1911

The invitation to preside over the Agricultural Sub-Section on this occasion naturally gave me great pleasure, but after accepting it I have felt embarrassment in a considerable degree. The motto of the great Society which has been responsible for so much progress in agricultural affairs in this country very clearly expresses the subject of our deliberations in the words "Practice with Science", and to be competent to address you, a man should be well conversant with both. But even if agriculture is allowed to include horticulture, as may perhaps be generally conceded, I am sadly conscious that my special qualifications are much weaker than you have a right to demand of a President.

The aspects of agriculture from which it offers hopeful lines for scientific attack are, in the main, three: Physiological, Pathological, and Genetic. All are closely interrelated, and for successful dealing with the problems of any one of these departments of research, knowledge of the results attained in the others is now almost indispensable. I myself can claim personal acquaintance with the third or genetic group alone, and therefore in considering how science is to be applied to the practical operations of agriculture, I must necessarily choose it as the more special subject of this address. I know very well that wider experience of those other branches of agricultural science or practical agriculture would give to my remarks a weight to which they cannot now pretend.

Before, however, proceeding to these topics of special consideration, I have thought it not unfitting to say something of a more general nature as to the scope of an applied science, such as that to which we here are devoted. We are witnessing a very remarkable outburst of activity in the promotion of science in its application to agriculture. Public bodies distributed throughout this country and our possessions are organising various enterprises with that object. Agricultural research is now everywhere admitted a proper subject for University support and direction.

With the institution of the Development Grant a national subsidy is provided on a considerable scale in England for the first time.

At such a moment the scope of this applied science and the conditions under which it may most successfully be advanced are prominent matters of consideration in the minds of most of us. We hope great things from these new ventures. We are, however, by no means the first to embark upon them. Many of the other great nations have already made enormous efforts in the same direction. We have their experience for a guide.

Now, it is not in dispute that wherever agricultural science has been properly organised valuable results have been attained, some of very high importance indeed; yet with full appreciation of these achievements, it is possible to ask whether the whole outcome might not have been greater still. In the course of recent years I have come a good deal into contact with those who in various countries are taking part in such work, and I have been struck with the unanimity that they have shewn in their comments on the conditions imposed upon them. Those who receive large numbers of agricultural bulletins purporting to give the results of practical trials and researches will, I feel sure, agree with me that with certain notable exceptions they form on the whole dull reading. True they are in many cases written for farmers and growers in special districts, rather than for the general scientific reader, but I have sometimes asked myself whether those farmers get much more out of this literature than I do. I doubt it greatly. Nevertheless, to the production of these things much labour and expense have been devoted. I am sure, and I believe that most of those engaged in these productions themselves feel, that the effort might have been much better applied elsewhere. Work of this unnecessary kind is done, of course, to satisfy a public opinion which is supposed to demand rapid returns for outlay, and to prefer immediate apparent results, however trivial, to the long delay which is the almost inevitable accompaniment of any serious production. For my own part, I greatly doubt whether in this estimate present public opinion has been rightly gauged. Enlightenment as to the objects, methods, and conditions of scientific research is proceeding at a rapid rate. I am quite sure, for example, that no organisation of agricultural research now to be inaugurated under the Development Commission will be subjected to the conditions laid down in 1887 when the Experimental Stations of the United States were established. For them it is decreed in Sect. 4 of the Act of Establishment:

That bulletins or reports of progress shall be published at said stations at least once in three months, one copy of which shall be sent to each newspaper in the States or Territories in which they are respectively located, and to such individuals actually engaged in farming as may request the same and as far as the means of the station will permit.

It would be difficult to draft a condition more unfavourable to the primary purpose of the Act, which was "to conduct original researches or verify experiments on the physiology of plants and animals". I can scarcely suppose the most prolific discoverer should be invited to deliver himself more than once a year. Not only does such a rule compel premature publication—that nuisance of modern scientific life—but it puts the investigator into a wrong attitude towards his work. He will do best if he forget the public and the newspaper of his State or Territory for long periods, and should only return to them when, after repeated verification, he is quite certain he has something to report.

In this I am sure the best scientific opinion of all countries would be agreed. If it is true that the public really demand continual scraps of results, and cannot trust the investigators to pursue research in a reasonable way, then the public should be plainly given to understand that the time for inaugurating researches in the public's name has not arrived. Men of science have in some degree themselves to blame if the outer world has been in any mistake on these points. It cannot be too widely known that in all sciences, whether pure or applied, research is nearly always a very slow process, uncertain in production, and full of disappointments. This is true, even in the new industries, chemical and electrical, for instance, where the whole industry has been built up from the beginning on a basis developed entirely by scientific method and by the accumulation of precise knowledge. Much more must any material advance be slow in the case of an ancient art like agriculture, where practice represents the casual experience of untold ages and accurate investigation is of yesterday. Problems moreover relating to unorganised matter are in their nature simpler than those concerned with the properties of living things, a region in which accurate knowledge is more difficult to attain. Here the research of the present day can aspire no higher than to lay the foundation on which the following generations will build. When this is realised it will at once be perceived that both those who are engaged in agricultural research and those who are charged with the supervision and control of these researches must be prepared to exercise a large measure of patience.

The applicable science must be created before it can be

applied. It is with the discovery and development of such science that agricultural research will for long enough best occupy its energies. Sometimes, truly, there come moments when a series of obvious improvements in practice can at once be introduced, but this happens only when the penetrative genius of a Pasteur or a Mendel has worked out the way into a new region of knowledge, and returns with a treasure that all can use. Given the knowledge it will soon enough become applied.

I am not advocating work in the clouds. In all that is attempted we must stick near to the facts. Though the methods of research and of thought must be strict and academic, it is in the farm and the garden that they must be applied. If inspiration is to be found anywhere it will be there. The investigator will do well to work

> As if his highest plot
> To plant the bergamot.

It is only in the closest familiarity with phenomena that we can attain to that perception of their orderly relations, which is the beginning of discovery.

To the creation of applicable science the very highest gifts and training are well devoted. In a foreign country an eminent man of science was speaking to me of a common friend, and he said that as our friend's qualifications were not of the first rank he would have to join the agricultural side of the university. I have heard remarks of similar disparagement at home. Now, whether from the standpoint of agriculture or pure science, I can imagine no policy more stupid and short-sighted.

The man who devotes his life to applied science should be made to feel that he is in the main stream of scientific progress. If he is not, both his work and science at large will suffer. The opportunities of discovery are so few that we cannot afford to miss any, and it is to the man of trained mind who is in contact with the phenomena of a great applied science that such opportunities are most often given. Through his hands pass precious material, the outcome sometimes of years of effort and design. To tell him that he must not pursue that inquiry further because he cannot foresee a direct and immediate application of the knowledge he would acquire, is, I believe almost always, a course detrimental to the real interests of the applied science. I could name specific instances where in other countries thoroughly competent and zealous investigators have by the short-sightedness of superior officials been thus debarred from following to their conclusion researches of great value and novelty.

In this country where the Development Commission will presumably for many years be the main instigator and controller of agricultural research, the constitution of the Advisory Board, on which science is largely represented, forms a guarantee that broader counsels will prevail, and it is to be hoped that not merely this inception of the work, but its future administration also will be guided in the same spirit. So long as a train of inquiry continues to extend, and new knowledge, that most precious commodity, is coming in, the enterprise will not be in vain and it will be usually worth while to pursue it.

The relative value of the different parts of knowledge in their application to industry is almost impossible to estimate, and a line of work should not be abandoned until it leads to a dead end, or is lost in a desert of detail.

We have, not only abroad, but also happily in this country, several private firms engaged in various industries—I may mention especially metallurgy, pharmacy, and brewing—who have set an admirable example in this matter, instituting researches of a costly and elaborate nature, practically unlimited in scope, connected with the subjects of their several activities, conscious that it is only by men in close touch with the operations of the industry that the discoveries can be made, and well assured that they themselves will not go unrewarded.

Let us on our part beware of giving false hopes. We know no haemony "of sovran use against all enchantments, mildew blast, or damp". Those who are wise among us do not even seek it yet. Why should we not take the farmer and gardener into our fullest confidence and tell them this? I read lately a newspaper interview with a fruit farmer who was being questioned as to the success of his undertaking, and spoke of the pests and difficulties with which he had had to contend. He was asked whether the Board of Agriculture and the scientific authorities were not able to help him. He replied that they had done what they could, that they had recommended first one thing and then another, and he had formed the opinion that they were only in an experimental stage. He was perfectly right, and he would hardly have been wrong had he said that in these things science is only approaching the experimental stage. This should be notorious. There is nothing to extenuate. To affect otherwise would be unworthy of the dignity of science.

Those who have the means of informing the public mind on the state of agricultural science should make clear that though something can be done to help the practical man already, the chief realisation of the hopes of that science is still very far away, and that it can only be reached by long and strenuous effort, expended in many various directions, most of which must seem to the uninitiated mere profitless wandering. So only will the confidence of the laity be permanently assured towards research.

Nowhere is the need for wide views of our problems more evident than in the study of plant diseases. Hitherto this side of agriculture and of horticulture, though full of possibilities for the introduction of scientific method, has been examined only in the crudest and most empirical fashion. To name the disease, to burn the affected plants, and to ply the crop with all the sprays and washes in succession ought not to be regarded as the utmost that science can attempt. There is at the present time hardly any comprehensive study of the morbid physiology of plants comparable with that which has been so greatly developed in application to animals. The nature of the resistance to disease characteristic of so many varieties, and the modes by which it may be ensured, offers a most attractive field for research, but it is one in which the advance must be made by the development of pure science, and those who engage in it must be prepared for a long period of labour without ostensible practical results. It has seemed to me that the most likely method of attack is here, as often, an indirect one. We should probably do best if we left the direct and special needs of agriculture for a time out of account, and enlisted the services of pathologists trained in the study of disease as it affects man and animals, a science already developed and far advanced towards success. Such a man, if he were to devote himself to the investigation of the same problems in the case of plants could, I am convinced, make discoveries which would not merely advance the theory of disease resistance in general very greatly, but would much promote the invention of rational and successful treatment.

As regards the application of Genetics to practice, the case is not very different. When I go to the Temple Show or to a great exhibition of live stock, my first feeling is one of admiration and deep humility. Where all is so splendidly done and results so imposing are already attained, is it not mere impertinence to suppose that any advice we are able to give is likely to be of value?

But as soon as one enters into conversation with breeders, one finds that almost all have before them some ideal to which they have not yet attained, operations to perform that they would fain do with greater ease and certainty, and that, as a matter of fact, they *are* looking to scientific research as a possible source of the greater knowledge which they require. Can we, without presumption, declare that genetic science is now able to assist these inquirers? In certain selected cases it undoubtedly can—and I will say, moreover, that if the practical men and we students could combine our respective experiences into one head, these cases would already be numerous. On the other hand, it is equally clear that in a great range of examples practice is so far ahead that science can scarcely hope in finite time even to represent what has been done, still less to better the performance. We cannot hope to improve the Southdown sheep for its own districts, to take a second off the trotting record, to increase the flavour of the muscat of Alexandria, or to excel the orange and pink of the rose Juliet. Nothing that we know could have made it easier to produce the Rambler roses, or even to evoke the latest novelties in sweet peas, though it may be claimed that the genetic system of the sweet pea is, as things go, fairly well understood. To do any of these things would require a control of events so lawless and rare that for ages they must probably remain classed as accidents. On the other hand, the modes by which combinations can be made, and by which new forms can be fixed, are through Mendelian analysis and the recent developments of genetic science now reasonably clear, and with that knowledge much of the breeder's work is greatly simplified. This part of the subject is so well understood that I need scarcely do more than allude to it.

A simple and interesting example is furnished by the work which Mr H. M. Leake is carrying out in the case of cotton in India. The cottons of fine quality grown in India are monopodial in habit, and are consequently late in flowering. In the United Provinces a comparatively early flowering form is required, as otherwise there is not time for the fruits to ripen. The early varieties are sympodial in habit, and the primary apex does not become a flower. Hitherto no sympodial form with cotton of high quality has existed, but Mr Leake has now made the combination needed, and has fixed a variety with high-class cotton and the sympodial habit, which is suitable for cultivation in the United Provinces. Until genetic physiology was developed by Mendelian analysis, it is safe to say that a practical achievement of this kind could not have been made with rapidity or certainty. The research was planned on broad lines. In the course of it much light was obtained on the genetics of cotton, and features of interest were discovered which considerably advance our knowledge of Heredity in several important respects. This work forms an admirable

illustration of that simultaneous progress both towards the
solution of a complex physiological problem and also towards
the successful attainment of an economic object which should
be the constant aim of agricultural research.

Necessarily it follows that such assistance as genetics can at
present give is applicable more to the case of plants and
animals which can be treated as annuals than to creatures of
slower generation. Yet this already is a large area of operations.
One of the greatest advances to be claimed for the work is that
it should induce raisers of seed crops especially to take more
hopeful views of their absolute purification than have hitherto
prevailed. It is at present accepted as part of the natural
perversity of things that most high-class seed crops must throw
"rogues", or that at the best the elimination of these waste
plants can only be attained by great labour extended over a
vast period of time. Conceivably that view is correct, but no
one acquainted with modern genetic science can believe it
without most cogent proof. Far more probably we should
regard these rogues either as the product of a few definite
individuals in the crop, or even as chance impurities brought
in by accidental mixture. In either case they can presumably
be got rid of. I may even go further and express a doubt
whether that degeneration which is vaguely supposed to be
attendant on all seed crops is a physiological reality. Degeneration may perhaps affect plants like the potato which are
continually multiplied asexually, though the fact has never been
proved satisfactorily. Moreover it is not in question that races
of plants taken into unsuitable climates do degenerate rapidly
from uncertain causes, but that is quite another matter.

The first question is to determine whether a given "rogue" has
in it any factor which is *dominant* to the corresponding character
in the typical plants of the crop. If it has, then we may feel
considerable confidence that these rogues have been introduced by accidental mixture. The only alternative, indeed, is
cross-fertilisation with some distinct variety possessing the
dominant, or crossing within the limits of the typical plants
themselves occurring in such a way that complementary factors
have been brought together. This last is a comparatively
infrequent phenomenon, and need not be considered till more
probable hypotheses have been disposed of. If the rogues are
first crosses the fact can be immediately proved by sowing their
seeds, for segregation will then be evident. For example, a
truly round seed is occasionally, though very rarely, found on
varieties of pea which have wrinkled seeds. I have three times

seen such seeds on my own plants. A few more were kindly given me by Mr Arthur Sutton, and I have also received a few from M. Philippe de Vilmorin—to both of whom I am indebted for most helpful assistance and advice. Of these abnormal or unexpected seeds some died without germinating, but all which did germinate in due course produced the normal mixture of round and wrinkled, proving that a cross had occurred. Cross-fertilisation in culinary peas is excessively rare, but it is certainly sometimes effected, doubtless by the leaf-cutter bee (*Megachile*) or a humble-bee visiting flowers in which for some reason the pollen has been inoperative. But in peas crossing is assuredly not the source of the ordinary rogues. These plants have a very peculiar conformation, being tall and straggling, with long internodes, small leaves, and small flowers, which together give them a curious wild look. When one compares them with the typical cultivated plants which have a more luxuriant habit, it seems difficult to suppose that the rogue can really be recessive in such a type. True, we cannot say definitely *a priori* that any one character is dominant to another, but old preconceptions are so strong that without actual evidence we always incline to think of the wilder and more primitive characteristics as dominants. Nevertheless, from such observations as I have been able to make I cannot find any valid reason for doubting that the rogues are really recessives to the type. One feature in particular is quite inconsistent with the belief that these rogues are in any proper sense degenerative returns to a wild type, for in several examples the rogues have *pointed* pods like the cultivated sorts from which they have presumably been derived. All the more primitive kinds have the dominant stump-ended pod. If the rogues had the stump pods they would fall in the class of dominants, but they have no single quality which can be declared to be certainly dominant to the type, and I see no reason why they may not be actually recessives to it after all. Whether this is the true account or not we shall know for certain next year. Mr Sutton has given me a quantity of material which we are now investigating at the John Innes Horticultural Institution, and by sowing the seed of a great number of individual plants separately I anticipate that we shall prove the rogue-throwers to be a class apart. The pure types then separately saved should, according to expectation, remain rogue-free, unless further sporting or fresh contamination occurs. If it prove that the long and attenuated rogues are really recessive to the shorter and more robust type, the case will be one of much physiological significance, but I believe a parallel already exists in the case of wheats, for among certain crosses bred by Professor Biffen, some curious spelt-like plants occurred among the derivatives from such robust wheats as Rivet and Red Fife.

There is another large and important class of cases to which similar considerations apply. I refer to the bolting or running to seed of crops grown as biennials, especially root crops. It has hitherto been universally supposed that the loss due to this cause, amounting in Sugar Beet as it frequently does to 5 or even more per cent., is not preventable.[1] This may prove to be the truth, but I think it is not impossible that the bolters can be wholly, or almost wholly, eliminated by the application of proper breeding methods. In this particular example I know that season and conditions of cultivation count for a good deal in promoting or checking the tendency to run to seed, nevertheless one can scarcely witness the sharp distinction between the annual and biennial forms without suspecting that genetic composition is largely responsible. If it proves to be so, we shall have another remarkable illustration of the direct applicability of knowledge gained from a purely academic source. "Let not him that putteth his armour on boast him as he that putteth it off", and I am quite alive to the many obstacles which may lie between the conception of an idea and its realisation. One thing, however, is certain, that we have now the power to formulate rightly the question which the breeder is to put to Nature; and this power and the whole apparatus by which he can obtain an answer to his question—in whatever sense that answer may be given—has been derived from experiments designed with the immediate object of investigating that scholastic and seemingly barren problem, "What is a species"? If Mendel's eight years' work had been done in an agricultural school supported by public money, I can imagine much shaking of heads on the County Council governing that institution, and yet it is no longer in dispute that he provided the one bit of solid discovery upon which all breeding practice will henceforth be based.

Everywhere the same need for accurate knowledge is apparent. I suppose horse-breeding is an art which has by the application of common-sense and great experience been carried to about as high a point of perfection as any. Yet even here I have seen a mistake made which is obvious to anyone accustomed to analytical breeding. Among a number of stallions

[1] A strain free from bolting was subsequently produced by him at Merton. C. B. B.

provided at great expense to improve the breed of horses in a certain district was one which was shewn me as something of a curiosity. This particular animal had been bred by one of the provided stallions out of an indifferent country mare. It had been kept as an unusually good-looking colt, and was now travelling the country as a breeding stallion, under the highest auspices. I thought to myself that if such a practice is sanctioned by breeding acumen and common-sense, science is not after all so very ambitious if she aspires to do rather better. The breeder has continually to remind himself that it is not what the animal or plant *looks* that matters, but what it *is*. Analysis has taught us to realise, first, that each animal and plant is a double structure, and next that the appearance may shew only half its composition.

With respect to the inheritance of many physiological qualities of divers kinds we have made at least a beginning of knowledge, but there is one class of phenomena as yet almost untouched. This is the miscellaneous group of attributes which are usually measured in terms of size, fertility, yield, and the like. This group of characters has more than common significance to the practical man. Analysis of them can nevertheless only become possible when pure science has progressed far beyond the point yet reached.

I know few lines of pure research more attractive and at the same time more likely to lead to economic results than an investigation of the nature of variation in size of the whole organism or of its parts. By what factors is it caused? By what steps does it proceed? By what limitations is it beset? In illustration of the application of these questions I may refer to a variety of topics that have been lately brought to my notice. In the case of merino sheep I have been asked by an Australian breeder whether it is possible to combine the optimum length of wool with the optimum fineness and the right degree of crimping. I have to reply that absolutely nothing is yet known for certain as to the physiological factors determining the length or the fineness of wool. The crimping of the fibres is an expression of the fact that each particular hair is curved, and if free and untwisted would form a corkscrew spiral, but as to the genetics of curly hair even in man very little is yet known. But leaving the question of curl on one side, we have in regard to the length and fineness of wool, a problem which genetic experiment ought to be able to solve. Note that in it, as in almost all problems of the "yield" of any product of farm or garden, two distinct elements are concerned—the one is *size*, and the other is *number*. The length of the hair is determined by the rate of excretion and length of the period of activity of the hair follicles, but the fineness is determined by the number of follicles in unit area. Now analogy is never a safe guide, but I think if we had before us the results of really critical experiments on the genetics of size and number of multiple organs in any animal or even any plant, we might not wholly be at a loss in dealing with this important problem.

A somewhat similar question comes from South Africa. Is it possible to combine the qualities of a strain of ostriches which has extra long plumes with those of another strain which has its plumes extra lustrous? I have not been able fully to satisfy myself upon what the lustre depends, but I incline to think it is an expression of fineness of fibre, which again is probably a consequence of the smallness and increased number of the excreting cells, somewhat as the fineness of wool is a consequence of the increased number and smallness of the excreting follicles.

Again the question arises in regard to flax, how should a strain be bred which shall combine the maximum length with maximum fineness of fibre? The element of number comes in here, not merely with regard to the number of fibres in a stem, but also in two other considerations: first, that the plant should not tiller at the base, and, secondly, that the decussation of the flowering branches should be postponed to the highest possible level.

Now in this problem of the flax,[1] and not impossibly in the others I have named, we have questions which can in all likelihood be solved in a form which will be of general, if not of universal, application to a host of other cognate questions. By good luck the required type of flax may be struck at once, in which case it may be fixed by ordinary Mendelian analysis, but if the problem is investigated by accurate methods on a large scale, the results may shew the way into some of those general problems of size and number which make a great part of the fundamental mystery of growth.

I see no reason why these things should remain inscrutable. There is indeed a little light already. We are well acquainted with a few examples in which the genetic behaviour of these properties is fairly definite. We have examples in which, when two varieties differing in number of divisions are crossed, the lower number dominates—or, in other words, that the

---

[1] A few years later, as a result of such experimental work, he produced at Merton a strain of flax in which the maximum length of fibre was fixed. C. B. B.

increased number is a consequence of the removal of a factor which prevents or inhibits particular divisions, so that they do not take place. It is likely that in so far as the increased productivity of a domesticated form as compared with its wild original depends on more frequent division, the increase is due to loss of inhibiting factors. How far may this reasoning be extended? Again we know that in several plants—peas, sweet peas, *Antirrhinum*, and certain wheats—a tall variety differs in that respect from a dwarf in possessing one more factor. It would be an extraordinarily valuable addition to knowledge if we could ascertain exactly how this factor operates, how much of its action is due to linear repetition, and how much to actual extension of individual parts. The analysis of the plants of intermediate size has never been properly attempted, but would be full of interest and have innumerable bearings on other cases in animals and plants, some of much economic importance.

That in all such examples the objective phenomena we see are primarily the consequence of the interaction of genetic factors is almost certain. The lay mind is at first disposed, as always, to attribute such distinctions to anything rather than to a specific cause which is invisible. An appeal to differences in conditions—which a moment's reflection shews to be either imaginary or altogether independent—or to those vague influences invoked under the name of Selection, silently postponing any laborious analysis of the nature of the material selected, repels curiosity for a time, and is lifted as a veil before the actual phenomena; and so even critical intelligences may for an indefinite time be satisfied that there is no specific problem to be investigated, in the same facile way that, till a few years ago, we were all content with the belief that malarial fevers could be referred to any damp exhalations in the atmosphere, or that in suppuration the body was discharging its natural humours. In the economics of breeding, a thousand such phenomena are similarly waiting for analysis and reference to their specific causes. What, for instance, is self-sterility? The phenomenon is very widely spread among plants, and is far commoner than most people suppose who have not specially looked for it. Why is it that the pollen of an individual in these plants fails to fertilise the ova of the same individual? Asexual multiplication seems in no way to affect the case. The American experimenters are doubtless right in attributing the failure of large plantations of a single variety of apples or of pears in a high degree to this cause. Sometimes, as Mr W. O. Backhouse has found in his work on plums at the John Innes Horticultural

Institution, the behaviour of the varieties is most definite and specific. He carefully self-fertilised a number of varieties, excluding casual pollination, and found that while some sorts, for example, Victoria, Czar, and Early Transparent set practically every fruit self-pollinated, others including several (perhaps all) Greengages, Early Orleans, and Sultan do not set a single fruit without pollination from some other variety. Dr Erwin Baur has found indications that self-sterility in *Antirrhinum* may be a Mendelian recessive, but whether this important suggestion be confirmed or not, the subject is worth the most minute study in all its bearings. The treatment of this problem well illustrates the proper scope of an applied science. The economic value of an exact determination of the empirical facts is obvious, but it should be the ambition of anyone engaging in such a research to penetrate further. If we can grasp the *rationale* of self-sterility we open a new chapter in the study of life. It may contain the solution of the question, What is an individual?—no mere metaphysical conundrum, but a physiological problem of fundamental significance.

What, again, is the meaning of that wonderful increase in size or in "yield" which so often follows on a first cross? We are no longer content, as Victorian teleology was, to call it a "beneficial" effect and pass on. The fact has long been known and made use of in breeding stock for the meat market, and of late years the practice has also been introduced in raising table poultry. Mr G. N. Collins,[1] of the U.S. Department of Agriculture, has recently proposed with much reason that it might be applied in the case of maize. The cross is easy to make on a commercial scale, and the gain in yield is striking, the increase ranging as high as 95 per cent. These figures sound extravagant, but from what I have frequently seen in peas and sweet peas I am prepared for even greater increase. But what *is* this increase? How much of it is due to change in number of parts, how much to transference of differentiation or homoeosis, as I have called it—leaf buds becoming flower buds, for instance—and how much to actual increase in size of parts? To answer these questions would be to make an addition to human knowledge of incalculably great significance.

Then we have the further question, How and why does the increase disappear in subsequent generations? The very uniformity of the cross-breds between pure strains must be taken as an indication that the phenomenon is orderly. Its subsidence is probably orderly also. Shull has advocated the most natural

[1] *Bureau of Plant Industry, Bulletin No.* 191, 1910.

view that heterozygosis is the exciting cause, and that with the gradual return to the homozygous state the effects pass off. I quite think this may be a part of the explanation, but I feel difficulties, which need not here be detailed, in accepting this as a complete account. Some of the effect we may probably also attribute to the combination of complementary factors; but whether heterozygosis, or complementary action, is at work, our experience of cross-breeding in general makes it practically certain that genetic factors of special classes only can have these properties, and no pains should be spared in identifying them. It is not impossible that such identification would throw light on the nature of cell division and of that meristic process by which the repeated organs of living things are constituted, and I have much confidence that in the course of the analysis discoveries will be made bearing directly both on the general theory of Heredity and on the practical industry of breeding.

In the application of science to the arts of agriculture, chemistry, the foundation of sciences, very properly and inevitably came first, while breeding remained under the unchallenged control of simple common-sense alone. The science of genetics is so young that when we speak of what it also can do we must still for the most part ask for a long credit; but I think that if there is full co-operation between the practical breeder and the scientific experimenter, we shall be able to redeem our bonds at no remotely distant date. In the mysterious properties of the living bodies of plants and animals there is an engine capable of wonders scarcely yet suspected, waiting only for the constructive government of the human mind. Even in the seemingly rigorous tests and trials which have been applied to living material apparently homogeneous, it is not doubtful that error has often come in by reason of the individual genetic heterogeneity of the plants and animals chosen. A batch of fruit trees may be all of the same variety, but the stocks on which the variety was grafted have hitherto been almost always seminally distinct individuals, each with its own powers of luxuriance or restriction, their own root systems, and properties so diverse that only in experiments on a colossal scale can this diversity be supposed to be levelled down. Even in a closely bred strain of cattle, though all may agree in their "points", there may still be great genetic diversity in powers of assimilation and rapidity of attaining maturity, by which irregularities by no means negligible are introduced. The range of powers which organic variation and genetic composition can confer is so vast as to override great dissimilarities in the conditions of cultivation. This truth is familiar to every raiser and grower, who knows it in the form that the first necessity is for him to get the right breed and the right variety for his work. If he has a wheat of poor yield, no amount of attention to cultivation or manuring will give him a good crop. An animal that is a bad doer will remain so in the finest pasture. All praise and gratitude to the student of the conditions of life, for he can do, and has done, much for agriculture, but the breeder can do even more.

When more than fifteen years ago the proposal to found a school of agriculture in Cambridge was being debated, much was said of the importance of the chemistry of soils, of researches into the physiological value of food-stuffs, and of other matters then already prominent on the scientific horizon. I remember then interpolating with an appeal for some study of the physiology of breeding, which I urged should find a place in the curriculum, and I pointed out that the improvement in the strains of plants and animals had done at least as much—more, I really meant—to advance agriculture than had been accomplished by other means. My advice found little favour, and I was taken to task afterwards by a prominent advocate of the new school for raising a side issue. Breeding was a purely empirical affair. Common-sense and selection comprised the whole business, and physiology flew at higher game. I am, nevertheless, happy now to reflect that of the work which is making the Cambridge School of Agriculture a force for progress in the agricultural world the remarkable researches and results of my late colleague, Professor Biffen, based as they have been on modern discoveries in the pure sciences of breeding, occupy a high and greatly honoured place.

In conclusion I would sound once more the note with which I began. If we are to progress fast there must be no separation made between pure and applied science. The practical man with his wide knowledge of specific natural facts, and the scientific student ever seeking to find the hard general truths which the diversity of Nature hides—truths out of which any lasting structure of progress must be built—have everything to gain from free interchange of experience and ideas. To ensure this community of purpose those who are engaged in scientific work should continually strive to make their aims and methods known at large, neither exaggerating their confidence nor concealing their misgivings,

Till the world is wrought
To sympathy with hopes and fears it heeded not.

PRESIDENTIAL ADDRESS TO THE BRITISH ASSOCIATION, AUSTRALIA

(a) Melbourne Meeting. 1914

The outstanding feature of this Meeting must be the fact that we are here—in Australia. It is the function of a President to tell the Association of advances in science, to speak of the universal rather than of the particular or the temporary. There will be other opportunities of expressing the thoughts which this event must excite in the dullest heart, but it is right that my first words should take account of those achievements of organisation and those acts of national generosity by which it has come to pass that we are assembled in this country. Let us, too, on this occasion, remember that all the effort, and all the goodwill, that binds Australia to Britain would have been powerless to bring about such a result had it not been for those advances in science which have given man a control of the forces of Nature. For we are here by virtue of the feats of genius of individual men of science, giant variations from the common level of our species; and since I am going soon to speak of the significance of individual variation, I cannot introduce that subject better than by calling to remembrance the line of pioneers in chemistry, in physics, and in engineering, by the working of whose rare—or, if you will, abnormal—intellects a meeting of the British Association on this side of the globe has been made physically possible.

I have next to refer to the loss within the year of Sir David Gill, a former President of this Association, himself one of the outstanding great. His greatness lay in the power of making big foundations. He built up the Cape Observatory; he organised international geodesy; he conceived and carried through the plans for the photography of the whole sky, a work in which Australia is bearing a conspicuous part. Astronomical observation is now organised on an international scale, and of this great scheme Gill was the heart and soul. His labours have ensured a base from which others will proceed to discovery otherwise impossible. His name will be long remembered with veneration and gratitude.

As the subject of the Addresses which I am to deliver here and in Sydney I take *Heredity*. I shall attempt to give the essence of the discoveries made by Mendelian or analytical methods of study, and I shall ask you to contemplate the deductions which these physiological facts suggest in application both to evolutionary theory at large and to the special case of the natural history of human society.

Recognition of the significance of Heredity is modern. The term itself in its scientific sense is no older than Herbert Spencer. Animals and plants are formed as pieces of living material split from the body of the parent organisms. Their powers and faculties are fixed in their physiological origin. They are the consequence of a genetic process, and yet it is only lately that this genetic process has become the subject of systematic research and experiment. The curiosity of naturalists has of course always been attracted to such problems; but that accurate knowledge of genetics is of paramount importance in any attempt to understand the nature of living things has only been realised quite lately even by naturalists, and with casual exceptions the laity still know nothing of the matter. Historians debate the past of the human species, and statesmen order its present or profess to guide its future as if the animal Man, the unit of their calculations, with his vast diversity of powers, were a homogeneous material, which can be multiplied like shot.

The reason for this neglect lies in ignorance and misunderstanding of the nature of Variation; for not until the fact of congenital diversity is grasped, with all that it imports, does knowledge of the system of hereditary transmission stand out as a primary necessity in the construction of any theory of Evolution, or any scheme of human polity.

The first full perception of the significance of variation we owe to Darwin. The present generation of evolutionists realises perhaps more fully than did the scientific world in the last century that the theory of Evolution had occupied the thoughts of many and found acceptance with not a few before ever the *Origin* appeared. We have come also to the conviction that the principle of Natural Selection cannot have been the chief factor in delimiting the species of animals and plants, such as we now with fuller knowledge see them actually to be. We are even more sceptical as to the validity of that appeal to changes in the conditions of life as direct causes of modification, upon which latterly at all events Darwin laid much emphasis. But that he was the first to provide a body of fact demonstrating the variability of living things, whatever be its causation, can never be questioned.

There are some older collections of evidence, chiefly the work of the French school, especially of Godron[1]—and I would mention also the almost forgotten essay of Wollaston[2]—these however are only fragments in comparison. Darwin regarded variability as a property inherent in living things, and eventually we must consider whether this conception is well founded; but postponing that inquiry for the present, we may declare that with him began a general recognition of Variation as a phenomenon widely occurring in Nature.

If a population consists of members which are not alike but differentiated, how will their characteristics be distributed among their offspring? This is the problem which the modern student of heredity sets out to investigate. Formerly it was hoped that by the simple inspection of embryological processes the modes of heredity might be ascertained, the actual mechanism by which the offspring is formed from the body of the parent. In that endeavour a noble pile of evidence has been accumulated. All that can be made visible by existing methods has been seen, but we come little if at all nearer to the central mystery. We see nothing that we can analyse further—nothing that can be translated into terms less inscrutable than the physiological events themselves. Not only does embryology give no direct aid, but the failure of cytology is, so far as I can judge, equally complete. The chromosomes of nearly related creatures may be utterly different both in number, size, and form. Only one piece of evidence encourages the old hope that a connection might be traceable between the visible characteristics of the body and those of the chromosomes. I refer of course to the accessory chromosome, which in many animals distinguishes the spermatozoon about to form a female in fertilisation. Even it however cannot be claimed as the cause of sexual differentiation, for it may be paired in forms closely allied to those in which it is unpaired or accessory. The distinction may be present or wanting, like any other secondary sexual character. Indeed, so long as no one can shew consistent distinctions between the cytological characters of somatic tissues in the same individual we can scarcely expect to perceive such distinctions between the chromosomes of the various types.

For these methods of attack we now substitute another, less ambitious, perhaps, because less comprehensive, but not less

[1] *De l'Espèce et des Races dans les Êtres Organisés*, 1859.
[2] *On the Variation of Species*, 1856.

direct. If we cannot see how a fowl by its egg and its sperm gives rise to a chicken or how a sweet pea from its ovule and its pollen grain produces another sweet pea, we at least can watch the system by which the differences between the various kinds of fowls or between the various kinds of sweet peas are distributed among the offspring. By thus breaking the main problem up into its parts we give ourselves fresh chances. This analytical study we call Mendelian because Mendel was the first to apply it. To be sure, he did not approach the problem by any such line of reasoning as I have sketched. His object was to determine the genetic definiteness of species; but though in his writings he makes no mention of inheritance it is clear that he had the extension in view. By cross-breeding he combined the characters of varieties in mongrel individuals and set himself to see how these characters would be distributed among the individuals of subsequent generations. Until he began this analysis nothing but the vaguest answers to such a question had been attempted. The existence of any orderly system of descent was never even suspected. In their manifold complexity human characteristics seemed to follow no obvious system, and the fact was taken as a fair sample of the working of heredity.

Misconception was especially brought in by describing descent in terms of "blood". The common speech uses expressions such as "consanguinity", "pure-blooded", "half-blood" and the like, which call up a misleading picture to the mind. Blood is in some respects a fluid, and thus it is supposed that this fluid can be both quantitatively and qualitatively diluted with other bloods, just as treacle can be diluted with water. Blood in primitive physiology being the peculiar vehicle of life, at once its essence and its corporeal abode, these ideas of dilution and compounding of characters in the commingling of bloods inevitably suggest that the ingredients of the mixture once combined are inseparable, that they can be brought together in any relative amounts, and in short that in heredity we are concerned mainly with a quantitative problem. Truer notions of genetic physiology are given by the Hebrew expression "seed". If we speak of a man as "of the blood-royal" we think at once of plebeian dilution, and we wonder how much of the royal fluid is likely to be "in his veins"; but if we say he is "of the seed of Abraham" we feel something of the permanence and indestructibility of that germ which can be divided and scattered among all nations, but remains recognisable in type and characteristics after 4000 years.

I knew a breeder who had a chest containing bottles of

coloured liquids by which he used to illustrate the relationships of his dogs, pouring from one to another and titrating
them quantitatively to illustrate their pedigrees. Galton was
beset by the same kind of mistake when he promulgated
his "Law of Ancestral Heredity". With modern research all
this has been cleared away. The allotment of characteristics
among offspring is not accomplished by the exudation of drops
of a tincture representing the sum of the characteristics of the
parent organism, but by a process of *cell division*, in which
numbers of these characters, or rather the elements upon which
they depend, are sorted out among the resulting germ cells in
an orderly fashion. What these elements, or *factors* as we call
them, are we do not know. That they are in some way directly
transmitted by the material of the ovum and of the spermatozoon is obvious, but it seems to me unlikely that they are in any
simple or literal sense material particles. I suspect rather that
their properties depend on some phenomenon of arrangement.
However that may be, analytical breeding proves that it is
according to the distribution of these genetic factors, to use
a non-committal term, that the characters of the offspring are
decided. The first business of experimental genetics is to
determine their number and interactions, and then to make
an analysis of the various types of life.

Now the ordinary genealogical trees, such as those which the
stud-books provide in the case of the domestic animals, or the
Heralds' College provides in the case of man, tell nothing of
all this. Such methods of depicting descent cannot even shew
the one thing they are devised to shew—"purity of blood".
For at last we know the physiological meaning of that expression. An organism is pure-bred when it has been formed
by the union in fertilisation of two germ cells which are alike
in the factors they bear; and since the factors for the several
characteristics are independent of each other, this question of
purity must be separately considered for each of them. A man,
for example, may be pure-bred in respect of his musical ability
and cross-bred in respect of the colour of his eyes or the shape
of his mouth. Though we know nothing of the essential nature
of these factors, we know a good deal of their powers. They
may confer height, colour, shape, instincts, powers both of mind
and body; indeed, so many of the attributes which animals and
plants possess that we feel justified in the expectation that with
continued analysis they will be proved to be responsible for
most if not all of the differences by which the varying individuals of any species are distinguished from each other. I will

not assert that the greater differences which characterise distinct species are due generally to such independent factors, but that is the conclusion to which the available evidence points. All this is now so well understood, and has been so often demonstrated and expounded, that details of evidence are now superfluous.

But for the benefit of those who are unfamiliar with such work let me briefly epitomize its main features and consequences. Since genetic factors are definite things, either present in or absent from any germ cell, the individual may be either "pure-bred" for any particular factor, or its absence, if he is constituted by the union of two germ cells both possessing or both destitute of that factor. If the individual is thus pure, all his germ cells will in that respect be identical, for they are simply bits of the similar germ cells which united in fertilisation to produce the parent organism. We thus reach the essential principle, that an organism cannot pass on to offspring a factor which it did not itself receive in fertilisation. Parents, therefore, which are both destitute of a given factor can only produce offspring equally destitute of it; and, on the contrary, parents both pure-bred for the presence of a factor produce offspring equally pure-bred for its presence. Whereas the germ cells of the pure-bred are all alike, those of the cross-bred, which results from the union of dissimilar germ cells, are mixed in character. Each positive factor segregates from its negative opposite, so that some germ cells carry the factor and some do not. Once the factors have been identified by their effects, the average composition of the several kinds of families formed from the various matings can be predicted.

Only those who have themselves witnessed the fixed operations of these simple rules can feel their full significance. We come to look behind the simulacrum of the individual body and we endeavour to disintegrate its features into the genetic elements by whose union the body was formed. Set out in cold general phrases such discoveries may seem remote from ordinary life. Become familiar with them and you will find your outlook on the world has changed. Watch the effects of segregation among the living things with which you have to do—plants, fowls, dogs, horses, that mixed concourse of humanity we call the English race, your friends' children, your own children, yourself—and however firmly imagination be restrained to the bounds of the known and the proved, you will feel something of that range of insight into Nature which Mendelism has begun to give. The question is often asked whether there are not also in operation systems of descent quite other than those contemplated by the Mendelian rules. I myself have expected such discoveries, but hitherto none have been plainly demonstrated. It is true we are often puzzled by the failure of a parental type to reappear in its completeness after a cross—the merino sheep or the fantail pigeon, for example. These exceptions may still be plausibly ascribed to the interference of a multitude of factors, a suggestion not easy to disprove; though it seems to me equally likely that segregation has been in reality imperfect. Of the descent of quantitative characters we still know practically nothing. These and hosts of difficult cases remain almost untouched. In particular the discovery of E. Baur, and the evidence of Winkler in regard to his "graft hybrids", both shewing that the sub-epidermal layer of a plant—the layer from which the germ cells are derived—may bear exclusively the characters of a part only of the soma, give hints of curious complications, and suggest that in plants at least the inter-relations between soma and gamete may be far less simple than we have supposed. Nevertheless, speaking generally, we see nothing to indicate that qualitative characters descend, whether in plants or animals, according to systems which are incapable of factorial representation.

The body of evidence accumulated by this method of analysis is now very large, and is still growing fast by the labours of many workers. Progress is also beginning along many novel and curious lines. The details are too technical for inclusion here. Suffice it to say that not only have we proof that segregation affects a vast range of characteristics, but in the course of our analysis phenomena of most unexpected kinds have been encountered. Some of these things twenty years ago must have seemed inconceivable. For example, the two sets of sex organs, male and female, of the same plant may not be carrying the same characteristics; in some animals characteristics, quite independent of sex, may be distributed solely or predominantly to one sex; in certain species the male may be breeding true to its own type, while the female is permanently mongrel, throwing off eggs of a distinct variety in addition to those of its own type; characteristics, essentially independent, may be associated in special combinations which are largely retained in the next generation, so that among the grandchildren there is numerical preponderance of those combinations which existed in the grandparents—a discovery which introduces us to a new phenomenon of polarity in the organism. We are accustomed to the fact that the fertilised egg has a polarity, a front and hind end for example; but we have now to recognise that it, or the primitive germinal cells formed from it, may have another polarity shewn in the groupings of the parental elements. I am entirely sceptical as to the occurrence of segregation solely in the maturation of the germ cells,[1] preferring at present to regard it as a special case of that patchwork condition we see in so many plants. These mosaics may break up, emitting bud sports at various cell divisions, and I suspect that the great regularity seen in the F_2 ratios of the cereals, for example, is a consequence of very late segregation, whereas the excessive irregularity found in other cases may be taken to indicate that segregation can happen at earlier stages of differentiation.

The paradoxical descent of colour-blindness and other sex-limited conditions—formerly regarded as an inscrutable caprice of Nature—has been represented with approximate correctness, and we already know something as to the way, or, perhaps, I should say ways, in which the determination of sex is accomplished in some of the forms of life—though, I hasten to add, we have no inkling as to any method by which that determination may be influenced or directed. It is obvious that such discoveries have bearings on most of the problems, whether theoretical or practical, in which animals and plants are concerned. Permanence or change of type, perfection of type, purity or mixture of race, "racial development", the succession of forms, from being vague phrases expressing matters of degree, are now seen to be capable of acquiring physiological meanings, already to some extent assigned with precision. For the naturalist—and it is to him that I am especially addressing myself to-day—these things are chiefly significant as relating to the history of organic beings—the theory of Evolution, to use our modern name. They have, as I shall endeavour to shew in my second address to be given in Sydney, an immediate reference to the conduct of human society.

I suppose that everyone is familiar in outline with the theory of the Origin of Species which Darwin promulgated. Through the last fifty years this theme of the Natural Selection of favoured races has been developed and expounded in writings innumerable. Favoured races certainly can replace others. The argument is sound, but we are doubtful of its value. For

---

[1] The fact that in certain plants the male and female organs respectively carry distinct factors may be quoted as almost decisively negativing the suggestion that segregation is confined to the reduction-division.

us that debate stands adjourned. We go to Darwin for his incomparable collection of facts. We would fain emulate his scholarship, his width and his power of exposition, but to us he speaks no more with philosophical authority. We read his scheme of Evolution as we would those of Lucretius or of Lamarck, delighting in their simplicity and their courage. The practical and experimental study of Variation and Heredity has not merely opened a new field; it has given a new point of view and new standards of criticism. Naturalists may still be found expounding teleological systems[1] which would have delighted Dr Pangloss himself, but at the present time few are misled. The student of genetics knows that the time for the development of theory is not yet. He would rather stick to the seed pan and the incubator.

In face of what we now know of the distribution of variability in Nature the scope claimed for Natural Selection in determining the fixity of species must be greatly reduced. The doctrine of the survival of the fittest is undeniable so long as it is applied to the organism as a whole, but to attempt by this principle to find value in all definiteness of parts and functions, and in the name of science to see fitness everywhere is mere eighteenth-century optimism. Yet it was in application to the parts, to the details of specific difference, to the spots on the peacock's tail, to the colouring of an orchid flower, and hosts of such examples, that the potency of Natural Selection was urged with the strongest emphasis. Shorn of these pretensions the doctrine of the survival of favoured races is a truism, helping scarcely at all to account for the diversity of species. Tolerance plays perhaps as considerable a part. By these admissions almost the last shred of that teleological fustian with which Victorian philosophy loved to clothe the theory of Evolu-

---

[1] I take the following from the abstract of a recent Croonian Lecture "On the Origin of Mammals" delivered to the Royal Society: "In Upper Triassic times the larger Cynodonts preyed upon the large Anomodont, *Kannemeyeria*, and carried on their existence so long as these Anomodonts survived, but died out with them about the end of the Trias or in Rhaetic times. The small Cynodonts, having neither small Anomodonts nor small Cotylosaurs to feed on, were forced to hunt the very active long-limbed Thecodonts. The greatly increased activity brought about that series of changes which formed the mammals—the flexible skin with hair, the four-chambered heart and warm blood, the loose jaw with teeth for mastication, an increased development of tactile sensation and a great increase of cerebrum. Not improbably the attacks of the newly-evolved Cynodont or mammalian type brought about a corresponding evolution in the Pseudosuchian Thecodonts which ultimately resulted in the formation of Dinosaurs and Birds" (Broom, R., *Proc. Roy. Soc.* B, LXXXVII, 88).

tion is destroyed. Those who would proclaim that whatever is is right will be wise henceforth to base this faith frankly on the impregnable rock of superstition and to abstain from direct appeals to natural fact.

My predecessor said last year that in physics the age is one of rapid progress and profound scepticism. In at least as high a degree this is true of biology, and as a chief characteristic of modern evolutionary thought we must confess also to a deep but irksome humility in presence of the great vital problems. Every theory of Evolution must be such as to accord with the facts of physics and chemistry, a primary necessity to which our predecessors paid small heed. For them the unknown was a rich mine of possibilities on which they could freely draw. For us it is rather an impenetrable mountain out of which the truth can be chipped in rare and isolated fragments. Of the physics and chemistry of life we know next to nothing. Somehow the characters of living things are bound up in properties of colloids, and are largely determined by the chemical powers of enzymes, but the study of these classes of matter has only just begun. Living things are found by a simple experiment to have powers undreamt of, and who knows what may be behind?

Naturally we turn aside from generalities. It is no time to discuss the origin of the Mollusca or of Dicotyledons, while we are not even sure how it came to pass that *Primula obconica* has in twenty-five years produced its abundant new forms almost under our eyes. Knowledge of Heredity has so reacted on our conceptions of Variation that very competent men are even denying that Variation in the old sense is a genuine occurrence at all. Variation is postulated as the basis of all evolutionary change. Do we then as a matter of fact find in the world about us variations occurring of such a kind as to warrant faith in a contemporary progressive Evolution? Till lately most of us would have said "yes" without misgiving. We should have pointed, as Darwin did, to the immense range of diversity seen in many wild species, so commonly that the difficulty is to define the types themselves. Still more conclusive seemed the profusion of forms in the various domesticated animals and plants, most of them incapable of existing even for a generation in the wild state, and therefore fixed unquestionably by human selection. These, at least, for certain, are new forms, often distinct enough to pass for species, which have arisen by variation. But when analysis is applied to this mass of variation the matter wears a different aspect. Closely examined,

what is the "variability" of wild species? What is the natural
fact which is denoted by the statement that a given species
exhibits much variation? Generally one of two things: either
that the individuals collected in one locality differ among themselves; or perhaps more often that samples from separate
localities differ from each other. As direct evidence of Variation it is clearly to the first of these phenomena that we must
have recourse—the heterogeneity of a population breeding
together in one area. This heterogeneity may be in any degree,
ranging from slight differences that systematists would disregard, to a complex variability such as we find in some moths,
where there is an abundance of varieties so distinct that many
would be classified as specific forms but for the fact that all are
freely breeding together. Naturalists formerly supposed that
any of these varieties might be bred from any of the others.
Just as the reader of novels is prepared to find that any kind of
parents might have any kind of children in the course of the
story, so was the evolutionist ready to believe that any pair of
moths might produce any of the varieties included in the
species. Genetic analysis has disposed of all these mistakes. We
have no longer the smallest doubt that in all these examples
the varieties stand in a regular descending order, and that they
are simply terms in a series of combinations of factors separately
transmitted, of which each may be present or absent.

The appearance of contemporary variability proves to be an
illusion. Variation from step to step in the series must occur
either by the addition or by the loss of a factor. Now, of the
origin of new forms *by loss* there seems to me to be fairly clear
evidence, but of the *contemporary acquisition* of any new factor
I see no satisfactory proof, though I admit there are rare
examples which may be so interpreted. We are left with a
picture of Variation utterly different from that which we saw at
first. Variation now stands out as a definite physiological
event. We have done with the notion that Darwin came
latterly to favour, that large differences can arise by accumulation of small differences. Such small differences are often
mere ephemeral effects of conditions of life, and as such are not
transmissible; but even small differences, when truly genetic,
are factorial like the larger ones, and there is not the slightest
reason for supposing that they are capable of summation. As
to the origin or source of these positive separable factors, we are
without any indication or surmise. By their effects we know
them to be definite, as definite, say, as the organisms which
produce diseases; but how they arise and how they come to

take part in the composition of the living creature so that when
present they are treated in cell division as constituents of the
germs, we cannot conjecture.

It was a commonplace of evolutionary theory that at least
the domestic animals have been developed from a few wild
types. Their origin was supposed to present no difficulty. The
various races of fowl, for instance, all came from *Gallus bankiva*,
the Indian Jungle-fowl. So we are taught; but try to reconstruct the steps in their evolution and you realise your hopeless
ignorance. To be sure there are breeds, such as Black-red
Game and Brown Leghorns, which have the colours of the
Jungle-fowl, though they differ in shape and other respects.
As we know so little as yet of the genetics of shape, let us assume
that those transitions could be got over. Suppose, further, as
is probable, that the absence of the maternal instinct in the
Leghorn is due to loss of one factor which the Jungle-fowl
possesses. So far we are on fairly safe ground. But how about
White Leghorns? Their origin may seem easy to imagine, since
white varieties have often arisen in well-authenticated cases.
But the white of White Leghorns is not, as white in Nature
often is, due to the loss of the colour-elements, but to the action
of something which inhibits their expression. Whence did that
something come? The same question may be asked respecting
the heavy breeds, such as Malays or Indian Game. Each of
these is a separate introduction from the East. To suppose that
these, with their peculiar combs and close feathering, could
have been developed from pre-existing European breeds is very
difficult. On the other hand, there is no wild species now living
any more like them. We may, of course, postulate that there
was once such a species, now lost. That is quite conceivable,
though the suggestion is purely speculative. I might thus go
through the list of domesticated animals and plants of ancient
origin and again and again we should be driven to this suggestion, that many of their distinctive characters must have
been derived from some wild original now lost. Indeed to this
unsatisfying conclusion almost every careful writer on such
subjects is now reduced. If we turn to modern evidence the
case looks even worse. The new breeds of domestic animals
made in recent times are the carefully selected products of recombination of pre-existing breeds. Most of the new varieties of
cultivated plants are the outcome of deliberate crossing. There
is generally no doubt in the matter. We have pretty full
histories of these crosses in gladiolus, orchids, cineraria, begonia,
calceolaria, pelargonium, etc. A very few certainly arise from

a single origin. The sweet pea is the clearest case, and there are others which I should name with hesitation. The cyclamen is one of them, but we know that efforts to cross cyclamens were made early in the cultural history of the plant, and they may very well have been successful. Several plants for which single origins are alleged, such as the Chinese primrose, the dahlia, and tobacco, came to us in an already domesticated state, and their origins remain altogether mysterious. Formerly single origins were generally presumed, but at the present time numbers of the chief products of domestication, dogs, horses, cattle, sheep, poultry, wheat, oats, rice, plums, cherries, have in turn been accepted as "polyphyletic" or, in other words, derived from several distinct forms. The reason that has led to these judgments is that the distinctions between the chief varieties can be traced as far back as the evidence reaches, and that these distinctions are so great, so far transcending anything that we actually know Variation capable of effecting, that it seems pleasanter to postpone the difficulty, relegating the critical differentiation to some misty antiquity into which we shall not be asked to penetrate. For it need scarcely be said that this is mere procrastination. If the origin of a form under domestication is hard to imagine, it becomes no easier to conceive of such enormous deviations from type coming to pass in the wild state. Examine any two thoroughly distinct species which meet each other in their distribution, as, for instance, *Lychnis diurna* and *vespertina* do. In areas of overlap are many intermediate forms. These used to be taken to be transitional steps, and the specific distinctness of *vespertina* and *diurna* was on that account questioned. Once it is known that these supposed intergrades are merely mongrels between the two species the transition from one to the other is practically beyond our powers of imagination to conceive If both these can survive, why has their common parent perished? Why when they cross do they not reconstruct it instead of producing partially sterile hybrids? I take this example to shew how entirely the facts were formerly misinterpreted.

When once the idea of a true-breeding—or, as we say, homozygous—type is grasped, the problem of Variation becomes an insistent oppression. What can make such a type vary? We know, of course, one way by which novelty can be introduced—by crossing. Cross two well-marked varieties—for instance, of Chinese primula—each breeding true, and in the second generation by mere recombination of the various factors which the two parental types severally introduced, there will be a profusion of forms, utterly unlike each other, distinct also from the original parents. Many of these can be bred true, and if found wild would certainly be described as good species. Confronted by the difficulty I have put before you, and contemplating such amazing polymorphism in the second generation from a cross in *Antirrhinum*, Lotsy[1] has lately with great courage suggested to us that all variation may be due to such crossing. I do not disguise my sympathy with this effort. After the blind complacency of conventional evolutionists it is refreshing to meet so frank an acknowledgment of the hardness of the problem. Lotsy's utterance will at least do something to expose the artificiality of systematic zoology and botany. Whatever might or might not be revealed by experimental breeding, it is certain that without such tests we are merely guessing when we profess to distinguish specific limits and to declare that this is a species and that a variety. The only definable unit in classification is the homozygous form which breeds true. When we presume to say that such and such differences are trivial and such others valid, we are commonly embarking on a course for which there is no physiological warrant. Who could have foreseen that the Apple and Pear—so like each other that their botanical differences are evasive—could not be crossed together, though species of *Antirrhinum* so totally unlike each other as *majus* and *molle* can be hybridised, as Baur has shewn, without a sign of impaired fertility? Jordan was perfectly right. The true-breeding forms which he distinguished in such multitudes are real entities, though the great systematists, dispensing with such laborious analysis, have pooled them into arbitrary Linnean species, for the convenience of collectors and for the simplification of catalogues. Such pragmatical considerations may mean much in the museum, but with them the student of the physiology of variation has nothing to do. These "little species", finely cut, true-breeding, and innumerable mongrels between them, are what he finds when he examines any so-called variable type. On analysis the semblance of variability disappears, and the illusion is shewn to be due to segregation and recombination of series of factors on pre-determined lines. As soon as the "little species" are separated out they are found to be fixed. In face of such a result we may well ask with Lotsy, is there such a thing as spontaneous variation anywhere? His answer is that there is not.

Abandoning the attempt to shew that positive factors can be

[1] Lotsy, J. P., "La Théorie du Croisement", *Arch. Néerland. Sc. Exact. Nat.* Ser. III, B II, 1914, 178.

added to the original stock, we have further to confess that we cannot often actually prove variation by loss of factors to be a real phenomenon. Lotsy doubts whether even this phenomenon occurs. The sole source of variation, in his view, is crossing. But here I think he is on unsafe ground. When a well-established variety like "Crimson King" primula, bred by Messrs Sutton in thousands of individuals, gives off, as it did a few years since, a salmon-coloured variety, "Coral King", we might claim this as a genuine example of variation by loss. The new variety is a simple recessive. It differs from "Crimson King" only in one respect, the loss of a single colour-factor, and, of course, bred true from its origin. To account for the appearance of such a new form by any process of crossing is exceedingly difficult. From the nature of the case there can have been no cross since "Crimson King" was established, and hence the salmon must have been concealed as a recessive from the first origin of that variety, even when it was represented by very few individuals, probably only by a single one. Surely, if any of these had been heterozygous for salmon this recessive could hardly have failed to appear during the process of self-fertilisation by which the stock would be multiplied, even though that "selfing" may not have been strictly carried out. Examples like this seem to me practically conclusive.[1] They can be challenged, but not, I think, successfully. Then again in regard to those variations in number and division of parts which we call meristic, the reference of these to original cross-breeding is surely barred by the circumstances in which they often occur. There remain also the rare examples mentioned already in which a single wild origin may with much confidence be assumed. In spite of repeated trials, no one has yet succeeded in crossing the sweet pea with any other leguminous species. We know that early in its cultivated history it produced at least two marked varieties which I can only conceive of as spontaneously arising, though, no doubt, the profusion of forms we now have was made by the crossing of those original varieties. I mention the sweet pea thus prominently for another reason, that it introduces us to another though subsidiary form of variation, which may be described as a *fractionation* of factors. Some of my Mendelian colleagues have spoken of genetic factors as permanent and indestructible. Relative permanence in a sense they have, for they commonly come out unchanged after segregation. But

---

[1] The numerous and most interesting "mutations" recorded by Professor T. H. Morgan and his colleagues in the fly, *Drosophila*, may also be cited as unexceptionable cases.

I am satisfied that they may occasionally undergo a quantitative disintegration, with the consequence that varieties are produced intermediate between the integral varieties from which they were derived. These disintegrated conditions I have spoken of as subtraction—or reduction—stages. For example, the Picotee sweet pea, with its purple edges, can surely be nothing but a condition produced by the factor which ordinarily makes the fully purple flower, quantitatively diminished. The pied animal, such as the Dutch rabbit, must similarly be regarded as the result of partial defect of the chromogen from which the pigment is formed, or conceivably of the factor which effects its oxidation. On such lines I think we may with great confidence interpret all those intergrading forms which breed true and are not produced by factorial interference.

It is to be inferred that these fractional degradations are the consequence of irregularities in segregation. We constantly see irregularities in the ordinary meristic processes, and in the distribution of somatic differentiation. We are familiar with half segments, with imperfect twinning, with leaves partially petaloid, with petals partially sepaloid. All these are evidences of departures from the normal regularity in the rhythms of repetition, or in those waves of differentiation by which the qualities are sorted out among the parts of the body. Similarly, when in segregation the qualities are sorted out among the germ cells in certain critical cell divisions, we cannot expect these differentiating divisions to be exempt from the imperfections and irregularities which are found in all the grosser divisions that we can observe. If I am right, we shall find evidence of these irregularities in the association of unconformable numbers with the appearance of the novelties which I have called fractional. In passing let us note how the history of the sweet pea belies those ideas of a continuous evolution with which we had formerly to contend. The big varieties came first. The little ones have arisen later, as I suggest by fractionation. Presented with a collection of modern sweet peas how prettily would the devotees of Continuity have arranged them in a graduated series, shewing how every intergrade could be found, passing from the full colour of the wild Sicilian species in one direction to white, in the other to the deep purple of "Black Prince", though happily we know these two to be among the earliest to have appeared.

Having in view these and other considerations which might be developed, I feel no reasonable doubt that though we may have to forgo a claim to variations by addition of factors, yet variation both by loss of factors and by fractionation of factors is a genuine phenomenon of contemporary nature. If then we have to dispense, as seems likely, with any addition from without we must begin seriously to consider whether the course of Evolution can at all reasonably be represented as an unpacking of an original complex which contained within itself the whole range of diversity which living things present. I do not suggest that we should come to a judgment as to what is or is not probable in these respects. As I have said already, this is no time for devising theories of Evolution, and I propound none. But as we have got to recognise that there has been an Evolution, that somehow or other the forms of life have arisen from fewer forms, we may as well see whether we are limited to the old view that evolutionary progress is from the simple to the complex, and whether after all it is conceivable that the process was the other way about. When the facts of genetic discovery become familiarly known to biologists, and cease to be the preoccupation of a few, as they still are, many and long discussions must inevitably arise on the question, and I offer these remarks to prepare the ground. I ask you simply to open your minds to this possibility. It involves a certain effort. We have to reverse our habitual modes of thought. At first it may seem rank absurdity to suppose that the primordial form or forms of protoplasm could have contained complexity enough to produce the divers types of life. But is it easier to imagine that these powers could have been conveyed by extrinsic additions? Of what nature could these additions be? Additions of material cannot surely be in question. We are told that salts of iron in the soil may turn a pink hydrangea blue. The iron cannot be passed on to the next generation. How can the iron multiply itself? The power to assimilate the iron is all that can be transmitted. A disease-producing organism like the *pébrine* of silkworms can in a very few cases be passed on through the germ cells. Such an organism can multiply and can produce its characteristic effects in the next generation. But it does not become part of the invaded host, and we cannot conceive it taking part in the geometrically ordered processes of segregation. These illustrations may seem too gross; but what refinement will meet the requirements of the problem, that the thing introduced must be, as the living organism itself is, capable of multiplication and of subordinating itself in a definite system of segregation? That which is conferred in variation must rather itself be a change, not of material, but of arrangement, or of motion. The invocation of additions extrinsic to the organism does not seriously help us to imagine how the power to change can be conferred, and if it proves that hope in that direction must be abandoned, I think we lose very little. By the re-arrangement of a very moderate number of things we soon reach a number of possibilities practically infinite.

That primordial life may have been of small dimensions need not disturb us. Quantity is of no account in these considera-tions. Shakespeare once existed as a speck of protoplasm not so big as a small pin's head. To this nothing was added that would not equally well have served to build up a baboon or a rat. Let us consider how far we can get by the process of removal of what we call "epistatic" factors, in other words those that control, mask, or suppress underlying powers and faculties. I have spoken of the vast range of colours exhibited by modern sweet peas. There is no question that these have been derived from the one wild bi-colour form by a process of successive removals. When the vast range of form, size, and flavour to be found among the cultivated apples is considered it seems difficult to suppose that all this variety is hidden in the wild crab-apple. I cannot positively assert that this is so, but I think all familiar with Mendelian analysis would agree with me that it is probable, and that the wild crab contains pre-sumably inhibiting elements which the cultivated kinds have lost. The legend that the seedlings of cultivated apples become crabs is often repeated. After many inquiries among the raisers of apple seedlings I have never found an authentic case —once only even an alleged case, and this on inquiry proved to be unfounded. I have confidence that the artistic gifts of mankind will prove to be due not to something added to the make-up of an ordinary man, but to the absence of factors which in the normal person inhibit the development of these gifts. They are almost beyond doubt to be looked upon as *releases* of powers normally suppressed. The instrument is there, but it is "stopped down". The scents of flowers or fruits, the finely repeated divisions that give its quality to the wool of the Merino, or in an analogous case the multiplicity of quills to the tail of the fantail pigeon, are in all probability other examples of such releases. You may ask what guides us in the discrimi-nation of the positive factors and how we can satisfy ourselves that the appearance of a quality is due to loss. It must be con-ceded that in these determinations we have as yet recourse only to the effects of dominance. When the tall pea is crossed with the dwarf, since the offspring is tall we say that the tall parent passed a factor into the cross-bred which makes it tall. The pure tall parent had two doses of this factor; the dwarf had none; and since the cross-bred is tall we say that one dose of the dominant tallness is enough to give the full height. The reasoning seems unanswerable. But the commoner result of crossing is the production of a form intermediate between the two pure parental types. In such examples we see clearly enough that the full parental characteristics can only appear when they are homozygous—formed from similar germ cells, and that one dose is insufficient to produce either effect fully. When this is so we can never be sure which side is positive and which negative. Since, then, when dominance is incomplete we find ourselves in this difficulty, we perceive that the amount of the effect is our only criterion in distinguishing the positive from the negative, and when we return even to the example of the tall and dwarf peas the matter is not so certain as it seemed. Professor Cockerell lately found among thousands of yellow sunflowers one which was partly red. By breeding he raised from this a form wholly red. Evidently the yellow and the wholly red are the pure forms, and the partially red is the heterozygote. We may then say that the yellow is YY with two doses of a positive factor which inhibits the development of pigment; the red is yy, with no dose of the inhibitor; and the partially red are Yy, with only one dose of it. But we might be tempted to think the red was a positive characteristic, and invert the expressions, representing the red as RR, the partly red as Rr, and the yellow as rr. According as we adopt the one or the other system of expression we shall interpret the evolutionary change as one of loss or as one of addition. May we not interpret the other apparent new dominants in the same way? The white dominant in the fowl or in the Chinese primula can inhibit colour. But may it not be that the original coloured fowl or primula had two doses of a factor which inhibited this inhibitor? The pepper moth, *Amphidasys betularia*, produced in England about 1840 a black variety, then a novelty, now common in certain areas, which behaves as a full dominant. The pure blacks are no blacker than the cross-bred. Though at first sight it seems that the black *must* have been something added, we can without absurdity suggest that the normal is the term in which two doses of inhibitor are present, and that in the absence of one of them the black appears.

In spite of seeming perversity, therefore, we have to admit that there is no evolutionary change which in the present state of our knowledge we can positively declare to be not due to loss. When this has been conceded it is natural to ask whether

the removal of inhibiting factors may not be invoked in alleviation of the necessity which has driven students of the domestic
breeds to refer their diversities to multiple origins. Something,
no doubt, is to be hoped for in that direction, but not until
much better and more extensive knowledge of what variation
by loss may effect in the living body can we have any real
assurance that this difficulty has been obviated. We should be
greatly helped by some indication as to whether the origin of
life has been single or multiple. Modern opinion is, perhaps,
inclining to the multiple theory, but we have no real evidence.
Indeed, the problem still stands outside the range of scientific
investigation, and when we hear the spontaneous formation
of formaldehyde mentioned as a possible first step in the origin
of life, we think of Harry Lauder in the character of a Glasgow
schoolboy pulling out his treasures from his pocket—"That's
a wassher—for makkin' motor cars"!

As the evidence stands at present all that can be safely added
in amplification of the evolutionary creed may be summed up
in the statement that variation occurs as a definite event often
producing a sensibly discontinuous result; that the succession
of varieties comes to pass by the elevation and establishment of
sporadic groups of individuals owing their origin to such
isolated events; and that the change which we see as a
nascent variation is often, perhaps always, one of loss. Modern
research lends not the smallest encouragement or sanction to
the view that gradual evolution occurs by the transformation
of masses of individuals, though that fancy has fixed itself on
popular imagination. The isolated events to which variation
is due are evidently changes in the germinal tissues, probably
in the manner in which they divide. It is likely that the
occurrence of these variations is wholly irregular, and as to
their causation we are absolutely without surmise or even
plausible speculation. Distinct types once arisen, no doubt a
profusion of the forms called species have been derived from
them by simple crossing and subsequent recombination. New
species may be now in course of creation by this means, but the
limits of the process are obviously narrow. On the other hand,
we see no changes in progress around us in the contemporary
world which we can imagine likely to culminate in the evolution of forms distinct in the larger sense. By intercrossing dogs,
jackals, and wolves new forms of these types can be made, some
of which may be species, but I see no reason to think that from
such material a fox could be bred in indefinite time, or that
dogs could be bred from foxes.

Whether science will hereafter discover that certain groups can by peculiarities in their genetic physiology be declared to have a prerogative quality justifying their recognition as species in the old sense, and that the differences of others are of such a subordinate degree that they may in contrast be termed varieties, further genetic research alone can shew. I myself anticipate that such a discovery will be made, but I cannot defend the opinion with positive conviction.

Somewhat reluctantly, and rather from a sense of duty, I have devoted most of this address to the evolutionary aspects of genetic research. We cannot keep these things out of our heads, though sometimes we wish we could. The outcome, as you will have seen, is negative, destroying much that till lately passed for gospel. Destruction may be useful, but it is a low kind of work. We are just about where Boyle was in the seventeenth century. We can dispose of Alchemy, but we cannot make more than a quasi-chemistry. We are awaiting our Priestley and our Mendeléeff. In truth it is not these wider aspects of genetics that are at present our chief concern. They will come in their time. The great advances of science are made like those of Evolution, not by imperceptible mass-improvement, but by the sporadic birth of penetrative genius. The journeymen follow after him, widening and clearing up, as we are doing along the track that Mendel found.

PRESIDENTIAL ADDRESS TO THE BRITISH ASSOCIATION, AUSTRALIA

(b) Sydney Meeting. 1914

At Melbourne I spoke of the new knowledge of the properties of living things which Mendelian analysis has brought
us. I indicated how these discoveries are affecting our outlook
on that old problem of natural history, the origin and nature of
Species, and the chief conclusion I drew was the negative one,
that, though we must hold to our faith in the Evolution of
Species, there is little evidence as to how it has come about,
and no clear proof that the process is continuing in any considerable degree at the present time. The thought uppermost
in our minds is that knowledge of the nature of life is altogether
too slender to warrant speculation on these fundamental
subjects. Did we presume to offer such speculations they would
have no more value than those which alchemists might have
made as to the nature of the elements. But though in regard
to these theoretical aspects we must confess to such deep
ignorance, enough has been learnt of the general course of
heredity within a single species to justify many practical
conclusions which cannot in the main be shaken. I propose
now to develop some of these conclusions in regard to our own
species, Man.

In my former Address I mentioned the condition of certain
animals and plants which are what we call "polymorphic".
Their populations consist of individuals of many types, though
they breed freely together with perfect fertility. In cases of this
kind which have been sufficiently investigated it has been found
that these distinctions—sometimes very great and affecting
most diverse features of organisation—are due to the presence
or absence of elements, or factors as we call them, which are
treated in heredity as separate entities. These factors and their
combinations produce the characteristics which we perceive.
No individual can acquire a particular characteristic unless
the requisite factors entered into the composition of that individual at fertilisation, being received either from the father
or from the mother or from both, and consequently no individual can pass on to his offspring positive characters which
he does not himself possess. Rules of this kind have already
been traced in operation in the human species; and though

I admit that an assumption of some magnitude is involved when we extend the application of the same system to human characteristics in general, yet the assumption is one which I believe we are fully justified in making. With little hesitation we can now declare that the potentialities and aptitudes, physical as well as mental, sex, colours, powers of work or invention, liability to diseases, possible duration of life, and the other features by which the members of a mixed population differ from each other, are determined from the moment of fertilisation; and by all that we know of heredity in the forms of life with which we can experiment we are compelled to believe that these qualities are in the main distributed on a factorial system. By changes in the outward conditions of life the expression of some of these powers and features may be excited or restrained. For the development of some an external opportunity is needed, and if that be withheld the character is never seen, any more than if the body be starved can the full height be attained; but such influences are superficial and do not alter the genetic constitution.

The factors which the individual receives from his parents and no others are those which he can transmit to his offspring; and if a factor was received from one parent only, not more than half the offspring, on an average, will inherit it. What is it that has so long prevented mankind from discovering such simple facts? Primarily the circumstance that as man must have *two* parents it is not possible quite easily to detect the contributions of each. The individual body is a *double* structure, whereas the germ cells are *single*. Two germ cells unite to produce each individual body, and the ingredients they respectively contribute interact in ways that leave the ultimate product a medley in which it is difficult to identify the several ingredients. When, however, their effects are conspicuous the task is by no means impossible. In part also even physiologists have been blinded by the survival of ancient and obscurantist conceptions of the nature of man by which they were discouraged from the application of any rigorous analysis. Medical literature still abounds with traces of these archaisms, and, indeed, it is only quite recently that prominent horse-breeders have come to see that the dam matters as much as the sire. For them, though vast pecuniary considerations were involved, the old "homunculus" theory was good enough. We were amazed at the notions of genetic physiology which Professor Baldwin Spencer encountered in his wonderful researches among the natives of Central Australia; but in truth, if we reflect that these problems have engaged the attention of civilised man for ages, the fact that he, with all his powers of recording and deduction, failed to discover any part of the Mendelian system is almost as amazing. The popular notion that any parents can have any kind of children within the racial limits is contrary to all experience, yet we have gravely entertained such ideas. As I have said elsewhere, the truth might have been found out at any period in the world's history if only pedigrees had been drawn the right way up. If, instead of exhibiting the successive pairs of progenitors who have contributed to the making of an ultimate individual, some one had had the idea of setting out the posterity of a single ancestor who possessed a marked feature such as the Habsburg lip, and shewing the transmission of this feature along some of the descending branches and the permanent loss of the feature in collaterals, the essential truth that heredity can be expressed in terms of presence and absence must have at once become apparent. For the descendant is not, as he appears in the conventional pedigree, a sort of pool into which each tributary ancestral stream had poured something, but rather a conglomerate of ingredient characters taken from his progenitors in such a way that some ingredients are represented and others are omitted.

Let me not, however, give the impression that the un-ravelling of such descents is easy. Even with fairly full details, which in the case of man are very rarely to be had, many complications occur, often preventing us from obtaining more than a rough general indication of the system of descent. The nature of these complications we partly understand from our experience of animals and plants, which are amenable to breeding under careful restrictions, and we know that they are mostly referable to various effects of interaction between factors by which the presence of some is masked.

Necessarily the clearest evidence of regularity in the in-heritance of human characteristics has been obtained in regard to the descent of marked abnormalities of structure and congenital diseases. Of the descent of ordinary distinctions such as are met with in the normal healthy population we know little for certain. Hurst's[1] evidence, that two parents both with light-coloured eyes—in the strict sense, meaning that no pigment is present on the front of the iris—do not have dark-eyed children, still stands almost alone in this respect. With

[1] Hurst, C. C., "On the Inheritance of Eye-colour in Man", *Proc. Roy. Soc.* B, LXXX, 1908, 85.

regard to the inheritance of other colour-characteristics some advance has been made, but everything points to the inference that the genetics of colour and many other features in man will prove exceptionally complex. There are, however, plenty of indications of system comparable with those which we trace in various animals and plants, and we are assured that to extend and clarify such evidence is only a matter of careful analysis. For the present, in asserting almost any general rules for human descent, we do right to make large reservations for possible exceptions. It is tantalising to have to wait, but of the ultimate result there can be no doubt.

I spoke of complications. Two of these are worth illustrating here, for probably both of them play a great part in human genetics. It was discovered by Nilsson-Ehle, in the course of experiments with certain wheats, that several factors having the same power may co-exist in the same individual. These cumulative factors do not necessarily produce a cumulative effect, for any one of them may suffice to give the full result. Just as the pure-bred tall pea with its two factors for tallness is no taller than the cross-bred with a single factor, so these wheats with three pairs of factors for red colour are no redder than the ordinary reds of the same family. Similar observations have been made by East and others. In some cases, as in the primulas studied by Gregory, the effect is cumulative. These results have been used with plausibility by Davenport and the American workers to elucidate the curious case of the mulatto. If the descent of colour in the cross between the negro and the white man followed the simplest rule, the offspring of two first-cross mulattos would be, on an average, one black : two mulattos : one white, but this is notoriously not so. Evidence of some segregation is fairly clear, and the deficiency of real whites may perhaps be accounted for on the hypothesis of cumulative factors, though by the nature of the case strict proof is not to be had. But at present I own to a preference for regarding such examples as instances of imperfect segregation. The series of germ cells produced by the cross-bred consists of some with no black, some with full black, and others with intermediate quantities of black. No statistical tests of the condition of the gametes in such cases exist, and it is likely that by choosing suitable crosses all sorts of conditions may be found, ranging from the simplest case of total segregation, in which there are only two forms of gametes, up to those in which there are all intermediates in various proportions. This at least is what general experience of hybrid products leads me to anticipate. Segregation is somehow effected by the rhythms of cell division, if such an expression may be permitted. In some cases the whole factor is so easily separated that it is swept out at once; in others it is so intermixed that gametes of all degrees of purity may result. That is admittedly a crude metaphor, but as yet we cannot substitute a better. Be all this as it may, there are many signs that in human heredity phenomena of this kind are common, whether they indicate a multiplicity of cumulative factors or imperfections in segregation. Such phenomena, however, in no way detract from the essential truths that segregation occurs, and that the organism cannot pass on a factor which it has not itself received.

In human heredity we have found some examples, and I believe that we shall find many more, in which the descent of factors is limited by sex. The classical instances are those of colour-blindness and haemophilia. Both these conditions occur with much greater frequency in males than in females. Of colour-blindness at least we know that the *sons* of the colour-blind man do not inherit it (unless the mother is a transmitter) and do not transmit it to their children of either sex. Some, probably all, of the daughters of the colour-blind father inherit the character, and though not themselves colour-blind, they transmit it to some (probably, on an average, half) of their offspring of both sexes. For since these normal-sighted women have only received the colour-blindness from one side of their parentage, only half their offspring, on an average, can inherit it. The sons who inherit the colour-blindness will be colour-blind, and the inheriting daughters become themselves again transmitters. Males with normal colour-vision, whatever their own parentage, do not have colour-blind descendants, unless they marry transmitting women. There are points still doubtful in the interpretation, but the critical fact is clear, that the germ cells of the colour-blind man are of two kinds: (i) those which do not carry on the affection and are destined to take part in the formation of sons; and (ii) those which do carry on the colour-blindness and are destined to form daughters. There is evidence that the ova also are similarly predestined to form one or other of the sexes, but to discuss the whole question of sex-determination is beyond my present scope. The descent of these sex-limited affections nevertheless calls for mention here, because it is an admirable illustration of factorial predestination. It moreover exemplifies that *parental polarity* of the zygote to which I alluded in my first Address, a phenomenon which we suspect to be at the bottom of various anomalies of heredity, and suggests that there may be truth in the popular notion that in some respects sons resemble their mothers and daughters their fathers.

As to the descent of hereditary diseases and malformations, however, we have abundant data for deciding that many are transmitted as dominants and a few as recessives. The most remarkable collection of these data is to be found in family histories of diseases of the eye. Neurology and dermatology have also contributed many very instructive pedigrees. In great measure the ophthalmological material was collected by Edward Nettleship, for whose death we so lately grieved. After retiring from practice as an oculist he devoted several years to this most laborious task. He was not content with hearsay evidence, but travelled incessantly, personally examining all accessible members of the families concerned, working in such a way that his pedigrees are models of orderly observation and recording. His zeal stimulated many younger men to take part in the work, and it will now go on, with the result that the systems of descent of all the common hereditary diseases of the eye will soon be known with approximate accuracy.

Give a little imagination to considering the chief deduction from this work. Technical details apart, and granting that we cannot wholly interpret the numerical results, sometimes noticeably more and sometimes fewer descendants of these patients being affected than Mendelian formulae would indicate, the expectation is that in the case of many diseases of the eye a large proportion of the children, grandchildren, and remoter descendants of the patients will be affected with the disease. Sometimes it is only defective sight that is transmitted; in other cases it is blindness, either from birth or coming on at some later age. The most striking example perhaps is that of a form of night-blindness still prevalent in a district near Montpellier, which has affected at least 130 persons, all descending from a single affected individual[1] who came into the country in the seventeenth century. The transmission is in every case through an affected parent, and no normal has been known to pass on the condition. Such an example well serves to illustrate the fixity of the rules of descent. Similar

[1] The first human descent proved to follow Mendelian rules was that of a serious malformation of the hand studied by Farabee in America. Drinkwater subsequently worked out pedigrees for the same malformation in England. After many attempts, he now tells me that he has succeeded in proving that the American family and one of his own had an abnormal ancestor in common, five generations ago.

instances might be recited relating to a great variety of other conditions, some trivial, others grave.

At various times it has been declared that men are born equal, and that the inequality is brought about by unequal opportunities. Acquaintance with the pedigrees of disease soon shews the fatuity of such fancies. The same conclusion, we may be sure, would result from the true representation of the descent of any human faculty. Never since Galton's publications can the matter have been in any doubt. At the time he began to study family histories even the broad significance of heredity was frequently denied, and resemblances to parents or ancestors were looked on as interesting curiosities. Inveighing against hereditary political institutions, Tom Paine remarks that the idea is as absurd as that of an "hereditary wise man", or an "hereditary mathematician", and to this day I suppose many people are not aware that he is saying anything more than commonly foolish. We, on the contrary, would feel it something of a puzzle if two parents, both mathematically gifted, had any children *not* mathematicians. Galton first demonstrated the overwhelming importance of these considerations, and had he not been misled, partly by the theory of pangenesis, but more by his mathematical instincts and training, which prompted him to apply statistical treatment rather than qualitative analysis, he might, not improbably, have discovered the essential facts of Mendelism.

It happens rarely that science has anything to offer to the common stock of ideas at once so comprehensive and so simple that the courses of our thoughts are changed. Contributions to the material progress of mankind are comparatively frequent. They result at once in application. Transit is quickened; communication is made easier; the food-supply is increased and population multiplied. By direct application to the breeding of animals and plants such results must even flow from Mendel's work. But I imagine the greatest practical change likely to ensue from modern genetic discovery will be a quickening of interest in the true nature of man and in the biology of races. I have spoken cautiously as to the evidence for the operation of any simple Mendelian system in the descent of human faculty; yet the certainty that systems which differ from the simpler schemes only in degree of complexity are at work in the distribution of characters among the human population cannot fail to influence our conceptions of life and of ethics, leading perhaps ultimately to modification of social usage. That change cannot but be in the main one of simplification. The eighteenth century made great pretence of a return to Nature, but it did not occur to those philosophers first to inquire what Nature is; and perhaps not even the patristic writings contain fantasies much further from physiological truth than those which the rationalists of the *Encyclopaedia* adopted as the basis of their social schemes. For men are so far from being born equal or similar that to the naturalist they stand as the very type of a polymorphic species. Even most of our local races consist of many distinct strains and individual types. From the population of any ordinary English town as many distinct human breeds could in a few generations be isolated as there are now breeds of dogs, and indeed such a population in its present state is much what the dogs of Europe would be in ten years' time but for the interference of the fanciers. Even as at present constituted, owing to the isolating effects of instinct, fashion, occupation, and social class, many incipient strains already exist.

In one respect civilised man differs from all other species of animal or plant in that, having prodigious and ever-increasing power over Nature, he invokes these powers for the preservation and maintenance of many of the inferior and all the defective members of his species. The inferior freely multiply, and the defective, if their defects be not so grave as to lead to their detention in prisons or asylums, multiply also without restraint. Heredity being strict in its action, the consequences are in civilised countries much what they would be in the kennels of the dog-breeder who continued to preserve all his puppies, good and bad: the proportion of defectives increases. The increase is so considerable that outside every great city there is a smaller town inhabited by defectives and those who wait on them. Round London we have a ring of such towns with some 30,000 inhabitants, of whom about 28,000 are defective, largely, though of course by no means entirely, bred from previous generations of defectives. Now, it is not for us to consider practical measures. As men of science we observe natural events and deduce conclusions from them. I may perhaps be allowed to say that the remedies proposed in America, in so far as they aim at the eugenic regulation of marriage on a comprehensive scale, strike me as devised without regard to the needs either of individuals or of a modern State. Undoubtedly if they decide to breed their population of one uniform puritan grey, they can do it in a few generations; but I doubt if timid respectability will make a nation happy, and I am sure that qualities of a different sort are needed if it is to compete with more vigorous and more varied communities. Everyone must have a preliminary sympathy with the aims of eugenists both abroad and at home. Their efforts at the least are doing something to discover and spread truth as to the physiological structure of society. The spirit of such organisations, however, almost of necessity suffers from a bias towards the accepted and the ordinary, and if they had power it would go hard with many ingredients of Society that could be ill-spared. I notice an ominous passage in which even Galton, the founder of Eugenics, feeling perhaps some twinge of his Quaker ancestry, remarks that "as the Bohemianism in the nature of our race is destined to perish, the sooner it goes, the happier for mankind".[1] It is not the eugenists who will give us what Plato has called divine releases from the common ways. If some fancier with the catholicity of Shakespeare would take us in hand, well and good; but I would not trust even Shakespeares, meeting as a committee. Let us remember that Beethoven's father was an habitual drunkard and that his mother died of consumption. From the genealogy of the patriarchs also we learn—what may very well be the truth—that the fathers of such as dwell in tents, and of all such as handle the harp or organ, and the instructor of every artificer in brass and iron—the founders, that is to say, of the arts and the sciences—came in direct descent from Cain, and not in the posterity of the irreproachable Seth, who is to us, as he probably was also in the narrow circle of his own contemporaries, what naturalists call a *nomen nudum*.

Genetic research will make it possible for a nation to elect by what sort of beings it will be represented not very many generations hence, much as a farmer can decide whether his byres shall be full of Shorthorns or Herefords. It will be very surprising indeed if some nation does not make trial of this new power. They may make awful mistakes, but I think they will try.

Whether we like it or not, extraordinary and far-reaching changes in public opinion are coming to pass. Man is just beginning to know himself for what he is—a rather long-lived animal, with great powers of enjoyment if he does not deliberately forgo them. Hitherto superstition and mythical ideas of sin have predominantly controlled these powers. Mysticism will not die out: for those strange fancies knowledge is no cure; but their forms may change, and mysticism as a force for the suppression of joy is happily losing its hold on the

[1] *Hereditary Genius*, 1869, p. 347.

modern world. As in the decay of earlier religions Ushabti dolls were substituted for human victims, so telepathy, necromancy, and other harmless toys take the place of eschatology and the inculcation of a ferocious moral code. Among the civilised races of Europe we are witnessing an emancipation from traditional control in thought, in art, and in conduct which is likely to have prolonged and wonderful influences. Returning to freer or, if you will, simpler conceptions of life and death, the coming generations are determined to get more out of this world than their forefathers did. Is it then to be supposed that when science puts into their hand means for the alleviation of suffering immeasurable, and for making this world a happier place, they will demur to using those powers? The intenser struggle between communities is only now beginning, and with the approaching exhaustion of that capital of energy stored in the earth before man began it must soon become still more fierce. In England some of our great-grandchildren will see the end of the easily accessible coal, and, failing some miraculous discovery of available energy, a wholesale reduction in population. There are races who have shewn themselves able at a word to throw off all tradition and take into their service every power that science has yet offered them. Can we expect that they, when they see how to rid themselves of the ever-increasing weight of a defective population, will hesitate? The time cannot be far distant when both individuals and communities will begin to think in terms of biological fact, and it behoves those who lead scientific thought carefully to consider whither action should tend. At present I ask you merely to observe the facts. The powers of science to preserve the defective are now enormous. Every year these powers increase. This course of action must reach a limit. To the deliberate intervention of civilisation for the preservation of inferior strains there must sooner or later come an end, and before long nations will realise the responsibility they have assumed in multiplying these "cankers of a calm world and a long peace".

The definitely feeble-minded we may with propriety restrain, as we are beginning to do even in England, and we may safely prevent unions in which both parties are defective, for the evidence shews that as a rule such marriages, though often prolific, commonly produce no normal children at all. The union of such social vermin we should no more permit than we would allow parasites to breed on our own bodies. Further than that in restraint of marriage we ought not to go, at least not yet. Something too may be done by a reform of medical ethics. Medical students are taught that it is their duty to prolong life at whatever cost in suffering. This may have been right when diagnosis was uncertain and interference usually of small effect; but deliberately to interfere now for the preservation of an infant so gravely diseased that it can never be happy or come to any good is very like wanton cruelty. In private few men defend such interference. Most who have seen these cases lingering on agree that the system is deplorable, but ask where can any line be drawn. The biologist would reply that in all ages such decisions have been made by civilised communities with fair success both in regard to crime and in the closely analogous case of lunacy. The real reason why these things are done is because the world collectively cherishes occult views of the nature of life, because the facts are realised by few, and because between the legal mind—to which society has become accustomed to defer—and the seeing eye, there is such physiological antithesis that hardly can they be combined in the same body. So soon as scientific knowledge becomes common property, views more reasonable and, I may add, more humane, are likely to prevail.

To all these great biological problems that modern society must sooner or later face there are many aspects besides the obvious ones. Infant mortality we are asked to lament without the slightest thought of what the world would be like if the majority of these infants were to survive. The decline in the birth-rate in countries already over-populated is often deplored, and we are told that a nation in which population is not rapidly increasing must be in a decline. The slightest acquaintance with biology, or even schoolboy natural history, shews that this inference may be entirely wrong, and that before such a question can be decided in one way or the other, hosts of considerations must be taken into account. In normal stable conditions population is stationary. The laity never appreciates, what is so clear to a biologist, that the last century and a quarter, corresponding with the great rise in population, has been an altogether exceptional period. To our species this period has been what its early years in Australia were to the rabbit. The exploitation of energy-capital of the earth in coal, development of the new countries, and the consequent pouring of food into Europe, the application of antiseptics, these are the things that have enabled the human population to increase. I do not doubt that if population were more evenly spread over the earth it might increase very much more; but the essential fact is that under any stable conditions a limit must be reached. A pair of wrens will bring off a dozen young every year, but each year you will find the same number of pairs in your garden. In England the limit beyond which under present conditions of distribution increase of population is a source of suffering rather than of happiness has been reached already. Younger communities living in territories largely vacant are very probably right in desiring and encouraging more population. Increase may, for some temporary reason, be essential to their prosperity. But those who live, as I do, among thousands of creatures in a state of semi-starvation will realise that too few is better than too many, and will acknowledge the wisdom of Ecclesiasticus who said "Desire not a multitude of unprofitable children".

But at least it is often urged that the decline in the birth-rate of the intelligent and successful sections of the population—I am speaking of the older communities—is to be regretted. Even this cannot be granted without qualification. As the biologist knows, differentiation is indispensable to progress. If population were homogeneous civilisation would stop. In every army the officers must be comparatively few. Consequently, if the upper strata of the community produce more children than will recruit their numbers some must fall into the lower strata and increase the pressure there. Statisticians tell us that an average of four children under present conditions is sufficient to keep the number constant, and as the expectation of life is steadily improving we may perhaps contemplate some diminution of that number without alarm.

In the study of history biological treatment is only beginning to be applied. For us the causes of the success and failure of races are physiological events, and the progress of man has depended upon a chain of these events, like those which have resulted in the "improvement" of the domesticated animals and plants. It is obvious, for example, that had the cereals never been domesticated cities could scarcely have existed. But we may go further, and say that in temperate countries of the Old World (having neither rice nor maize) populations concentrated in large cities have been made possible by the appearance of a "thrashable" wheat. The ears of the wild wheats break easily to pieces, and the grain remains in the thick husk. Such wheat can be used for food, but not readily. Ages before written history began, in some unknown place, plants, or more likely a plant, of wheat lost the dominant factor to which this brittleness is due, and the recessive, thrashable wheat resulted. Some man noticed this wonderful novelty, and it has been disseminated over the earth. The original variation may well have occurred once only, in a single germ cell.

So must it have been with Man. Translated into terms of factors, how has that progress in control of nature which we call civilisation been achieved? By the sporadic appearance of variations, mostly, perhaps all, consisting in a loss of elements, which inhibit the free working of the mind. The members of civilised communities, when they think about such things at all, imagine the process a gradual one, and that they themselves are active agents in it. Few, however, contribute anything but their labour; and except in so far as they have freedom to adopt and imitate, their physiological composition is that of an earlier order of beings. Annul the work of a few hundreds—I might almost say scores—of men, and on what plane of civilisation should we be? We should not have advanced beyond the mediaeval stage without printing, chemistry, steam, electricity, or surgery worthy the name. These things are the contributions of a few excessively rare minds. Galton reckoned those to whom the term "illustrious" might be applied as one in a million, but in that number he is, of course, reckoning men famous in ways which add nothing to universal progress. To improve by subordinate invention, to discover details missed, even to apply knowledge never before applied, all these things need genius in some degree, and are far beyond the powers of the average man of our race; but the true pioneer, the man whose penetration creates a new world, as did that of Newton and of Pasteur, is inconceivably rare. But for a few thousands of such men, we should perhaps be in the Palaeolithic era, knowing neither metals, writings, arithmetic, weaving, nor pottery.

In the history of Art the same is true, but with this remarkable difference, that not only are gifts of artistic creation very rare, but even the faculty of artistic enjoyment, not to speak of higher powers of appreciation, is not attained without variation from the common type. I am speaking, of course, of the non-Semitic races of modern Europe, among whom the power whether of making or enjoying works of art is confined to an insignificant number of individuals. Appreciation can in some degree be stimulated, but in our population there is no widespread physiological appetite for such things. When detached from the centres where they are made by others most of us pass our time in great contentment, making nothing that is beautiful, and quite unconscious of any deprivation. Musical taste is the most notable exception, for in certain races—for example, the Welsh and some of the Germans—it is almost universal. Otherwise artistic faculty is still sporadic in its occurrence. The case of music well illustrates the application of genetic analysis to human faculty. No one disputes that musical ability is congenital. In its fuller manifestation it demands sense of rhythm, ear, and special nervous and muscular powers. Each of these is separable and doubtless genetically distinct. Each is the consequence of a special departure from the common type. Teaching and external influences are powerless to evoke these faculties, though their development may be assisted. The only conceivable way in which the people of England, for example, could become a musical nation would be by the gradual rise in the proportional numbers of a musical strain or strains until the present type became so rare as to be negligible. It by no means follows that in any other respect the resulting population would be distinguishable from the present one. Difficulties of this kind beset the efforts of anthropologists to trace racial origins. It must continually be remembered that most characters are independently transmitted and capable of such recombination. In the light of Mendelian knowledge the discussion whether a race is pure or mixed loses almost all significance. A race is pure if it breeds pure and not otherwise. Historically we may know that a race like our own was, as a matter of fact, of mixed origin. But a character may have been introduced by a single individual, though subsequently it becomes common to the race. This is merely a variant on the familiar paradox that in the course of time if registration is accurate we shall all have the same surname. In the case of music, for instance, the gift, originally perhaps from a Welsh source, might permeate the nation, and the question would then arise whether the nation, so changed, was the English nation or not.

Such a problem is raised in a striking form by the population of modern Greece, and especially of Athens. The racial characteristics of the Athenian of the fifth century B.C. are vividly described by Galton in *Hereditary Genius*. The fact that in that period a population, numbering many thousands, should have existed, capable of following the great plays at a first hearing, revelling in subtleties of speech, and thrilling with passionate delight in beautiful things, is physiologically a most singular phenomenon. On the basis of the number of illustrious men produced by that age Galton estimated the average intelligence as at least two of his degrees above our own, differing from us as much as we do from the negro. A few generations later the display was over. The origin of that constellation of human genius which then blazed out is as yet beyond all biological analysis, but I think we are not altogether without suspicion of the sequence of the biological events. If I visit a poultry breeder who has a fine stock of thoroughbred game fowls breeding true, and ten years later—that is to say ten fowl-generations later—I go again and find scarcely a recognisable game fowl on the place, I know exactly what has happened. One or two birds of some other or of no breed must have strayed in and their progeny been left undestroyed. Now in Athens we have many indications that up to the beginning of the fifth century so long as the phratries and gentes were maintained in their integrity there was rather close endogamy, a condition giving the best chance of producing a homogeneous population. There was no lack of material from which intelligence and artistic power might be derived. Sporadically these qualities existed throughout the ancient Greek world from the dawn of history, and, for example, the vase painters, the makers of the Tanagra figurines, and the gem cutters were presumably pursuing family crafts, much as are the actor families[1] of England or the professorial families of Germany at the present day. How the intellectual strains should have acquired predominance we cannot tell, but in an in-breeding community homogeneity at least is not surprising. At the end of the sixth century came the "reforms" of Cleisthenes (507 B.C.), which sanctioned foreign marriages and admitted to citizenship a number not only of resident aliens but also of manumitted slaves. As Aristotle says, Cleisthenes legislated with the deliberate purpose of breaking up the phratries and gentes, in order that the various sections of the population might be mixed up as much as possible, and the old tribal associations abolished. The "reform" was probably a recognition and extension of a process already begun; but is it too much to suppose that we have here the effective beginning of a series of genetic changes which in a few generations so greatly altered the character of the people? Under Pericles the old law was restored (451 B.C.), but losses in the great wars led to further laxity in practice, and though at the end of the fifth century the strict rule was re-enacted that a citizen must be of citizen birth on both sides, the population by that time may well have become largely mongrelised.

[1] For tables of these families, see the Supplement to *Who's Who in the Theatre*.

Let me not be construed as arguing that mixture of races is an evil: far from it. A population like our own, indeed, owes much of its strength to the extreme diversity of its components, for they contribute a corresponding abundance of aptitudes. Everything turns on the nature of the ingredients brought in, and I am concerned solely with the observation that these genetic disturbances lead ultimately to great and usually unforeseen changes in the nature of the population.

Some experiments of this kind are going on at the present time, in the United States, for example, on a very large scale. Our grandchildren may live to see the characteristics of the American population entirely altered by the vast invasion of Italian and other South European elements. We may expect that the Eastern States, and especially New England, whose people still exhibit the fine Puritan qualities with their appropriate limitations, absorbing little of the alien elements, will before long be in feelings and aptitudes very notably differentiated from the rest. In Japan, also, with the abolition of the feudal system and the rise of commercialism, a change in population has begun which may be worthy of the attention of naturalists in that country. Till the revolution the Samurai almost always married within their own class, with the result, as I am informed, that the caste had fairly recognisable features. The changes of 1868 and the consequent impoverishment of the Samurai have brought about a beginning of disintegration which may not improbably have perceptible effects.

How many genetic vicissitudes has our own peerage undergone! Into the hard-fighting stock of mediaeval and Plantagenet times have successively been crossed the cunning shrewdness of Tudor statesmen and courtiers, the numerous contributions of Charles II and his concubines, reinforcing peculiar and persistent attributes which popular imagination especially regards as the characteristic of peers, ultimately the heroes of finance and industrialism. Definitely intellectual elements have been sporadically added, with rare exceptions, however, from the ranks of lawyers and politicians. To this aristocracy art, learning, and science have contributed sparse ingredients, but these mostly chosen for celibacy or childlessness. A remarkable body of men, nevertheless; with an average "horse-power", as Samuel Butler would have said, far exceeding that of any random sample of the middle class. If only man could be reproduced by budding what a simplification it would be! In vegetative reproduction heredity is usually complete. The Washington plum can be divided to produce as many identical individuals as are required. If, say, Washington, the statesman, or preferably King Solomon, could similarly have been propagated, all the nations of the earth could have been supplied with ideal rulers.

Historians commonly ascribe such changes as occurred in Athens, and will almost certainly come to pass in the United States, to conditions of life and especially to political institutions. These agencies, however, do little unless they are such as to change the breed. External changes may indeed give an opportunity to special strains, which then acquire ascendency. The industrial developments which began at the end of the eighteenth century, for instance, gave a chance to strains till then submerged, and their success involved the decay of most of the old aristocratic families. But the demagogue who would argue from the rise of the one and the fall of the other that the original relative positions were not justifiable altogether mistakes the facts.

Conditions give opportunities but cause no variations. For example, in Athens, to which I just referred, the universality of cultivated discernment could never have come to pass but for the institution of slavery which provided the opportunity, but slavery was in no sense a cause of that development, for many other populations have lived on slaves and remained altogether inconspicuous.

The long-standing controversy as to the relative importance of nature and nurture, to use Galton's "convenient jingle of words", is drawing to an end, and of the overwhelmingly greater significance of nature there is no longer any possibility of doubt. It may be well briefly to recapitulate the arguments on which naturalists rely in coming to this decision both as regards races and individuals. First as regards human individuals, there is the common experience that children of the same parents reared under conditions sensibly identical may develop quite differently, exhibiting in character and aptitudes a segregation just as great as in their colours or hair-forms. Conversely all the more marked aptitudes have at various times appeared and not rarely reached perfection in circumstances the least favourable for their development. Next, appeal can be made to the universal experience of the breeder, whether of animals or plants, that strain is absolutely essential, that though bad conditions may easily enough spoil a good strain, yet that under the best conditions a bad strain will never give a fine result. It is faith, not evidence, which encourages educationists and economists to hope so greatly in the ameliorating effects of the conditions of life. Let us consider what they can do and what they cannot. By reference to some sentences in a charming though pathetic book, *What Is, and What Might Be*, by Mr Edmond Holmes, which will be well known in the Educational Section, I may make the point of view of us naturalists clear. I take Mr Holmes's pronouncement partly because he is an enthusiastic believer in the efficacy of nurture as opposed to nature, and also because he illustrates his views by frequent appeals to biological analogies which help us to a common ground. Wheat badly cultivated will give a bad yield, though, as Mr Holmes truly says, wheat of the same strain in similar soil well cultivated may give a good harvest. But, having witnessed the success of a great natural teacher in helping unpromising peasant children to develop their natural powers, he gives us another botanical parallel. Assuming that the wild bullace is the origin of domesticated plums, he tells us that by cultivation the bullace can no doubt be improved so far as to become a better bullace, but by no means can the bullace be made to bear plums. All this is sound biology; but translating these facts into the human analogy, he declares that the work of the successful teacher shews that with man the facts are otherwise, and that the *average* rustic child, whose normal ideal is "bullacehood", can become the rare exception, developing to a stage corresponding with that of the plum. But the naturalist knows exactly where the parallel is at fault. For the wheat and the bullace are both breeding approximately true, whereas the human crop, like jute and various cottons, is in a state of polymorphic mixture. The population of many English villages may be compared with the crop which would result from sowing a bushel of kernels gathered mostly from the hedges, with an occasional few from an orchard. If anyone asks how it happens that there are any plum kernels in the sample at all, he may find the answer perhaps in spontaneous variation, but more probably in the appearance of a long-hidden recessive. For the want of that genetic variation, consisting probably, as I have argued, in loss of inhibiting factors, by which the plum arose from the wild form, neither food, nor education, nor hygiene can in any way atone. Many wild plants are half-starved through competition, and transferred to garden soil they grow much bigger; so good conditions might certainly enable the bullace population to develop beyond the stunted physical and mental stature they commonly attain, but plums they can never be. Modern statesmanship aims rightly at helping those who have got sown as wildings to come into their proper class; but let not any one suppose such a policy democratic in its ultimate effects, for no course of action can be more effective in strengthening the upper classes whilst weakening the lower.

In all practical schemes for social reform the congenital diversity, the essential polymorphism of all civilised communities must be recognised as a fundamental fact, and reformers should rather direct their efforts to facilitating and rectifying class distinctions than to any futile attempt to abolish them. The teaching of biology is perfectly clear. We are what we are by virtue of our differentiation. The value of civilisation has in all ages been doubted. Since, however, the first variations were not strangled in their birth, we are launched on that course of variability of which civilisation is the consequence. We cannot go back to homogeneity again, and differentiated we are likely to continue. For a period measures designed to create a spurious homogeneity may be applied. Such attempts will, I anticipate, be made when the present unstable social state reaches a climax of instability, which may not be long hence. Their effects can be but evanescent. The instability is due not to inequality, which is inherent and congenital, but rather to the fact that in periods of rapid change like the present, convection currents are set up such that the elements of the strata get intermixed and the apparent stratification corresponds only roughly with the genetic. In a few generations under uniform conditions these elements settle in their true levels once more.

In such equilibrium is content most surely to be expected. To the naturalist the broad lines of solution of the problems of social discontent are evident. They lie neither in vain dreams of a mystical and disintegrating equality, nor in the promotion of that malignant individualism which in older civilisations has threatened mortification of the humbler organs, but rather in a physiological co-ordination of the constituent parts of the social organism. The rewards of commerce are grossly out of proportion to those attainable by intellect or industry. Even regarded as compensation for a dull life, they far exceed the value of the services rendered to the community. Such disparity is an incident of the abnormally rapid growth of population and is quite indefensible as a permanent social condition. Nevertheless capital, distinguished as a provision for offspring, is an eugenic institution; and unless human instinct undergoes some profound and improbable variation, abolition of capital means the abolition

of effort; but as in the body the power of independent growth
of the parts is limited and subordinated to the whole, similarly
in the community we may limit the powers of capital, preserving
so much inequality of privilege as corresponds with physiological fact.

At every turn the student of political science is confronted
with problems that demand biological knowledge for their
solution. Most obviously is this true in regard to education,
the criminal law, and all those numerous branches of policy
and administration which are directly concerned with the
physiological capacities of mankind. Assumptions as to what
can be done and what cannot be done to modify individuals
and races have continually to be made, and the basis of fact on
which such decisions are founded can be drawn only from
biological study.

A knowledge of the facts of nature is not yet deemed an
essential part of the mental equipment of politicians; but as
the priest, who began in other ages as medicine man, has
been obliged to abandon the medical parts of his practice, so
will the future behold the schoolmaster, the magistrate, the
lawyer, and ultimately the statesman, compelled to share with
the naturalist those functions which are concerned with the
physiology of race.

THE METHODS AND SCOPE OF GENETICS

Inaugural Lecture delivered 23 October 1908
Cambridge

PREFATORY NOTE. The Professorship of Biology was founded in 1908 for a period of five years partly by the generosity of an anonymous benefactor, and partly by the University of Cambridge. The object of the endowment was the promotion of inquiries into the physiology of Heredity and Variation, a study now spoken of as Genetics.

It is now recognised that the progress of such inquiries will chiefly be accomplished by the application of experimental methods, especially those which Mendel's discovery has suggested. The purpose of this inaugural lecture is to describe the outlook over this field of research in a manner intelligible to students of other parts of knowledge.

The opportunity of addressing fellow-students pursuing lines of inquiry other than his own falls seldom to a scientific man. One of these rare opportunities is offered by the constitution of the Professorship to which I have had the honour to be called. That Professorship, though bearing the comprehensive title "of Biology", is founded with the understanding that the holder shall apply himself to a particular class of physiological problems, the study of which is denoted by the term Genetics. The term is new; and though the problems are among the oldest which have vexed the human mind, the modes by which they may be successfully attacked are also of modern invention. There is therefore a certain fitness in the employment of this occasion for the deliverance of a discourse explaining something of the aims of Genetics and of the methods by which we trust they may be reached.

You will be aware that the claims put forward in the name of Genetics are high, but I trust to be able to shew you that they are not high without reason. It is the ambition of every one who in youth devotes himself to the search for natural truth, that his work may be found somewhere in the main stream of progress. So long only as he keeps something of the limitless hope with which his voyage of discovery began, will his courage and his spirit last. The moment we most dread is one in which it may appear that, after all, our effort has been spent in exploring some petty tributary, or worse, a backwater of the great current. It is because genetic research is still pushing forward in the central undifferentiated trunk of biological science that we confess no guilt of presumption in declaring boldly that whatever difficulty may be in store for those who cast in their lot with us, they need fear no disillusionment or misgiving that their labour has been wasted on a paltry quest.

In research, as in all business of exploration, the stirring times come when a fresh region is suddenly unlocked by the discovery of a new key. Then conquest is easy and there are prizes for all. We are happy in that during our own time not a few such territories have been revealed to the vision of mankind. I do not dare to suggest that in magnitude or splendour the field of Genetics may be compared with that now being disclosed to the physicist or the astronomer; for the glory of the celestial is one and the glory of the terrestrial is another. But I will say that for once to the man of ordinary power who cannot venture into those heights beyond, Mendel's clue has shewn the way into a realm of nature which for surprising novelty and adventure is hardly to be excelled.

It is no hyperbolical figure that I use when I speak of Mendelian discovery leading us into a new world, the very existence of which was unsuspected before.

The road thither is simple and easy to follow. We start from a common fact, familiar to everyone, that all the ordinary animals and plants began their individual life by the union of two cells, the one male, the other female. Those cells are known as germ cells or *gametes*, that is to say, "marrying" cells.

Now obviously the diversity of form which is characteristic of the animal and plant world must be somehow represented in the gametes, since it is they which bring into each organism all that it contains. I am aware that there is interplay between the organism and the circumstances in which it grows up, and that opportunity given may bring out a potentiality which without that opportunity must have lain dormant. But while noting parenthetically that this question of opportunity has an importance, which some day it may be convenient to estimate, the one certain fact is that all the powers, physical and mental, that a living creature possesses were contributed by one or by both of the two germ cells which united in fertilisation to give it existence. The fact that *two* cells are concerned in the production of all the ordinary forms of life was discovered a long while ago, and has been part of the common stock of elementary knowledge of all educated persons for about half a century. The full consequences of this double nature seem nevertheless to have struck nobody before Mendel. Simple though the fact is, I have noticed that to many it is difficult to assimilate as a working idea. We are accustomed to think of a man, a butterfly, or an apple tree as each *one* thing. In order to understand the significance of Mendelism we must get thoroughly familiar with the fact that they are each *two* things, double throughout every part of their composition. There is perhaps no better exercise as a preparation for genetic research than to examine the people one meets in daily life and to try in a rough way to analyse them into the two assemblages of characters which are united in them. That we are assemblages or medleys of our parental characteristics is obvious. We all know that a man may have his father's hair, his mother's colour, his father's voice, his mother's insensibility to music, and so on, but that is not enough.

Such an analysis is true, inasmuch as the various characters *are* transmitted independently, but it misses the essential point. For in each of these respects the individual is double; and so to get a true picture of the composition of the individual we have to think how *each* of the two original gametes was provided in the matter of height, hair, colour, mathematical ability, nail-shape, and the other features that go to make the man we know. The contribution of each gamete in each respect has thus to be separately brought to account. If we could make a list of all the ingredients that go to form a man and could set out how he is constituted in respect of each of them, it would not suffice to give one column of values for these ingredients, but we must rule two columns, one for the ovum and one for the spermatozoon, which united in fertilisation to form that man, and in each column we must represent how that gamete was supplied in respect of each of the ingredients in our list. When the problem of heredity is thus represented we can hardly avoid discovering, by mere inspection, one of the chief conclusions to which genetic research has led. For it is obvious that the contributions of the male and female gametes may in respect of any of the ingredients be either the same, or different. In any case in which the contribution made by the two cells is the same, the resulting organism—in our example the man—is, as we call it, *pure-bred* for that ingredient, and in all respects in which the contribution from the two sides of the parentage is dissimilar the resulting organism is *cross-bred*.

To give an intelligible account of the next step in the analysis without having recourse to precise and technical language is not very easy.

We have got to the point of view from which we see the individual made up of a large number of distinct ingredients, contributed from two sources, and in respect of any of them he may have received two similar portions or two dissimilar portions. We shall not go far wrong if we extend and elaborate our illustration thus. Let us imagine the contents of a gamete as a fluid made by taking a drop from each of a definite number of bottles in a chest, containing tinctures of the several ingredients. There is one such chest from which the male gamete is to be made up, and a similar chest containing a corresponding set of bottles out of which the components of the female gamete are to be taken. But in either chest one or more of the bottles may be empty; then nothing goes in to represent that ingredient from that chest, and if corresponding bottles are empty in both chests, then the individual made on fertilisation by mixing the two collections of drops together does not contain the missing ingredient at all. It follows therefore that an individual may thus be "pure-bred", namely alike on both sides of his composition as regards each ingredient in one of two ways, either by having received the ingredient from the male chest and from the female, or in having received it from neither. Conversely in respect of any ingredient he may be "cross-bred", receiving the presence of it from one gamete and the absence of it from the other.

The second conception with which we have now to become thoroughly familiar is that of the individual as composed of what we call presences and absences of all the possible ingredients. It is the basis of all progress in genetic analysis. Let me give you two illustrations. A blue eye is due to the absence of a factor which forms pigment on the front of the iris. Two blue-eyed parents therefore, as Hurst has proved, do not have dark-eyed children. The dark eye is due to either a single or double dose of the factor missing from the blue eye. So dark-eyed persons may have families all dark-eyed, or families composed of a mixture of dark and light-eyed children in certain proportions which on the average are definite.

Two plants of *Oenothera* which I exhibit illustrate the same thing. One of them is the ordinary *Lamarckiana*. I bend its stem. It will not break, or only breaks with difficulty on account of the tough fibres it contains. The stem of the other, one of de Vries' famous mutations, snaps at once like short pastry, because it does not contain the factor for the formation of the fibres. Such plants may be sister plants produced by the self-fertilisation of one parent, but they are distinct in their composition and properties—and this distinction turns on the presence or absence of elements which are treated as definite entities when the germ cells are formed. When we speak of such qualities as the formation of pigment in an eye, or the development of fibres in a stem, as due to transmitted elements or factors, you will perhaps ask if we have formed any notion as to the actual nature of those factors. For my own part as regards that ulterior question I confess to a disposition to hold my fancy on a tight rein. It cannot be very long before we shall *know* what some of the factors are, and we may leave guessing till then. Meanwhile however there is no harm in admitting that several of them behave much as if they were ferments, and others as if they constructed the substances on which the ferments act. But we must not suppose for a moment that it is the ferment, or the objective substance, which is transmitted. The thing transmitted can only be the power or faculty to produce the ferment or the objective substance.

So far we have been considering the synthesis of the individual from ingredients brought into him by the two gametes. In the next step of our consideration we reverse the process, and examine how the ingredients of which he was originally compounded are distributed among the gametes that are eventually budded off from him.

Take first the case of the components in respect of which he is pure-bred. Expectation would naturally suggest that all the germ cells formed from him would be alike in respect of those ingredients, and observation shews, except in the rare cases of originating variations, the causation of which is still obscure, that this expectation is correct.

Hitherto though without experimental evidence no one could have been certain that the facts were as I have described them, yet there is nothing altogether contrary to common expectation. But when we proceed to ask how the germ cells will be constituted in the case of an individual who is cross-bred in some respect, containing that is to say, an ingredient from the one side of his parentage and not from the other, the answer is entirely contrary to all the preconceptions which either science or common-sense had formed about heredity. For we find definite experimental proof in nearly all the cases which have been examined, that the germ cells formed by such individuals do either contain or not contain a representation of the ingredient, just as the original gametes did or did not contain it.

If *both* parent gametes brought a certain quality in, then all the daughter gametes have it; if neither brought it in, then none of the daughter gametes have it. If it came in from one side and not from the other, then on an average in half the resulting gametes it will be present and from half it will be absent. This last phenomenon, which is called segregation, constitutes the essence of Mendel's discovery.

So recurring to the simile of the man as made by the mixing of tinctures, the process of redistribution of his characters among the germ cells may be represented as a sorting back of the tinctures again into a double row of bottles, a pair corresponding to each ingredient; and each of the germ cells as then made of a drop from one or other bottle of each pair: and in our model we may represent the phenomenon of segregation in a crude way by supposing that the bottles having no tincture in them, instead of being empty contained an inoperative fluid, say water, with which the tincture would not mix. When the new germ cells are formed, the two fluids instead of diluting each other simply separate again. It is this fact which entitles us to speak of the purity of germ cells. They are pure in the possession of an ingredient, or in not possessing it; and the ingredients, or factors, as we generally call them, are units because they are so treated in the process of formation of the new gametes, and because they come out of the process of segregation in the same condition as they went in at fertilisation.

As a consequence of these facts it follows that however complex may be the origin of two given parents the composition of the offspring they can produce is limited. There is only a limited number of types to be made by the possible recombinations of the parental ingredients, and the relative numbers in which each type will be represented are often predicable by very simple arithmetical rules.

For example, if neither parent possesses a certain factor at all, then none of the offspring will have it. If either parent has two doses of the factor then all the children will have it; and if either parent has one dose of the factor and the other has none, then on an average half the family will have it, and half be without it.

To know whether the parent possesses the factor or not may be difficult for reasons which will presently appear, but often it is quite easy and can be told at once, for there are many factors which cannot be present in the individual without manifesting their presence. I may illustrate the descent of such a factor by the case of a family possessing a peculiar form of night-blindness. The affected individuals marrying with those unaffected have a mixture of affected and unaffected children, but their unaffected children not having the responsible ingredient cannot pass it on.[1]

In such an observation two things are strikingly exemplified, (1) the fact of the permanence of the unit, and (2) the fact that a *mixture* of types in the family means that one or other parent is cross-bred in some respect, and is giving off gametes of more than one type.

The problem of heredity is thus a problem primarily analytical. We have to detect and enumerate the factors out of which the bodies of animals and plants are built up, and the laws of their distribution among the germ cells. All the processes of which I have spoken are accomplished by means of cell divisions, and in the one cell union which occurs in fertilisation. If we could watch the factors segregating from each other in cell division, or even if by microscopic examination we could recognise this multitudinous diversity of composition that must certainly exist among the germ cells of all ordinary individuals, the work of genetics would be much simpler than it is.

But so far no such direct method of observation has been discovered. In default we are obliged to examine the constitution of the germ cells by experimental breeding, so contrived that each mating shall test the composition of an individual in one or more chosen respects, and, so to speak, sample its germ cells by counting the number of each kind of offspring which it can produce. But cumbersome as this method must necessarily be, it enables us to put questions to Nature which never have been put before. She, it has been said, is an unwilling witness. Our questions must be shaped in such a way that the only possible answer is a direct "Yes" or a direct "No". By putting such questions we have received some astonishing answers which go far below the surface. Amazing though they be, they are nevertheless true; for though our witness may prevaricate, she cannot lie. Piecing these answers together, getting one hint from this experiment, and another from that, we begin little by little to reconstruct what is going on in that hidden world of gametes. As

---

[1] The investigation of this remarkable family was made originally by Cunier. The facts have been re-examined and the pedigree much extended by Nettleship. The numerical results are somewhat irregular, but it is especially interesting as being the largest pedigree of human disease or defect yet made. It contains 2121 persons, extending over ten generations. Of these persons, 135 are known to have been night-blind. In no single case was the peculiarity transmitted through an unaffected member. It should be mentioned that for night-blindness such a system of descent is peculiar. More usually it follows the scheme described for colour-blindness. It is not known wherein the peculiarity of this family consists.

we proceed, like our brethren in other sciences, we sometimes receive answers which seem inconsistent or even contradictory. But by degrees a sufficient body of evidence can be attained to shew what is the rule and what the exception. My purpose to-day must be to speak rather of the regular than of the irregular.

One clear exception I may mention. Castle finds that in a cross between the long-eared lop rabbit and a short-eared breed, ears of intermediate length are produced: and that these intermediates breed approximately true.

Exceptions in general must be discussed elsewhere. Nevertheless if I may throw out a word of counsel to beginners, it is: Treasure your exceptions! When there are none, the work gets so dull that no one cares to carry it further. Keep them always uncovered and in sight. Exceptions are like the rough brickwork of a growing building which tells that there is more to come and shews where the next construction is to be.

You will readily understand that the presentation here given of the phenomena is only the barest possible outline. Some of the details we may now fill in. For example, I have spoken of the characters of the organism, its colour, shape, and the like, as if they were due each to one ingredient or factor. Some of them are no doubt correctly so represented; but already we know numerous bodily features which need the concurrence of several factors to produce them. Nevertheless though the character only appears when all the complementary ingredients are together present, each of these severally and independently follows, as regards its transmission, the simple rules I have described.

This complementary action may be illustrated by some curious results that Mr Punnett and I have encountered when experimenting with the height of sweet peas. There are two dwarf varieties, one the prostrate *Cupid*, the other the half-dwarf or *Bush* sweet peas. Crossed together they give a cross-bred of full height. There is thus some element in the *Cupid* which when it meets the complementary element from the *Bush*, produces the characteristic length of the ordinary sweet pea. We may note in passing that such a fact demonstrates at once the nature of variation and reversion. The reversion occurs because the two factors that made the *height* of the old sweet pea again come together after being parted: and the variations by which each of the dwarfs came into existence must have taken place by the dropping out of one of these elements or of the other.

Conversely there are factors which by their presence can prevent or inhibit the development and appearance of others present and unperceived.

For example, all the factors for pigmentation may be present in a plant or an animal; but in addition there may be another factor present which keeps the individual white, or nearly so.

There are cases in which the action of the factors is superposed one on top of the other, and not until each factor is removed in turn can the effects of the underlying factors be perceived. So in the mouse if no other colour factor is present, the fur is chocolate. If the next factor in the series be there, it is black. If still another factor be added, it has the brownish grey of the common wild mouse. Conversely, by the variation which dropped out the top factor, a black mouse came into existence. By the loss of the black factor, the chocolate mouse was created, and for aught we can tell there may be still more possibilities hidden beneath.

In the disentanglement of the properties and interactions of these elementary factors, the science we must call to our aid is Physiological Chemistry. The relations of Genetics with the other branches of biology are close. Such work can only be conducted by those who have the good fortune to be able to count upon continual help and advice from specialists in the various branches of Zoology, Physiology, and Botany. Often we have questions with which only a cytologist can deal, and often it is the experience of a systematist we must invoke. The school of Genetics in Cambridge starts under happy auspices in that we are surrounded by colleagues qualified, and as we have often found, willing to give us such aid unstinted. But with Chemical Physiology we stand in an even closer relation; and from the little I have dared to say respecting the action and interaction of factors, it is evident that for their disentanglement there must one day be an intimate and enduring partnership arranged with the physiological chemists.

Now, as the whole of the elaborate process by which the various elements are apportioned among the gametes must be got through in a few cell divisions at most, and perhaps in one division only, it is not surprising that there is sometimes an interaction between factors that have quite distinct rôles to perform. These interactions are probably of several kinds. One, which I shall illustrate presently, is probably to be represented as a repulsion between two factors. As a consequence of its operations when the various factors are sorted out into the gametes, if the individual be cross-bred in respect of the *two* repelling factors, having received so to speak only a single dose of each, then the gametes are made up in such a way that each takes one or other of the two repelling factors, not both.

Mutual repulsions of this kind probably play a significant part in the phenomena of heredity. A single concrete case which Mr Punnett and I have been investigating for some years will illustrate several of these principles. We crossed together a pure white sweet pea having an erect standard, with another pure white sweet pea having a hooded standard. The result is, as you see, a purple flower with an erect standard. The colour comes from the concurrence of complementary elements. A dose of a certain ingredient from one parent meets a dose of another ingredient from the other parent and the two make pigment in the flower. From other experiments we know that the *purple* colour of the pigment is due to a dose of a third ingredient brought in from the hooded parent; and that in the absence of that blue factor, as we may call it, the flower would be red. The standard is erect because it contains a dose of the erectness factor from the erect parent, and the hooded parent can readily be proved to owe its peculiar shape to the absence of that element.

Our purple plant is thus cross-bred for four factors, containing only one dose of each.

We let it fertilise itself, and its offspring shew all the possible combinations of the four different factors and their absences which the genetic constitution of the plant can make.

Note that one of the combinations we expect to find is missing. There are white erect and white hooded—white because they are lacking one or other of the complementary ingredients necessary to the production of pigment. There are purple erect and purple hooded, of which the purple erect must perforce contain all the four factors, and the purple hooded must similarly contain all of them except that for erectness. But when we turn to the red class we are surprised to find that they are all erect, none hooded. One of the possible combinations is missing. If you examine this series of facts you will find there is only one possible interpretation: namely that the ingredient which turns the flower purple—alkalinity, perhaps we may call it—never goes into the same germ cell as the ingredient which makes the standard erect. There are plenty of ways of testing the truth of this interpretation. For example, it follows that the purple erects from such a family will in perpetuity have offspring 1 purple hooded : 2 purple erect : 1 red erect; also that all the white hooded crossed with pure reds will give purples, and so on. These experiments have been made and the result has in each case been conformable to expectation.

Between these two factors, the purpleness and the erectness of standard, some antagonism or repulsion must exist. In some way therefore the chemical and the geometrical phenomena of heredity must be inter-related.

Some one will say perhaps this is all very well as a scientific curiosity, but it has nothing to do with real life. The right answer to such criticism is of course the lofty one that science and its applications are distinct: that the investigator fixes his gaze solely on the search for truth and that his attention must not be distracted by trivialities of application. But while we make this answer and at least try to work in the spirit it proclaims, we know in our hearts that it is a counsel of perfection. I suspect that even the astronomer who at his spectroscope is analysing the composition of Vega or Capella has still an eye sometimes free for the affairs of this planet, and at least the fact that his discoveries may throw light on our destinies does not diminish his zeal in their pursuit. And surely to the study of Heredity, pre-eminently among all the sciences, we are looking for light on human destiny. To pretend otherwise would be mere hypocrisy. So while reserving the higher line of defence I will reply that again and again in our experimental work we come very near indeed to human affairs. Sometimes this is obvious enough. No practical dog-breeder or seedsman can see the results of Mendelian recombination without perceiving that here is a bit of knowledge he can immediately apply. No sociologist can examine the pedigrees illustrating the simple descent of a deformity or a congenital disease, and not see that the new knowledge gives a solid basis for practical action by which the composition of a race could be modified if society so chose. More than this: we know for certain in one case, from the work of Professor Biffen, that the power to resist a disease caused by the invasion of a pathogenic organism, wheat-rust, is due to the absence of one of the simple factors or ingredients of which I have spoken, and what we know to be true in that one case we are beginning to suspect to be true of resistance to certain other diseases. No pathologist can see such an experiment as this of Professor Biffen's without realising that here is a contribution of the first importance to the physiology of disease.

There is no lack of utility and direct application in the study of Genetics. I have alluded to some strictly practical results. If we want to raise mangolds that will not run to seed, or to breed a cow that will give more milk in less time, or milk with more butter and less water, we can turn to genetics with every hope that something can be done in these laudable directions. But here I would plead what I cannot but regard as a higher usefulness in our work. Genetic inquiry aims at providing knowledge that may bring, and I think will bring, certainty into a region of human affairs and concepts which might have been supposed reserved for ages to be the domain of the visionary. We have long known that it was believed by some that our powers and conduct were dependent on our physical composition, and that other schools have maintained that nurture not nature, to use Galton's antithesis, had a preponderating influence on our careers; but so soon as it becomes common knowledge—not a philosophical speculation, but a certainty—that liability to a disease, or the power of resisting its attack, addiction to a particular vice, or to superstition, is due to the presence or absence of a specific ingredient; and finally that these characteristics are transmitted to the offspring according to definite, predicable rules, then man's views of his own nature, his conceptions of justice, in short his whole outlook on the world, must be profoundly changed. Yet as regards the more tangible of these physical and mental characteristics there can be little doubt that before many years have passed the laws of their transmission will be expressible in simple formulae.

The blundering cruelty we call criminal justice will stand forth divested of natural sanction, a relic of the ferocious inventions of the savage. Well may such justice be portrayed as blind. Who shall say whether it is crime or punishment which has wrought the greater suffering in the world? We may live to know that to the keen satirical vision of Sam Butler on the pleasant mountains of Erewhon there was revealed a dispensation, not kinder only, but wiser than the terrific code which Moses delivered from the flames of Sinai.

If there are societies which refuse to apply the new knowledge, the fault will not lie with Genetics. I think it needs but little observation of the newer civilisations to foresee that *they* will apply every scrap of scientific knowledge which can help them, or seems to help them in the struggle, and I am good enough selectionist to know that in that day the fate of the recalcitrant communities is sealed.

The thrill of discovery is not dulled by a suspicion that the discovery can be applied. No harm is done to the investigator if he can resist the temptation to deviate from his aim. With rarest exceptions the discoveries which have formed the basis of physical progress have been made without any thought but for the gratification of curiosity. Of this there can be few examples more conspicuous than that which Mendel's work presents. Untroubled by any itch to make potatoes larger or bread cheaper, he set himself in the quiet of a cloister garden to find out the laws of hybridity, and so struck a mine of truth, inexhaustible in brilliancy and profit.

I will now suggest to you that it is by no means unlikely that even in an inquiry so remote as that which I just described in the case of the sweet pea, we may have the clue to a mystery which concerns us all in the closest possible way. I mean the problem of the physiological nature of sex. In speaking of the interpretation of sexual difference suggested by our experimental work as of some practical moment, I do not imply that as in the other instances I have given, the knowledge is likely to be of immediate use to our species; but only that if true it makes a contribution to the stock of human ideas which no one can regard as insignificant.

In the light of Mendelian knowledge, when a family consists of more than one type the fact means that the germ cells of one or other parent must certainly be of more than one kind. In the case of sex the members of the family are thus of two kinds, and the presumption is overwhelming that this distinction is due to a difference among the germ cells. Next, since for all practical purposes the numbers of the two sexes produced are approximately equal, sex exhibits the special case in which a family consists of two types represented in equal numbers, half being male, half female. But I called your attention to the fact that equality of types results when *one* parent was cross-bred in the character concerned, having received one dose only of the factor on which it depends. So we may feel fairly sure that the distinction between the sexes depends on the presence in one or other of them of an unpaired factor. This conclusion appears to me to follow so immediately on all that we have learnt of genetic physiology that with every confidence we may accept it as representing the actual fact.

The question which of the two sexes contains the unpaired factor is less easy to answer, but there are several converging lines of evidence which point to the deduction that in vertebrates at least, and in some other types, it is the female, and I feel little doubt that we shall succeed in proving that in them femaleness is a definite Mendelian factor absent from the male and following the ordinary Mendelian rules.

Before shewing you how the sweet pea phenomenon aids in this inquiry I must tell you of some other experimental results.

The first concerns the common currant moth, *Abraxas grossulariata*. It has a definite pale variety called *lacticolor*. With these two forms Doncaster has made a remarkable series of experiments. When he began, *lacticolor* was only known as a female form. This was crossed with the *grossulariata* male and gave *grossulariata* only, shewing that the male was pure to type. The hybrids bred together gave *grossulariata* males and females and *lacticolor* females only. But the hybrid males bred to *lacticolor* females produced all four combinations, *grossulariata* males and females, and *lacticolor* males and females. When the *lacticolor* males were bred to *grossulariata* females, whether hybrid, or wild from a district where *lacticolor* does not exist, the result was that all the males were *grossulariata* and all the females *lacticolor*! It is difficult to follow the course of such an experiment on once hearing and all I ask you to remember is first that there is a series of matings giving very curious distributions of the characters of type and variety among the two sexes. And then, what is perhaps the most singular fact of all, that the wild typical *grossulariata* female can when crossed with the *lacticolor* male produce all females *lacticolor*. This last fact can, we know, mean only one thing, namely that these wild females are in reality hybrids of *lacticolor*; though since the males are pure *grossulariata*, that fact would in the natural course of things never be revealed.

When we encounter such a series of phenomena as this, our business is to find a means of symbolical expression which will represent all the factors involved, and shew how each behaves in descent. Such a system or scheme we have at length discovered, and I incline to think that it must be the true one. If you study this case you will find that there are nine distinct kinds of matings that can be made between the variety, the type and the hybrid, and the scheme fits the whole group of results. It is based on two suppositions:

1. That the female is cross-bred, or as we call it heterozygous for femaleness factor, the male being without that factor. The eggs are thus each destined from the first to become either males or females, but as regards sex the spermatozoa are alike in being non-female.

2. That there is a repulsion between the femaleness factor and the *grossulariata* factor.

Such a repulsion between two factors we are justified in regarding as possible because we have had proof of the occurrence of a similar repulsion in the case of the two factors in the sweet pea.

If the case of this moth stood alone it would be interesting, but its importance is greatly increased by the fact that we know two cases in birds which are closely comparable. The simpler case to which alone I shall refer has been observed in the canary. Like the currant moth it has a kind of albino, called *Cinnamon*, and males of this variety when mated with ordinary dark green hen canaries produce dark males and *Cinnamons* which are always hens; while the green male and the *Cinnamon* hen produce nothing but greens of both sexes. This case, which has been experimentally studied by Miss Durham, offers a certain complication, but in its main outlines it is exactly like that of the moth, and the same interpretation is applicable to both.

The particular interpretation may be imperfect and even partially wrong; but that we are at last able to form a working idea of the course of such phenomena at all is a most encouraging fact. If we are right, as I am strongly inclined to believe, we get a glimpse of the significance of the popular idea that in certain respects daughters are apt to resemble their fathers and sons their mothers; a phenomenon which is certainly sometimes to be observed.

[There are several collateral indications that we are on the right track in our theory of the nature of sex. One of these, derived from the peculiar inheritance of colour-blindness, is especially interesting. That affection is common in men, rare in women. Men who are colour-blind can transmit the affection, but men who have normal vision cannot. Women however who are ostensibly normal may have colour-blind sons; and women who are colour-blind have, so far as we know, no sons who are not colour-blind.[1]

Mendelian analysis of these facts shews that colour-blindness is due, not, as might have been supposed, to the absence of something from the composition of the body, but to the presence of something which affects the sight. Just as nicotine poisoning can paralyse the colour sense, so may we conceive the development of a secretion in the body which has a similar action. The comparative exemption of the woman must therefore mean that there is in her a positive factor which counteracts the colour-blindness factor, and it is not improbable that the counteracting element is no other than the femaleness factor itself.][2]

---

[1] We have knowledge now of seven colour-blind women, having, in all, 17 sons who are all colour-blind. Most of these cases have been collected by Mr Nettleship.

[2] The author noted in pencil in his copy that these two paragraphs [ ] were *out of date: to be deleted or replaced*. This note was probably made after the appearance of the late Dr Doncaster's papers, *Journ. Genetics*, 1911, I, 377, and 1913, III, 11. For his revised views of the case see Address 17th Intern. Cong. Med. 1913 (*Sci. Papers*). C. B. B.

I think I have said enough to prove that after all, those curiosities collected from observation of sweet peas and canaries have no remote bearing on some very fascinating problems of human life.

Lastly I suppose it is self-evident that they have a bearing on the problem of Evolution. The facts of Heredity and Variation are the materials out of which all theories of Evolution are constructed. At last by genetic methods we are beginning to obtain such facts of unimpeachable quality, and free from the flaws that were inevitable in older collections. From a survey of these materials we see something of the changes which will have to be made in the orthodox edifice to admit of their incorporation, but he must be rash indeed who would now attempt a comprehensive reconstruction. The results of genetic research are so bewilderingly novel that we need time and an exhaustive study of their inter-relations before we can hope to see them in proper value and perspective. In all the discussions of the stability and fitness of species who ever contemplated the possibility of a wild species having one of its sexes permanently hybrid? When I spoke of adventures to be encountered in genetic research I was thinking of such astonishing discoveries as that.

There are others no less disconcerting. Who would have supposed it possible that the pollen cells of a plant could be all of one type, and its egg cells of two types? Yet Miss Saunders' experiments have provided definite proof that this is the condition of certain stocks, of which the pollen grains all bear doubleness, while the egg cells are some singles and some doubles. We cannot think yet of interpreting these complex phenomena in terms of a common plan. All that we know is that there is now open for our scrutiny a world of varied, orderly and specific physiological wonders into which we have as yet only peeped. To lay down positive propositions as to the origin and inter-relation of species in general, now, would be a task as fruitless as that of a chemist must have been who had tried to state the relationship of the elements before their properties had been investigated.

For the first time *Variation* and *Reversion* have a concrete, palpable meaning. Hitherto they have stood by in all evolutionary debates, convenient genii, ready to perform as little or as much as might be desired by the conjurer. That vaporous stage of their existence is over; and we see Variation shaping itself as a definite, physiological event, the addition or omission of one or more definite elements; and Reversion as that particular addition or subtraction which brings the total of the elements back to something it had been before in the history of the race.

The time for discussion of Evolution as a problem at large is closed. We face that problem now as one soluble by minute, critical analysis. Lord Acton in his inaugural lecture said that in the study of history we are at the beginning of the documentary age. No one will charge me with disrespect to the great name we commemorate this year, if I apply those words to the history of Evolution: Darwin, it was, who first shewed us that the species have a history that can be read at all. If in the new reading of that history, there be found departures from the text laid down in his first recension, it is not to his fearless spirit that they will bring dismay.

BIOLOGICAL FACT AND THE STRUCTURE OF SOCIETY

The Herbert Spencer Lecture. 28 February 1912
Oxford

There are signs that the civilised world is at length awakening to the fact that the knowledge needed for the right direction of social progress must be gained by biological observation and experiment. Such a turn in public opinion would, we may be sure, have been viewed by Herbert Spencer with exceptional interest and approval. The truth, so obvious to the naturalist, that man is an animal, subject to the same physical laws of development as other animals, is a doctrine he constantly expounded, and perhaps his teaching did more than that of any other philosopher towards helping men to see themselves as they really are, stripped of the sanctity with which superstition and ignorance have through all ages invested the human species.

Spencer not only contributed that great service, but I suppose that no one ever looked forward with serener confidence or a fuller optimism to the consequences which follow upon a recognition of these natural facts, to the possibility of a further evolution of our species, and to the certainty that by his own action the destiny of man may be controlled. It is natural therefore that in a lecture founded to commemorate his work we should examine the possibilities of biological discovery as applied to the constitution and future of human society.

Many causes have combined to give prominence at this moment to the biological aspects of Sociology. There exists a general perception on the part of the more intelligent that the present condition of the social structure in civilised states is one of extreme instability. The apprehension that changes of exceptional magnitude are impending is widely spread. In addition to these indefinite sensations of uneasiness, the minds of observant persons are becoming keenly alive to the fact that the unexampled changes in the conditions of human life, made possible by the applications of science, are likely to result in an alteration of the composition of the population. Owing to the control which civilised communities have acquired over the forces of nature the average human life has been materially lengthened, and we need no evidence beyond that of ordinary experience to shew that especially have the lives of those who are defective in mind or body been prolonged by application of these new powers on their behalf.

A general acquaintance with the idea of Evolution, in outline at least, has become universal. We are all habituated to the notion that the form of a society, like that of an individual, is a consequence of an evolutionary process. To that process experimental interference on an enormous scale is being applied, and it is inevitable that the community at large should be asking, not without anxiety, how far the outcome of these interferences with what have usually been regarded as natural forces will bring good or evil to the societies which attempt them. Within the last few years, moreover, mankind has suddenly begun to realise what heredity means. The deliberate interferences hitherto contemplated by economists have related to the distribution of wealth and opportunities of many kinds, the regulation of supply and demand, the creation or abolition of divers political institutions, and other measures of similar character. Though the effects of these devices are commonly described as profound, such measures are indirect, and to the mind of the naturalist most of them are essentially superficial. Every legislative encouragement given to one class and every repression of another has an effect on the future of the race. Exerted over long periods of time, these interferences must indeed influence the composition of a population; but with knowledge of the full meaning of the physiological process of heredity we perceive that man has it in his power to operate upon his species in a much more drastic way. In Spencer's time and long before, this fact was obvious to all who reflected on the matter. He himself in many passages alludes to these possibilities. In 1873, for example, he wrote:[1]

If anyone denies that children bear likenesses to their progenitors in character and capacity—if he holds that men whose parents and grandparents were habitual criminals, have tendencies as good as those of men whose parents and grandparents were industrious and upright, he may consistently hold that it matters not from what families in a society the successive generations descend. He may think it just as well if the most active, and capable, and prudent, and conscientious people die without issue; while many children are left by the reckless and dishonest. But whoever does not espouse so insane a proposition must admit that social arrangements which retard the multiplication of the mentally-best, and facilitate the multiplication of the mentally-worst, must be extremely injurious.

[1] *The Study of Sociology*, ed. 1908, p. 343.

In the period when these words were written practically nothing was known of heredity. Naturalists knew that in general offspring resemble their parents more or less, and that by selection for an indefinite number of generations types could be fixed so as to breed approximately true. That there was a vast province of exact and readily ascertainable knowledge, fraught with immeasurable practical consequence to mankind, hidden behind the word *heredity* had occurred to scarcely a single mind.

Many were perfectly aware of the importance of heredity. All upholders of evolutionary doctrines, both those who preceded Darwin and those who followed him, were familiar with the fact that change of type came about through the inheritance of modification. In many admirable and striking works the late Francis Galton had endeavoured to direct attention to the practical significance of heredity. He had shewn also that the descent of characters could be partially expressed in a system, which, though erroneous in fundamental conception, still gives an approximately correct representation of several of the phenomena.

But the discovery of Mendelian analysis, though as yet imperfectly developed, opens up a new world of physiology. Expressed in the briefest possible way the essence of the Mendelian principle is not difficult to grasp. It may be conveyed in the statement that organisms may be regarded as composed to a great extent of separate factors, by virtue of which they possess their various characters or attributes. These factors are detachable, and may be recombined in various ways. It thus becomes possible to institute a factorial analysis of an individual.

How far such analysis can be carried we do not yet know, but we have the certainty that it extends far, and ample indications that we should probably be right in supposing that it covers most of the features, whether of mind or body, which distinguish the various members of a mixed population like that of which we form a part. From such a representation we pass to the obvious conclusion that an individual parent is unable to pass on to offspring a factor which he or she does not possess.

Just as various features or characteristics may be due to the *presence* of the corresponding factor, so we have to recognise that other attributes appear only in the *absence* of certain factors. Moreover, since those individuals only which are possessed of the factors can pass them on to their offspring, so the offspring of those that are destitute of these elements do not acquire them in subsequent generations but continue to perpetuate the type which exists by reason of the deficiency. You will readily understand that in practice the analysis and detection of these factors is a difficult matter. The difficulty arises especially from the very important fact that some of the ingredient-factors have the property of *inhibiting* or masking the effects of other factors, and that many features of bodily organisation are due to the *combination* and interaction of two or more ingredients, which alone might be present without producing any perceptible sign of their presence. Thus one flower may be white, because it is lacking in the element which produces colour; but another may be white though it has everything needed to give it colour, because it has in addition an element which suppresses the pigmentation. Again, colour in some plants is due to one factor, but in others it is developed only when two independent complementary factors are present, and either of these may be present alone in a flower which is perfectly white.

Such rules have been demonstrated in operation for an immense diversity of characteristics in both animals and plants in great variety. It should be explicitly stated, however, that in the case of the ordinary attributes of normal men we have as yet unimpeachable evidence of the manifestation of this system of descent for one set of characters only, namely the colour of the eyes.

There is nevertheless no reasonable doubt that the extension to the normal attributes of man is one which we are well entitled to make. For with the doubtful exception of certain features of quantity, size, and number, no characters of animals and plants which have been made subject to adequate experimental tests have hitherto proved incapable of being represented as governed by such a system. Moreover, if the evidence as to normal characteristics of man is defective—which in view of the extreme difficulty of applying accurate research to normal humanity is scarcely surprising—there is in respect of numerous human abnormalities abundant evidence that a factorial system of descent is followed.

To appreciate the full significance of these things one must have practical experience of breeding. I wish it were in my power to bring to the minds of such an audience as this some part of the emotion which the contemplation of this display of order can excite. Imagine a green-house stage full of a miscellaneous collection of varieties of some plant, such as the Chinese primula, with all their varied shapes of leaves and flowers. Their colours also seem at first sight to range through an endless series of tints, of magenta, crimson, pink, and blue. By appropriate treatment we have it in our power to determine that in three generations at the most the offspring of these plants, the generation in fact which will then replace them and represent them, shall be entirely of one type only, or of two types, or three types in any required proportion. By choosing which parents shall leave offspring we can decide how the species shall be represented on our stages with a certainty almost as great as if the selection were made from plants already grown. And similarly for fowls and many other forms of life. Write *man* for primula or fowl, and the stage of the world for that of the greenhouse, and I believe that with a few generations of experimental breeding we should acquire the power similarly to determine how the varieties of men should be represented in the generations that succeed.

At a cattle show I look at the splendid animals in their pens, ranged breed by breed, and I look at the farmers and the sightseers passing by them in procession. They too are of manifold types, men from all parts of the country, often shewing the characteristics of their race plain and easy to recognise—big men, little men, men who fill out or "mature" early, as they say in the meat-market—spare men, that the farmers would call "bad doers", tame men, vicious men, sharp and dull, dark and fair—shepherds, stockmen, grooms, butchers, and salesmen. Could they too be arranged breed by breed in pens? A few most certainly could. (We might make a pen of shepherds and we should not often put in a groom by mistake.) Why could not all be sorted into breeds? The answer is obvious: because they are the offspring of matings made almost at random—and for no more recondite reason.

Many are disposed to imagine that the conditions of life play a great part in producing the diversity of such a mixed assemblage, but the more we learn of biological fact the less do we find much evidential ground for that opinion. The conditions of life provide opportunity for the development of characters, but they cannot increase the original endowment. If the right opportunity be withheld the characteristic does not appear. If the stout man had been starved from his birth, obviously his disposition to stoutness might have remained unknown; but the spare man, like the razor-back pigs of the Southern States, will not fatten though he take five meals a day. And so for qualities that may be regarded as more subtle. A muscat grape will produce its aromatic flavour if it have sun and a suitable soil— the pretentious Gros Colmar, with its fruits half as large again, is not worth eating, though it be fostered with all the gardener's skill. These qualities are, as we say, *genetic*, given to the creature at its birth, brought into it on fertilisation by one, or by the other, or by both of the cells which united to produce it. That the conclusions to which experimental studies of animals and plants have led us apply also on the whole to the descent of human faculty can be doubted by no one who has studied the evidence.

If any one is not already convinced he should refer to the accumulated proofs which Galton so successfully collected, especially in *Hereditary Genius*. Let him study any biographical records of human achievement or conduct—such as a dictionary of painters or of musicians—and observe the perpetual recurrence of the same names in groups of two, three, or more; or the geographical distribution of illegitimacy, shewing as it does the maintenance of "local custom" and morals under divers conditions of occupation, soil, and climate; or the pedigrees collected in medical literature shewing the descent of disease; or if he look no further than the distribution of qualities among the families known to himself he will be forced to the admission that, though the circumstances among which a man is born or thrust have some influence in the development and direction of his powers, yet the total contribution which circumstance makes to achievement is of that subordinate kind which is adequately described by the word "opportunity".

Men do not gather grapes of thorns or figs of thistles; and what is so clear for the budding branches of the plant would be no less obviously true of the branching generations of our species, were it not for the fact that we each of us come from the union of *two* cells derived from two parents. The fact that we arise by this sexual process throws, however, but a thin veil of obscurity over the laws of descent, and it is interesting to notice that if only human or any other pedigree-tables had been arranged to be read *downwards instead of upwards*, the essential fact of Mendelian segregation must have been long ago discovered in regard to many characteristics. Genealogists have been accustomed to make a table of descent as a fan, with its apex in the individual whose origin they wish to display and its base widening as far as possible into the ancestry, the parental stock of each ancestor being represented by a pair. But to shew how a character really descends we require the table constructed with its apex in one original individual who possessed the character, and from that apex to exhibit the devolution of that character among the diverging branches of his posterity. As the usual purpose of the genealogist has been to contribute either to political history or to family pride, rather than to natural knowledge, his mind has consequently been set on a demonstration rather of the origin and antiquity of his hero's qualities than of their distribution or absence among the collaterals.

I do not propose on this occasion to adduce facts in support of the general proposition that human genetic physiology follows in the main systems similar to those discovered in animals, but rather, assuming that this truth is admitted, to examine some phenomena of social physiology as they appear in the light of this knowledge.

May I clear myself at once of a possible misunderstanding? You will think, perhaps, that I am about to advocate interference by the State, or by public opinion, with the ordinary practices and habits of our society. There may be some who think that the English would be happier if their marriages were arranged at Westminster instead of, as hitherto, in Heaven. I am not of that opinion, nor can I suppose that the constructive proposals even of the less advanced Eugenists would be seriously supported by anyone who realised how slender is our present knowledge of the details of the genetic processes in their application to man. Before science can claim to have any positive guidance to offer, numbers of untouched problems must be solved. We need first some outline of an analysis of human characters, to know which are due to the presence of positive factors and which are due to their absence; how and in respect of what qualities the still mysterious phenomenon of sex causes departures from the simpler rules of descent, and many other data which will occur readily enough to those who are familiar with these inquiries. It is almost certain, for instance, that some qualities are transmitted differently according as they are possessed by the mother or by the father, and it is by no means improbable that various forms of conspicuous talent are among their number. It should be borne in mind that we do not yet know even which females among mankind correspond to which males. In man sexual differentiation is generally strongly marked. The case is almost like that of poultry. If a breeder ignorant of the breeds of poultry were asked to sort a miscellaneous assemblage of cocks and hens into pairs according to breed, he would often be quite at a loss to know what a given male type looked like when represented in a hen, and conversely. He would thus make many mistakes even when dealing with pure breeds; and in man, as individuals pure-bred in any respect are very rare, the operation would be far more difficult. For these and other reasons I am entirely opposed to the views of those who would subsidise the families of parents passed as unexceptionable. Galton, I know, contemplated some such possibility; but if we picture to ourselves the kind of persons who would infallibly be chosen as examples of "civic worth"— the term lately used to denote their virtues—the prospect is not very attractive. We need not for the present fear any scarcity of that class, and I think we may be content to postpone schemes for their multiplication.

As regards practical interference there is nevertheless one perfectly clear line of action which we may be agreed to take— the segregation of the hopelessly unfit. I need not argue this point. When it is realised that two parents, both of gravely defective or feeble mind, in the usual acceptance of that term, *do not have any normal children at all*, save perhaps in some very rare cases, and that the offspring of even one such parent mated to a normal generally contain a proportion of defectives,[1] no one can doubt that the right and most humane policy is to restrain them from breeding, and I suppose the principle of the Act now before Parliament for the institution of such a policy will have general approval. Under our present system the State exerts all the powers which science has developed for the preservation of such persons from their birth, most of whom would otherwise perish early. Brought to maturity their destiny is not difficult to imagine. However ignorant we may be as to the several ingredients which are required to compose a stable society, or of the proportions in which they are severally desirable, we are safe in preventing these creatures from reproducing themselves. Some of the more advanced of the American States are already going further, and even such a representative of older ideas as the State of New Jersey is, I am informed, introducing the practice of sterilising criminals of special classes. That appears to me the very utmost length to which it is safe to extend legislative interference of this kind, until social physiology has been much more fully explored.

Beyond that if there is authority to go, it is not drawn from genetic science. If a person who is born with cataract, or develops cataract very early in childhood, has children, it is almost certain that half those children will inherit the cataract, with varying degrees of blindness. The prospect is more or less the same for several other defects. Nevertheless, though from

---

[1] From such pedigrees as I have seen I should nevertheless hesitate to describe feeble-mindedness as a simple Mendelian recessive. It is possibly due to an absence of some factor or factors; but there is strong evidence that the usual result of a mating between normal and feeble-minded parents is a proportion of feeble-minded children, and it is difficult to suppose that most ostensibly normal persons are heterozygous in this respect. See especially H. H. Goddard, *Amer. Breeder's Magazine*, I, 1910, 172.

these causes many remain grievous burdens to their families, or
to the public funds, and though they could probably be eliminated after a few generations without difficulty by legislative
interference, that would be a very dangerous course. They are
not necessarily useless persons, nor are their own lives necessarily
miserable. There are many healthy and active types which are
a far greater nuisance to their neighbours and reproduce themselves with equal exactitude. Possibly, on a ballot, few of us
would be encouraged to perpetuate our likenesses! We all have
grave defects, not least those who contribute much to the happiness of the world. The monogamous pigeons sitting on the
barn roof perhaps are scandalised at the polygamy of the fowls
in the yard. Such decadence, they hold, is disgusting and should
be stopped. The fowls no doubt would reply that they may be
polygamous and even polyandrous, but as for decadence, they
at least do not limit their families to two. Such degeneracy is
race-suicide and they think it should be punished. And so the
debate might continue.

Seriously, let us remember that a polymorphic and mongrel
population like ours descends from many tributary streams.
We are made of fragments of divers races, all in their degree
contributing their special aptitudes, their special deficiencies,
their particular virtues and vices, and their multifarious notions
of right and wrong. Many of us have, for instance, the monogamous instinct as strong as pigeons, and many of both sexes
have it no more than fowls. Why should some be ambitious to
make all think or act alike? It is much better that we should
be of many sorts, saints, nondescripts, and sinners. Posterity is
likely to discover that to eliminate sinners there is only one way
—that which St Paul pointed to us when he wrote that "where
no law is, there is no transgression". Science knows nothing of
sin save by its evil consequence. In all reverence she inverts
the ancient saying and proclaims that the sting of Sin is Death.
It is not the tyrannical and capricious interference of a half-
informed majority which can safely mould or purify a population, but rather that simplification of instinct for which we ever
hope, which fuller knowledge alone can make possible. As
science strengthens our hold on nature, more and more will
man be able to annul the evil consequence of sin. Little by
little the law will lapse into oblivion, and sins which it created
will be sins no more.

The great and noble work which genetic science can do for
humanity at the present time is to bring men to take more true,
more simple, and, if so inexact a word can be used intelligibly,

more natural views of themselves and of each other. With fuller
knowledge of the physiology of races, and of the intimate relation between the physiological composition of the individual
and his vital possibilities, all the problems of social organisation
shew new aspects, and the vision is cleared of the fancies with
which subjective ingenuity has overlaid the facts. How hard it
is to realise the polymorphism of man! Think of the varieties
which the word denotes, merely in its application to one small
society such as ours, and of the natural, genetic distinctions
which differentiate us into types and strains—acrobats, actors,
artists, clergy, farmers, labourers, lawyers, mechanics, musicians,
poets, sailors, men of science, servants, soldiers, and tradesmen.
Think of the diversity of their experience of life. How few of
these could have changed parts with each other. Many of these
types are, even in present conditions, almost differentiated into
distinct strains. In no wild species, not even among the ants, so
often quoted, do we find any polymorphism approaching to this. I
never cease to marvel that the more divergent castes of civilised
humanity are capable of inter-breeding and of producing fertile
offspring from their crosses. Nothing but this paradoxical fact
prevents us from regarding many classes even of Englishmen as
distinct species in the full sense of the term. In a strident passage
the acute Cobbett long ago expanded this conclusion:

I am quite satisfied, that there are as many *sorts* of men as there are
of dogs.... It cannot be *education* alone that makes the amazing difference we see. Besides, we see men of the very same rank and riches and
education differing as widely as the pointer does from the pug. The
name, *man*, is common to all the sorts, and hence arises very great
mischief. What confusion must there be in rural affairs, if there were
no names whereby to distinguish hounds, greyhounds, pointers,
spaniels, terriers, and sheep-dogs from each other! And what pretty
work, if, without regard to the *sorts* of dogs, men were to attempt to
employ them! Yet this is done in the case of *men*! A man is always a
man; and without the least regard as to the sort, they are promiscuously
placed in all kinds of situations....What would be said of the 'Squire
who should take a fox-hound out to find partridges for him to shoot
at? Yet would this be *more* absurd than to set a man to law-making
who was manifestly formed for the express purpose of sweeping the
streets or digging out sewers?[1]

The problem which confronts the political philosopher is to
find a system by which these differentiated elements may combine together to form a co-ordinated community, while each

[1] W. Cobbett, *Rural Rides*, ed. 1853, p. 291.

element remains substantially contented with its lot. To discuss this mighty problem in its full scope I have neither qualification nor desire. All that I can venture to contribute are some reflections which must come often to the minds of naturalists who contemplate the facts. They may be familiar enough to those who engage in the study of human affairs, but I have noticed that among those natural divisions between the sorts of men to which I just referred there are few more marked than that which usually separates students of natural knowledge from those who care nothing for it; and with rare exception you will find that publicists of the various denominations are almost always in this latter group.[1] Legislators, nevertheless, whether they know it or not, are engaged in a practical experiment with living things of a peculiarly intricate kind.

Many features of social phenomena evidently wear to the legislator aspects entirely different from those which they present to us. Lately, for example, the nation has been debating the virtual abolition of the hereditary Chamber—obviously a problem to the solution of which biological data are essential. I did not see in the public utterance of any statesman an allusion even to this aspect of the matter. Yet such data are neither very difficult to collect nor to interpret.

Let us think of the criminal law and consider how a system can satisfy the legislator which to the naturalist is stupid and infamously cruel. Just now I spoke of the polymorphism of mankind. No one trained in biology is ignorant of that phenomenon. True we realise it now as we never did before the study of heredity had developed, and I doubt not that before many years are past genetic research will have successfully represented the varying compositions of many at least of the more aberrant types of men by irrefutable analysis. If we have not yet these exact expressions, none of us doubt they can be found. Yet "in the sight of the law", as the phrase goes, all men are equal! Are they equal in the sight of any one less blind than Justice? We do not find them equal in the out-patient room, in the school, at the recruiting *dépôt*—why in the court of law? If a lawyer cares to know how criminal procedure looks to

---

[1] Mr Canning did not learn till late in life that tadpoles turn into frogs, and thought that a schoolboy who gave him that information was fooling him. Mr Gladstone believed that twenty-eight was the normal total for the human teeth. Portentous ignorance of this kind is common among historians and legislators. In itself perhaps a trifle, it is a symptom of detachment from the actual world so complete as to disqualify a man from safely exercising high functions of statesmanship, demanding, as they must, a discernment which can only come from wide knowledge of natural fact.

biologists, let him read the sentence pronounced in *Erewhon*[1] by the judge on the prisoner convicted "of the great crime of labouring under pulmonary consumption". After expressing the pain he felt at having to pass a severe sentence on one who was yet young, and had otherwise excellent prospects, he continued:

You were convicted of aggravated bronchitis last year: and I find that though you are now only twenty-three years old, you have been imprisoned on no less than fourteen occasions for illnesses of a more or less hateful character; in fact, it is not too much to say that you have spent the greater part of your life in jail. It is all very well of you to say that you came of unhealthy parents, and had a severe accident in your childhood which permanently undermined your constitution; excuses such as these are the ordinary refuge of the criminal; but they cannot for one moment be listened to by the ear of justice. I am not here to enter upon curious metaphysical questions as to the origin of this or that—questions to which there would be no end were their introduction once tolerated, and which would result in throwing the only guilt on the tissues of the primordial cell, or on the elementary gases.... I do not hesitate therefore to sentence you to imprisonment, with hard labour, for the rest of your miserable existence.

A humane man—a lawyer too—after witnessing such a scene, not in "Erewhon" but in London, said once to me that he did think the judge might have noticed that the prisoner's head was a different shape from anybody else's in the court. The sickening cruelty of the courts is, I am happy to think, abating somewhat, but there will be no radical improvement until the functions of the administrator of criminal justice are recognised as in the main medical. The criminal may be and often is hopeless; but if his case be one for treatment, let us treat it with the only remedies capable of doing any good. Give him occupation, distraction, change of thoughts, if it be possible. These, and not solitary confinement, are the treatment we should prescribe for ourselves when we fear temptation.

Take the two converse aspects of the question of population. Infant mortality is conventionally regarded by both statesmen and philanthropists as deplorable, without further inquiry.[2]

---

[1] *Erewhon*, by Samuel Butler, 1872, p. 96.

[2] Such an infatuation does this idea become even with statistical experts, that I find so careful a writer as Dr Newsholme saying without qualification "that each member of the population, when the balance between expense of subsistence and wages earned through life is worked out, represents enormous wealth". This passage is introduced with the words "It has been already pointed out"; but even in the place where the subject is more fully treated and Farr's calculations are given, the only reservation overtly made is for the aged. Dr Newsholme of course means that on an average of the

Do they consider from what prospect most of these infants are delivered? Would it be better that they should be preserved to fill the workhouse infirmaries?

Other public men profess indignation against the practice, almost universal among the more intelligent and more provident classes in civilised countries, of limiting their families to two or three children. Have these patriots estimated what the pressure upon the resources of the country would be if we mostly had six to ten children, as our parents had? The naturalist knows that a great part of the population of this country ought not to exist at all under present conditions of distribution. To add greatly to the number even of the able and thrifty will not diminish the proportion of the unfit or lighten the strain. What would be thought of a breeder who tried to keep all his stock? He wants no more than he can do well; otherwise his stock and he too will soon be ruined. The distinction which Malthus drew between "a redundant population and one actually great" is sound, biologically as well as in economics. It is not the *maximum* number but the *optimum* number, having regard to the means of distribution, that it should be the endeavour of social organisation to secure. To spread a layer of human protoplasm of the greatest possible thickness over the earth—the implied ambition of many publicists—in the light of natural knowledge is seen to be reckless folly. We need not more of the fit, but fewer of the unfit. A high death-rate is often associated with a high birth-rate, but happily a low birth-rate and a low death-rate are quite compatible with each other.

In the gloom which shrouds the future of civilised communities there is one fact which gives encouragement and hope, the decline in the birth-rate, associated as it now is with a decline in the death-rate also.

To most writers on these questions continual increase of the population of a country is regarded as the normal condition of things. This proposition is explicitly stated, for example by Rümelin,[1] in one of the leading text-books. The naturalist knows,

population there is a balance of profit, and on an average of wage-earners a high profit, not that "each member of the population...represents enormous wealth". Yet that section of the population whose value is negative should be constantly and explicitly mentioned; for there is nothing to shew that a reduction in total population is incompatible with an equal or even greater profit on the whole. (See Newsholme, A., *Elements of Vital Statistics*, 1889, pp. 69 and 14.)

[1] "...so erscheint es nicht nur als empirische Tatsache sondern als die Ordnung der Natur, dass die Geburten in jeder menschlichen Gesellschaft einen Ueberschuss über die Todesfälle ergeben, somit die *stetige Zunahme einer*

however, that such a phenomenon can be but ephemeral. He is accustomed to take longer views of the life of a species. In nature the numbers of a species can only increase when it is taking up fresh means of subsistence, in consequence of variation or otherwise. Parasites increase when they invade a new host. The rabbits increased when they invaded Australia, as did the sparrows in America. The population of this country increased very slowly till the latter half of the eighteenth century, when it began to rise sharply, but it was in the first third of the nineteenth century that the rate of increase became alarming, culminating in the misery of the 'forties.[1]

No one can doubt that the new means of subsistence which made this rise in population possible was the energy latent in the coalfields. Nor have we to look far for the variation which enabled man to begin thus to devour the capital of the earth; and I suppose the coincidence of the first quick rise in population with the activities of that remarkable mutation, James Watt, needs no special emphasis of interpretation. Sir William Ramsay estimates that the coal of this country will be exhausted in 175 years, and in his opinion it is in the highest degree improbable that any comparable source of energy will become available.[2] He limited his remarks to this country; but though there is no reliable means of estimating the coal in the earth as a whole, it is probable that within some period which is short as biology counts time, our species will be once more limited to the energy-income of the earth. We are in fact passing through a phase which is quite exceptional in the history of a species—exceptionally favourable if you will—and it is in a decline in the birth-rate that the most promising omen exists for the happiness of future generations.

Bevölkerung als die Normale, der Stillstand oder Rückgang stets als etwas Naturwidriges, als eine krankhafte, durch ausserordentliche Umstände begründete Störung zu gelten hat" (Rümelin, in Schönberg's Handb. Polit. Oekon. 1890, I, 772).

[1] Sir A. Alison, The Principles of Population, 1840, I, 520: "It is in the midst of this prodigious manufacturing population that the human race advances with alarming rapidity, and shoals of human beings are ushered into the world without any adequate provision existing for their comfortable maintenance. Such is the improvidence, the recklessness, and the profligacy which characterise the great bulk of the urban population in all the great cities of the empire, that the rate of increase bears no proportion to the permanent demand for labour: but mankind go on multiplying, as in the Irish hovels, with hardly any other limit than that arising from the physical inability in the one sex to procreate, and in the other to bear children".

[2] Presidential Address to British Association, Portsmouth, 1911. This estimate followed that of the Coal Commission in excluding coal below 4000 feet, which, if included, would prolong the period for perhaps a century.

Professor Marshall, discussing not the consequences of the exhaustion of coal, but another phase of the population question, remarks: "It remains true that unless the checks on the growth of population in force at the end of the nineteenth century are on the whole increased (they are certain to change their form in places that are as yet imperfectly civilised) it will be impossible for the habits of comfort prevailing in Western Europe to spread themselves over the whole world and maintain themselves for many hundred years". In a note to this passage he estimates that if the present rate of increase of human population continue till the year 2400 "the population will then be 1000 for every mile of fairly fertile land: and so far as we can foresee now, the diet of such a population must needs be in the main vegetarian".[1]

And now regarding the central problem of social structure, the conditions of stability in the relations of the human classes to each other and to the State, has biological science any counsel of value to give? Is there any observation that naturalists have made, knowledge acquired, or principles perceived in their study of the manifold forms of life, which in this period of grave anxiety they dare to offer as a contribution to political philosophy? Let us examine the physiological aspects of that problem. Upon the data there is now an agreement almost universal. Society consists of differentiated elements, unlike in tastes, faculties, sex, health, and ability of every kind. Some are strong, most are weak. If this complexity of civilisation—the indispensable condition of evolutionary progress—is to continue, such differentiation, or some state approaching it, must be preserved. How then, in an age when knowledge is cheap and all know how the rest live, is any general content to be secured? Let us turn to the familiar comparison in which the community is likened to an organism with differentiated parts. The comparison is as old as Menenius Agrippa, or at least as Plutarch. It was one, too, which Herbert Spencer especially delighted to develop. Note next that to the biologist this presentation of the phenomenon is not a mere analogy but often a description of fact. The comparative anatomist cannot always draw a clear distinction between a compound organism with differentiated parts and a social organism with differentiated members.

I lay stress on this aspect of the social problem because I have seen several times of late the claim put forward that the teaching of biological science sanctions a system of freest competition for the means of subsistence between individuals, under which the

[1] Alfred Marshall, *Principles of Economics*, 3rd ed. 1895, p. 259.

fittest will survive and the less fit tend to extinction. That may conceivably be a true inference applicable to forms which, like thrushes, live independent lives, but so soon as social organisation begins, the competition is between societies and not between individuals. Just as the body needs its humbler organs, so a community needs its lower grades, and just as the body decays if even the humblest organs starve, so it is necessary for society adequately to ensure the maintenance of all its constituent members so long as they are contributing to its support. The simple hydroids, such as *Hydra*,[1] live alone, and no doubt compete freely against each other; but hydroids which remain united as compound forms have to let the food circulate among those degraded components which never even develop mouths, and all their lives function as tentacles. A body all muscle would be as helpless as a nation of Sandows; nor would a nation of Newtons live much longer than a brain removed from the skull.

From these considerations we may draw a conclusion that some elements of the doctrines vaguely described as socialism are consistent with, and indeed are essential to, stability. Society would do well to restrain competition between its parts so far as to ensure proper food and leisure for the lower grades of producers. How that restraint is to be effected is a question for the practical economist. Some such measures of restraint we have already enacted: on the whole with good results. Spencer, as everyone knows, protested with vehemence against this legislation, but I have never been able to comprehend the biological grounds on which he based his protest. For if society is in reality an organism, society must apply restraints on the undue growth of its parts analogous to that co-ordinating mechanism which controls the growth of organs in the body.

Apart also from actual restraint by civil authority, there is happily hope of some effective restraint by change in public feeling.

Formerly, cruelty to domesticated animals was defended on the principle that "a man may do what he likes with his own". Civilised humanity no longer recognises that defence; and slowly, even in our dealings with the weaker members of our own species, change in public feeling has begun to act in restraint of oppression.

Motive for individual exertion must nevertheless be preserved. It could be dispensed with only in a community in which the component members were in *complete* co-ordination, as the

[1] MS footnote by W. B.: "or substitute Lucernaria".

organs of the healthy body are. The only instinct in our race
which is sufficiently universal to supply this motive is the desire
to accumulate property, generally as a provision for offspring.
Other instincts, such as emulation, the altruistic emotions, or
the mere love of activity, may all be strongly developed in some,
but they are permanent in very few individuals. They are apt
to weaken after adolescence, and to disappear as middle age
supervenes. But for the institution of property the fibre of the
whole community, as at present physiologically constituted,
would slacken, and decay must immediately begin. Yet, admitting the principle that if life held no prizes no one would
compete, might we not prohibit prizes of such magnitude as
to jeopardise the stability of the community? To fix an upper
limit on accumulation would not greatly discourage effort, for
people will play hard, though the stakes be limited.

Socialism is a state that Nature knows well and has sometimes approved. Yet consider how this approval has been won.
Hive bees, for example, are socialists: the individual worker
amasses no property for herself. They defend their hive. Every
individual bee that stings you dies in a few hours. But the
success of this socialism is founded in the instinctive, almost
reflex, devotion of the bees.

Among us we have individuals who develop such feelings for
a few years in early youth, and lose them later. A few possess
these instincts all their lives. They sacrifice themselves, and but
too often others also in their course. Such casual devotion is no
base on which to form a social system. All permanent and stable
change of institutions is founded in the physiological variation
of instinct. In mankind we know a mysterious variation which
we call change in fashion or in public opinion.[1] It is to such a
variation that constructive socialism must look for its foundation. This is but a slender hope; for that "public opinion"
must take the form of an instinctive, mystic devotion to society,
not merely a passion to enjoy the fruits of other men's labour.
Of socialistic public opinion in that fuller sense we see few
signs.

Observe, too, how even the bees behave under sore temptation. Those who have witnessed the phenomenon of "robbing"
are not likely to forget the experience. If in August or September, when the honey-flow is failing, the bee-keeper drops a
comb near his apiary, he knows what to expect. The bees find

___

[1] In half serious mood he would sometimes maintain that could we
account for fashion, we should be near to unravelling the problem of
species. C. B. B.

this honey undefended, easy to seize. They become instantly demoralised. They fight for it at random, stinging and tearing each other to pieces. They charge promiscuously into their neighbours' hives and indescribable pandemonium begins. After such a scene the ground is littered with dead bees in hundreds, and in the bottom of a hive I have seen a layer of bodies an inch or more thick. Such is the instability of instinct even in the great prototype of socialism, and can we hope that the sight of undefended property would not similarly, in time of scarcity, upset the stability of a socialist State?

But there is still another side to the problem. If Nature gives some clear guidance as to the distribution of the means of life, her teaching is even clearer as to the distribution of political power. Socialistic she may sometimes be, but democratic she is not. Turn once more to the physiological facts. "All men are equal", say certain philosophers. "That is not true", replies the naturalist. "Proceed, then, as if it were", urges the statesman, and upon that course we have started. Founded in natural falsehood, the principle of equal rights is at length bearing fruits inevitable, though long deferred. The gift of equal power did not at first disturb the stability of society. Even the able seldom receive a new idea after they are grown up; for the dull mass that process is then impossible.[1] A generation passes and

[1] Herbert Spencer, in many of his strictures on the failure of legislation to achieve its avowed object, makes far too little allowance for the long latent period which often elapses before results appear. Commenting on the fact that laws rarely produce as much direct effect as was expected, and always produce indirect effects (which is all perfectly true), he proceeds to the following illustration, which at the present date reads somewhat naïvely: "It is so even with fundamental changes: witness the two we have seen in the constitution of our House of Commons. Both advocates and opponents of the first Reform Bill anticipated that the middle classes would select as representatives many of their own body. But both were wrong. The class-quality of the House of Commons remained very much what it was before. While, however, the immediate and special result looked for did not appear, there were vaster remote and general results, foreseen by no one. So, too, with the recent change. We had eloquently-uttered warnings that delegates from the working-classes would swamp the House of Commons; and nearly everyone expected that, at any rate, a sprinkling of working-class members would be chosen. Again all were wrong" (*The Study of Sociology*, ed. 1908, p. 270). So again he speaks with great contempt of the legislative efforts to suppress diseases among cattle, which (partly no doubt by the development of greater physiological knowledge) have now been very effective in most cases, and completely successful in many. In 1873 he wrote (*The Study of Sociology*, p. 164): "Since 1848 there have been seven Acts of Parliament bearing the general titles of Contagious Diseases (Animals) Acts. Measures to 'stamp out', as the phrase goes, this or that disease have been called for as imperative. Measures have been passed, and then, expectation not having been

their children, who learnt of it when young, become aware of the new power, with the consequences we are about to witness.

Of abstract rights, biology knows little: of equal rights, nothing. Philosophers have conceived men born with rights as they are with livers or with spleens. Perhaps they are; but since all those birthrights which can be expressed in terms of health or powers of mind or body are unequal, we find it difficult to suppose that there is some other kind of rights which we possess equally. Some would reply that *equal opportunity* is the right of all. But what use is equal opportunity to those who cannot use the opportunity equally? Either we must waste our strength in creating opportunities for those who cannot profit by them, or by aiming at the lower grades of mankind we deny to the rest the only opportunities which will enable them to develop.

All these familiar ideas will acquire new meaning in the light of the new knowledge of the definite composition of individuals; and it would be well, perhaps, if those who are now contemplating a great extension of equal political power to still lower grades of our population would consider how such a proposal reads when translated into physiological terms.

The political reformer claims to raise the standard of a population by thus providing opportunity in ameliorating the conditions of life, and it is worth noting the sense in which his claim is physiologically justified. The gardener by pricking out his seedlings gives them a chance of developing. Left crowded in the seed-pan, none, or very few, will become decent plants. The few successful, if there are any, may owe their success to their special qualities, but more often than not it is determined by mere accident of position near the tally, or against the edge of the pan, where they get most water or light. The botanist knows too that wild plants growing in the competition of a turf or amongst brushwood are usually half-starved. Set out, clear of their kind, or of weeds, many of them can grow to twice the size. So with the crowded masses of humanity. They may, so to speak, be "potted on". Given hygienic conditions and better opportunities, they may develop into decent specimens, but they will not turn into better kinds. In the new countries the

fulfilled, amended measures have been passed, and then re-amended measures; so that of late no session has gone by without a bill to cure evils which previous bills tried to cure, but did not. Notwithstanding the keen interest felt by the ruling classes in the success of these measures, they have succeeded so ill, that the 'foot-and-mouth disease' has not been 'stamped out', has not even been kept in check, but during the past year has spread alarmingly in various parts of the kingdom".

consequences of this process of planting out can be seen on a very large scale. The emigrants prosper. They are well fed. Except in a few large cities slums do not exist. All can develop; and if we do not expect what the gardener calls "important novelties" the result is admirable.

It is upon mutational novelties, definite favourable variations, that all progress in civilisation and in the control of natural forces must depend. How will *they* fare in a socialistic community? What stimulus is left to tempt them to exert their powers? In the born discoverer the instinct to find out natural truth is a strong passion, and those who have that feeling will gratify it, just as the artist or the poet works when rejected by the market; but those who invent applications of discoveries are generally thinking of patent rights, and if none are to be had, they may take life more easily. Is it not certain that all the forces of the community will be invoked against men of extra power? They will be treated as a disturbing nuisance. The progress of modification of a race composed of independent individuals can proceed by variation of individuals, but in a community organised on the principle of equality—if it can be imagined—an individual variation of any magnitude will be either without result or must produce immediate disorganisation and disruption.

The ideals therefore of socialism and of democracy are incompatible with each other, and the incompatibility will appear when the period of destruction is over. It is strange that the two words are so commonly associated. "Social democracy" denotes not one ideal but two. In order that the socialist community should succeed it must have but one mind, as the bees apparently have, not the uncoordinated resultant of all individual minds, which is the ideal of democracy. Until these two coincide, not occasionally only but in some permanent fashion, destruction may proceed but construction cannot begin.

The essential difference between the ideals of democracy and those which biological observation teaches us to be sound, is this: democracy regards class distinction as evil; we perceive it to be essential. It is the heterogeneity of modern man which has given him his control of the forces of nature. The maintenance of that heterogeneity, that differentiation of members, is a condition of progress. The aim of social reform must be not to abolish class, but to provide that each individual shall so far as possible get into the right class and stay there, and usually his children after him. Men rise from below and fall from above, and the fact is sometimes appealed to as evidence that such vicissitudes are a normal and wholesome phenomenon. The naturalist sees that the convection currents to which such displacements are due must indicate special kinds of disturbance. These disturbances are mainly due to interbreeding between the social grades, and between sections of the population formerly isolated. Such rapid social diffusion must mean either that much original variation is happening, or that extraordinary changes are affecting the conditions of life. There is no doubt that in the case of our own age *both* phenomena can be recognised, but the human variations in mental power are the primary factor, and they have created the disturbance in the conditions of life. Just as the numbers of the population tend always to reach an equilibrium in which births balance deaths, so do the differentiated elements of the population tend always to find their particular level, near which they would stop till the mass is again disturbed.

The fact that families or individuals rose into prominence or dropped into obscurity when the great industrial development of this country began, does not prove that the strains from which they came ought previously and in differing circumstances to have been in different relative positions. In various circumstances various qualities are required for success. It would be useful to illustrate this by actual examples discussed from the biological standpoint, but it will be sufficient to say that as we have come to recognise that evolutionary change proceeds not by fluctuations in the characters of the mass, but by the predominance of sporadic and special strains possessing definite characteristics, so in a society may previously existing types find their opportunity in the supervention of new social conditions.

When King David said, "I have been young, and now am old: and yet saw I never the righteous forsaken, nor his seed begging their bread", thus asserting the permanence and heritability of success, he is thought by some to shew himself singularly inobservant. But I doubt whether in the Middle Ages, or in any other epoch when conditions were comparatively uniform over long periods of time, he would have been regarded as saying anything contrary to general experience.

However that may be, he is declaring what *ought* to be true in an ideal State. We have abolished the Middle Age conception of the State as composed of classes permanently graded, with the ladder of lords rising from the *minuti homines* below to the king on his throne, and yet to such stratification, after each successive disturbance, society tends to return.

But those *minuti homines*, how are they to be contented, for is it not the duty and the desire of all to content them? The first and greatest step towards such contentment is taken when the grades find their right places. At such a time as the present much of the intensity of discontent is due to the fact that some are at the bottom who should be higher, while some are high who should be lower. For time is of the essence of the process; and two generations have scarcely passed since the great changes began. Then, strange as it may seem, content is not so very rare after all. There is a discontent which is caused not because something is withheld from us, but because we resent our own inferiority; for that there is no cure. Decent food and lodging, however, go far to satisfy *minuti homines* in general. Very early most of us accept the truth of Schumann's aphorism, that if everyone were determined to play first fiddle no orchestra could be got together.

As a boy in Cambridge I learnt that if a man got a first class he might be happy; if he got a second class he would be unhappy; if he got a third class, nothing but misery and a colonial life awaited him. When we grow older we unlearn these simple propositions, and we find that happiness is in many cases compatible with weekly wages and even with a pass degree. Some will have more than others. As in the body the heart is arranged so that the best blood goes to the head, so must and ought it to be with society.

Whatever is doubtful, this much I think is certain, that we are fast nearing one of those great secular changes through which history occasionally passes. The present social order is too unstable to last much longer, and he must be callous who greatly desires that it should. What will emerge from the approaching histolysis no one can predict. Let us hope, something better: and to this end may those upon whom devolves the duty of rearing that new organism, which is to grow from the dissolved tissues of society, be guided in their treatment, like physicians of the modern age, not by nostrum merely, but by the facts of natural physiology.

SCIENCE AND NATIONALITY

Presidential Address delivered at the Inaugural Meeting
of the Yorkshire Science Association

Edinburgh Review, 1919[1]

The position of science in its relation to the conduct and
policy of nations is a theme which has been in the thoughts of
most of us during these sad years. The end of the first act has
come but the tragedy may soon begin again. We claim the
proud title of scientific men, *Makers,* that is to say, of knowledge. In old times mankind was wont to turn to priests and
lawgivers for counsel. We are witnessing the ruin to which a
world professing the ideals of religion and law may come. Those
ideals claim to have made the world we see. The counsels of
science are as yet untried. Can the makers of natural knowledge help where the rest have failed? That is a question we
may well consider in this partial respite from horror which may
perhaps be brief.

A great cry has gone up in all the land; and not in our land
alone, but through all the earth, for is there a house where there
is not one dead? Caught in the wheels of a hideous destiny the
young men of the nations and the innocent boys have been torn
to pieces. The shattered victims from whom kindly death has
turned aside wring our hearts in every public place. They went
at the high call of Duty. The altar upon which they bled bears
the glorious names of Patriotism and Duty. From the enemy
cities and from their quiet villages has poured another stream
of youth to perish at the self-same shrine, calling alike on
Patriotism and Duty, with, we must believe, an equal devotion.
So it has been from the beginning; must it continue so to the
end?

We, scientific men, know that the life of one creature is the
death of another. Trained from our earliest years to face
natural fact, knowing with a precision and vividness, which the
layman seldom attains, that every animal and plant and every
race and strain of living things holds its place on the earth by
power and power alone, it is not we who cherish dreams of
equal rights and universal peace. Like other men we may seek
peace and ensue it, but in our hearts we know that never shall

[1] This Address is now restored as nearly as possible to its original form.
C. B. B.

true peace be found. With sincerity we may strive in the earlier words of the precept to eschew evil and do good, but not only can peace, absolute and whole, never be attained, it can never even be approached. The form of the struggle and its rapidity can alone be changed. Among mankind race may exterminate race by fire and sword, by drink and disease, or by the slower cruelties of competition in its countless disguises. The struggle is the same and will continue in the same perennial course while men increase and multiply upon the earth. When therefore we hear schemes for universal peace earnestly put forward, we are not misled, recognising, as enthusiasts commonly do not, that though wars may cease, struggle and competition go on.

With the cessation of war we are told that each nation will be free to develop, in President Wilson's phrase, "unhindered, un-threatened, unafraid". To the biologist the substance behind these words is illusion. And yet which of us that has human imagination and feeling, would not seek to abolish war? We *know* that war brings to men agony in volume and intensity that the secular sufferings of peace prolonged through the ages do not equal; that in war the devil that each of us hides in his heart is loosed in frenzy, driving honest men to commit crimes atrocious and unutterable, of which in sober health they scarcely know the names. The "calm, kind years, exacting their account of pain" we must all bear. The destiny of man is hard and he must fulfil it. Nature has laid on him her curse; he need not summon fiends from hell to multiply torture in its execution.

In time of war truth must be suppressed or garbled; history rewritten; the standards of candour and generosity suspended. To glorify our own virtues and to hide any that our enemy may have, conduct by which we used to recognise the cad, is no longer indecent, but, as it seems, incumbent on us all. No nation in modern times admits that it fights for an ignoble motive. Duty and Liberty draw armies forward, but there are divinities with titles less noble whose sinister forms hover in the rear. Sympathy for the oppressed seldom culminates in war but for considerations which commercial and geological maps reveal.

Those are foul accompaniments of war of which all are conscious. There are others which men like ourselves know to be scarcely less hateful. War is the very opportunity of evil. The common and the base rise; thought, art, science, the high things of the mind are put aside forgotten. The continuity of learning is threatened. The closing of the public museums and galleries in England fitly typified the general contempt of the community for the intellectual life. Surely men of our craft have cause to be haters of war.

But in what I have to say I am no advocate of the pacifist creed. The duty of self-defence is one which no Government can decline. I have never doubted that such a duty fell upon our Government in 1914. For the climax then reached the world had been long in preparation. To have averted that catastrophe the policy of nations must have changed its course long years ago.

I am not sanguine that it is in the power of the rulers and leaders of men, be they kings, statesmen, philosophers or men of science, to avert similar catastrophes in the future. Nevertheless there are courses which may be tried. The statement is sometimes made that from the principles of biology it is evident that wars must recur; that this deduction follows from scientific premisses, chiefly the fact that population increases more rapidly than the means of subsistence, and therefore that all attempts to abolish war must be in vain. Such pronouncements display a certain confusion of thought. Recognising, as we all do, that the struggle for existence must go on, that it is not in the power of any individuals or communities by mutual agreement or statesmanship to contract themselves out of that struggle, or in any way to protect themselves from it, I see nothing in what we know of biological fact which justifies the assertion that the struggle must periodically take the special form of war. War is the acute and spasmodic conflict between nations, or more correctly between *States*; a phase indeed of the struggle for existence, but not an essential manifestation of that phenomenon.

For example, since war is the conflict between States, it could be abolished in the limiting case in which the world were stably governed as a single State. Scientific men, whose studies necessarily have given them a sense of perspective in regard to time and space, will not agree that the conception of the world as a single State or unit is chimerical or incapable of realisation. Indeed of the possible future developments of civilisation, we may regard the idea of a World-State as one of the least improbable.

There are several ways in which this unity may come to pass. One method by which it might be attained has been already prominent in the public mind. It might be reached by the assumption of the whole terrestrial power by a single State. For an approach to universal peace so established and maintained, the Roman Empire offers an obvious parallel. Notice further that in a contracted world the chances of stability of a State conterminous with the globe are vastly increased. Holding the laboratories, arsenals and electrical communication, a resolute, unscrupulous and dominant State, having once firmly grasped the hegemony, might quite conceivably enforce peace for an indefinite period, perhaps as long as coal lasts. That some such dream as this was in the minds of an important section of the German people is fairly certain.

Is there no other mode in which the permanent cessation of war can be imagined?

The League of Nations is offered as a preventive. That is a lawyer's remedy. Confidence in that remedy is, I fear, strongest in those whose acquaintance with racial physiology is least. It is not among those who, like ourselves, are accustomed to rate the soundness of ideas according to their consonance with natural fact, that enthusiasm for the League of Nations will be found. Philosophers have declared that men are born equal. The naturalist knows that statement to be untrue. Whether we measure the bodily or the intellectual powers of men we find that the inequality is extreme. Moreover we know that the progress of civilisation has resulted solely from the work of the exceptional men. The rest merely copy and labour. By civilisation I mean, here as always, not necessarily a social ideal, but progress in man's control over Nature. As between individuals, so between nations, there is similar inequality. When even in Europe we observe that there are teeming populations which have scarcely made any significant contribution to art, learning or science throughout their history, we perfectly appreciate the meaning of that fact. We know how to deal also with the exceptions which believers in equality may adduce. If I see a red mangold in a field of " Prizewinners ", I know that its colour came from a red variety, such as Golden Tankard or Golden Globe, and that the plant is probably only a first cross. We are prepared therefore for the analogous fact that Copernicus, classified as a Polish astronomer, had a German mother, and that the father of Chopin, a Polish composer, was a Lorrainer. When we find proletariat nations producing genius of the highest kinds it will be time to reckon them our equals, but we are about as likely to see that time as we are to see a blonde Japanese. There are proletariat races as there are proletariat families and their redemption does not lie in statesmanship. The unequal distribution of illustrious men among the nations is a biological fact. France, Great Britain, Italy, Germany, and some smaller groups have since the revival of learning contributed many men of the magnitude we have now in mind. Some have excelled more in special arts or sciences, as for instance, in painting, music, literature, astronomy, chemistry and physics, biology or engineering, but in a wide view of these manifold excellences there is no obvious disparity to be noted between those nations. Besides the "illustrious" persons, each of the nations I have named has produced a vast number of men who have reached far above the degree which Galton defined as "eminence", seeing that the value of their work has been admitted universally. If they are reckoned, the inferiority of the proletariat nations becomes still more manifest; but since the performances of the eminent cannot always be distinguished from imitation, we are on safer ground in forgoing reference to them. The argument indeed founded on the distribution of the illustrious is so overwhelmingly convincing that further testimony is superfluous. If we needed other indications of the relative powers of the nations we find it in the fact that where the distributions of races overlap, the same races from which the "illustrious" and "eminent" are derived, become masters and foremen, while the proletariat races remain labourers and factory hands.

It appears to me from these considerations unlikely that nations dissimilar in intellectual attainments can co-operate in a League on equal terms. We have spoken only of the nations in Europe. The nations of Asia and Africa must also be represented in the League. Disparities still more fantastically great will then be introduced. For there again we find the same disproportionate performances, China and Japan with a record in art—not to speak of science or letters—that Europe has seldom equalled and never surpassed, side by side with hordes that have not only produced nothing memorable of their own, but are scarcely even capable of successful imitation. But all have, we are told, "equal rights". That they may have, but if it be so, there the equality is likely to end. Democracy, the system which confers equal political power on individuals, in defiance of genetic inequality, may, by forgoing that material progress which we know as civilisation, produce a phase of spurious equilibrium, the equilibrium of chaos and disruption, but the natural instability caused by the fact of physiological inequality is not unlikely to produce, as heretofore, its recurrent effects.

The contemplated League is a democracy of nations, and every consideration, drawn whether from science or commonsense, which raises doubt as to the supreme virtue of democracy in the single State, applies with special force to the case of a democracy of nations. Democracy is the combination of the mediocre and inferior to restrain the more able. As Flinders Petrie in a brilliant essay[1] lately shewed, democracy has been the final phase of each great civilisation. It may be an inevitable part of the physiology of nations that this phase should be reached, but the period of democracy is one of decay.

> Young boys and girls
> Are level now with men; the odds is gone,
> And there is nothing left remarkable
> Beneath the visiting moon.

In such a lethargy, approaching absolute zero of the mind, after the proscriptions—that *Ausrottung der Besten* as Seeck well called this ancient Bolshevism—the Roman Empire died.

Admittedly however the conditions of the modern world are different. In the older epochs each new period of civilisation began by invasion from without. In the modern world, contracted in time and space, such new invasion can scarcely occur; and could a democracy of nations be indeed firmly established, civilisation may perhaps enter on a period of decay from which there will be no revival until coal is exhausted.

But in truth there is little cause to dread that when the pressure of a common enemy is removed, the nations will restrain themselves for long by such a League. From the moment the League is constituted national consciousness will be still further stimulated, a condition little likely to promote peace. Seeing that the delegates to the League are not simply to be drawn from all nations but are to be sent into conference expressly as *representatives* of their several nations, the first consequence of the formation of the League must be to make prominent and to perpetuate that special form of the struggle for existence which arises between national units, of which war is the direct and immediate result. By the nature of the case men of spirit and ambition will be sent to this Parliament of the world, and quarrels will be bred, points of honour raised and exacerbation ensured, which but for the existence of such a court might perhaps have slept or died away.

Wars in the modern world are made by the awakening of national sentiment. National spirit has ebbed and flowed. After a period of abatement during which men's thoughts had turned rather to the common interests of nations than to their separate ambitions, the evil spirit of nationality has swept through the world.

Like all movement in human affairs, that spirit is roused from

---

[1] *The Revolutions of Civilisation*, Harper Bros. 1912.

quiescence by the force of individual minds, in various countries, acting and re-acting on each other. Among many such influences I will mention two that have been conspicuous. In the year 1834 were born two historians, Treitschke in Germany and Seeley in England, to whose writings perhaps more than to any other concrete incident may be traced that inflammation of national spirit which brought about the consequences we have been witnessing. We may note in passing the curious circumstance that in *The Expansion of England*, published in 1883, there should be never a hint or suspicion that the imperial sentiment which Seeley strove to excite would lead to a struggle with Germany. Russia is often named as an obvious source of danger ahead, and the possibility of friction with France is not forgotten. To those who are young there seems perhaps some natural, predestined fitness in the present grouping of the nations. They should remember that only thirty-five years ago national friendships and hatreds were very differently apportioned. So transient are those emotions which now appear inevitable and permanent! Upon Seeley in this country followed Mr Chamberlain, and upon Treitschke in Germany followed the ex-Kaiser with, of course, a host of others, fosterers of ambition and hatred.

Men of science, whose calling familiarises them with epochs, whose measure of achievement transcends the reckonings of statesmen, judge greatness by another scale. We have our units, and the commonalty have theirs. Seldom even are the two estimates commutable. For us a man is great not according as he has succeeded in influencing the ephemeral destinies of some artificial group on whom the lawyer has conferred the title of a State, but rather as he has extended thought or penetrated new provinces of knowledge. We speak sometimes of science, art and letters as an international domain. More truly we should think of them as *extra-national*.

Could the thoughts of man be turned, if only a little, away from the national towards the *extra*-national things, in such a change would lie the best hope of the world. Ultimately, as Lord Robert Cecil lately said, even if a League of Nations is in being, public opinion is the only force which can be invoked for the execution of its decisions. Public opinion can only be altered from above. It is for us men of science and our brethren in the arts and letters to lead. Public opinion has strange laws of growth, little understood as yet. Under influences which would repay analysis, the last fifty years have been a period of deterioration and of lowering in public ideals. The ascent may be less easy. Yet it is comforting to reflect that in private manners we have in the same period advanced. It *was* public opinion which abolished slavery in the West Indies overcoming vast financial interest. It might be impossible for Wilberforce and Clarkson to carry such a reform at the present day, but it *is* public opinion that has suppressed public cruelty to animals and children during our own lifetimes, and when a few years ago men suddenly gave up the habit of spitting in railway carriages, they changed their manner not for fear of a forty-shilling fine but at the bidding of fashion. The force of imitation, once developed, binds all but the rare exceptional men, and normal man, being almost incapable of independent thought, once bound by fashion is as powerless to escape as the hypnotised subject to resist the commands of the mesmerist. If the powers of the school, the church and the press were exerted to interest the world in extra-national things as they have been exerted to inflame national ambitions and to glorify war, the public opinion we long for may yet be created.

We have seen something of the resistless power of propaganda, the greatest political invention of our age. Those who control the press and the cinema can now in a few years inoculate the mass with any requisite opinion, whether poisonous or prophylactic. Democratic governments, intending a course of action, do not in modern conditions offer that measure directly. They prepare the way for a while by propaganda, judiciously exhibiting selected materials, arranged and timed to appear so as to produce a desired effect upon the passive minds of their populations, who presently find themselves thinking what they were meant to think, as they imagine, of their own mere motion. If the national leaders are sincere in their professed desire to abolish war they must proceed by this propagandist method, and they might then perhaps be successful. Simply to inveigh against war is as futile as to set up a League of Nations. The European races have no primary or instinctive dislike of war, nor is it probable that they can acquire that instinct. Though the horrors of war be depicted with the super-human force of Goya—beside whose terrific record contemporary cartoons are merely insipid—public opinion is not in any way disturbed or deflected. To form a public opinion unfavourable to war other and more positive images must be presented. The public mind must be turned to other thoughts and especially to objects which contribute to establish a sense of perspective and proportion.

Our children learn of kings and of battles, but of the stupendous extra-national events that have made the world, almost nothing. The names of the great pioneers, the discoverers, and the things they have done, of what races they were, and how, though separated by nationality, each has built on the work of the rest, these are the things that history should teach. The historian confesses that Newton was a greater man than Harley, but he still chooses Harley as the central figure of the reign of Anne.[1] Why do we suffer this abrogation of all value and proportion? It may be right to remember that 1848 was a year of political upheaval, but surely it is a date transcending in significance every other in the century, for in that year Pasteur made his first discovery of the properties of asymmetrical crystals out of which grew the whole science of bacteriology, modern surgery, very largely modern medicine, and other discoveries unrolling in endless series. Does someone say that the minds of common people cannot be made to see the size and meaning of such an event and that Waterloo makes a more moving picture? Low as may be our estimate of the common man, we can fairly rate him higher than that. We need the help of the writers and the poets. What epic theme of Titans or of blood can stand in grandeur beside the story of Promethean man, tearing their secrets from the elements, building, bit by bit, by his genius and toil the dazzling fabric of knowledge by which he shall surely scale the heavens? That is our *Paradise Regained*.

Wars may decide the destinies of nations; short of extermination they do not decide the destiny of *man*.

> The splendours of the firmament of time
> May be eclipsed but are extinguished not.

And just as science is an extra-national possession so is great art. Shakespeare, Beethoven, Rembrandt, Raphael, and the rest, the poets and the artists who have seen deepest into the heart of man, the makers of beauty, the creators of delight, the pioneers of emotion; in them shall all nations of the earth be blest. Their calm and mighty works soar eternally beyond the noises of temporal ambition, high above the plane on which the nations grapple. In their presence the voices of the partisans are hushed.

Who would not that his name should stand however low in this catalogue of the immortal? What honour can national pride offer that is fit to be compared with theirs? To put one's own nation above others for a while, awaiting its inevitable deposition at the hands of other patriots, is this really a high aim to set before our youth?

What is nationality that it should claim our devotion? That

[1] See *Expansion of England*, Lecture 1.

conception, primordially attaching to homogeneity of race, has plainly now no such exclusive meaning. We know how

> from a Mixture of all Kinds began
> That Het'rogeneous Thing an Englishman,

and yet "Your Roman-Saxon Danish-Norman English" are conspicuously a nation. If we travel through Provence we find a people with marked physical and mental characteristics, so well defined that we recognise a *méridional* immediately, wherever we meet him. But when we are in Normandy we are surrounded by a type as distinct as possible, distinct in appearance and with a wholly different outlook on life. Yet both are good Frenchmen, though the Norman has in him a good deal more of the Teuton, both of his virtues and his vices, than perhaps he cares to acknowledge.

If we inquire what does constitute a nation, we soon find that neither common racial origin, nor identity of language, or of religion, or of manners is essential. The most acceptable definition is probably that which declares that people compose a nation when they feel themselves to be a nation. But if we inquire how they come to share this feeling which has no necessary dependence on genetic relationship, on collocation in space, or common language, or common customs or beliefs, the answer is by no means obvious. National sentiment is a phenomenon of "polarity"—to use Tarde's ever-memorable expression—a compulsion under which a congeries of dissimilar elements can be controlled so as to point in one direction. It is in the power of those whom we call patriots, to give them their more noble title, thus to polarise the peoples. In their wake follow the journalist, the contractor and the manufacturer of armaments. The scene is set. Catch-words are chosen, insults bandied and the play begins.

Feelings closely analogous are not rarely aroused among purely chance association of persons. The boys of a school, of a particular boarding house, men forming the "class" of a University year, the crew of a vessel, owing their juxtaposition to the whims of guardians, the year of birth, or the vicissitudes of employment, are constantly liable to these blind emotions, often displaying them in their most violent forms.

Such instincts of coherence—for they are little more—have perhaps less of reasonable foundation than any others to which man is subject. They reach their highest development in men whose knowledge and experience is most limited, and are no more commendable than the instinct which compels a dog to fly at a stranger however amiable, in imaginary defence of a master however bad, a service he is equally ready to perform in a few weeks time for anyone else who may happen to buy or steal him.

It is possible to be convinced that our country is the best in the world without ever leaving our homes, without acquaintance with any language but our own, or reading anything but the newspapers, and some of the most unshakable judgments as to the respective merits of European countries are based on twenty years in India, reinforced by a few weeks in Swiss hotels or a visit to the Moulin Rouge.

Nationality therefore is a sentiment, picturesque and within limits laudable, yet in its essence accidental and ephemeral, capable of being turned to effects, often good so long as they last, but impermanent, adding nothing to *universal* good, and commonly a pretext for the grossest forms of selfishness and cruelty. Transitory as the fruits of patriotism must be, it is pathetic to observe that this force is especially invoked by the nations in their pitiful striving for terrestrial permanence and immortality.

This emotion acted on the German mind, I have been assured, in working up the passions of which the war was begotten. A year or two before the outbreak, a very patriotic but thoughtful German expressed this feeling to me. He said that wherever his countrymen settled they lost their language and national habits. Not only is this the case in England, Australia or America, but in Russia and even in Bohemia. If the process continued, he sighed, in a few generations there will be relatively little representing the German Tradition on earth, a thought he could not bring himself to face.

Surely one must have great confidence in the excellence of one's own ways to wish them permanently riveted upon the world. But however we may attempt to bind posterity, neither we nor our habits *will* persist even for periods which in the short view of the vulgar can be accounted long; and melancholy as the thought may be to the self-centred, we may be assured that in a few hundred years the language, the manners and customs, the ideals even, and all that we regard as distinctive of ourselves *will* have given place to others that we should not recognise as ours. Biology and history alike shew that all this transformation by mere lapse of time must come to pass even in our own lineal descendants, apart altogether from the process of racial replacement or infiltration by which in former ages more abrupt changes were effected.

Let us consider what we know of this phenomenon of lineal, physiological descent. It is pleasant to remember that we are the countrymen of Shakespeare or Newton, but when we base a claim to predominance in the world on such relationships, how much of that claim is substantial and how much idle vanity? With the discovery of genetic segregation the facts wear an aspect different from that in which men have been wont to see them. Where a few families of similar type have intermarried from time immemorial, a strain sensibly homogeneous may exist, and the members of the community may be so much alike that each generation does approximately represent the preceding in physical characteristics and aptitudes, so that they can feel a common pride in their successes and a common shame in their failures. Such homogeneous communities existed in many places up to the modern era of constant travel. I believe they still exist, for example, in some of the secluded valleys of Yorkshire. But nothing of the kind is now true of the population of any West European country as a whole. In various degrees the chief nations are mongrelised and we British are probably the most mongrel of all. I hasten to add that to this physiological fact I believe some of our national efficiency is largely due, partly because much differentiation of type means great variety of aptitudes, an essential to a large industrial community, but also because I am fairly sure, though the point cannot be developed here, that in special cases, cross-breeding even in mankind does contribute to vigour. The fact however which I wish more to emphasise is that by the workings of the phenomenon of genetic segregation a man's children may possess few of the transferable ingredients which characterised him, his grandchildren may possess none at all, and of his collaterals it is practically certain that few will contain so much of him that he need feel any personal satisfaction or humiliation in their performances.

When in these days a man claims that he is of a particular breed, say a Yorkshireman, he generally means little more than that his patronymic belongs to that race, or perhaps that for some generations his paternal ancestors lived in that part of the country. Unless he be of exclusive and immediate peasant origin it is improbable that he is in the genetic sense purely of one race. We are bi-parental, and since in each generation the mother's name is sunk, mixtures by intermarriage are soon forgotten; yet the female contributions to the genetic make-up are at least as significant as those of the male. For example, I have a name common on the borders of Lancashire and Yorkshire. My father's father's parents were peasants of that country, but I know of at least four other racial contributions. It is already not more than an even chance that I contain any given ingredient of my father's father's parents. If I lived in Yorkshire I should be called a Yorkshireman, but genetically I should be no more a Yorkshireman than my father was a Welshman because two of his maternal great-grandparents were Welsh. In all our urban populations, and in many of those of other countries, pedigrees at least as composite are so common as to be almost normal. In a very large number, moreover, elements definitely exotic occur.

Looked at coldly in the light of physiological knowledge, what is called the tie of blood is therefore in modern times exceedingly slender, and in all likelihood many of us contain no more of the elements that went to the making of Shakespeare and our heroes than the modern Greek contains of Zeus or Phoebus, despite the frequent alliances which those deities contracted with the daughters of men.

The old notion of the unity of a population has no physiological meaning in our modern world. It is framed on the legal figment of the undying Corporation, so prettily expressed by Sir John Davies:

> And so the ancestor and all his heires,
> Though they in number passe the Stars of Heaven,
> Are still but one; his forfeitures are theirs,
> And unto them are his advancements given.

That is a picture applicable to none of the polymorphic human societies.

The real and solid fact that does emerge from the study of pedigrees is, as I said earlier, that certain races have exhibited the sporadic property of being able to produce genius, while others have not done so at all or to a relatively small degree, and occasionally in their crosses.

The substitution of true ideas of heredity and of the biological structure of societies must therefore contribute something, if ever so little, to weaken the conviction that national sentiment is a proper and natural extension of normal fraternal affection. Confusion between pride of race and national pride is not unnatural in the ignorant and has been constantly fostered by publicists; but when once the two are clearly distinguished, we realise that slender as may be the grounds for pride of race in any given individual, those for pride of nationality are illusory. Moreover a well-informed public opinion will recognise that such pride of race as is derived from the production of the higher types of genius is shared by many races.

I look for no rapid change. It has taken more than two generations to bring national spirit to its present heat. The creation of an extra-national public opinion will take longer.

Observe that the suggestion here made appeals to definite human instincts—the love of imitation and the craving for approval—which have again and again prevailed in competition with more elemental forms of selfishness. The struggle for existence goes on. Men will remain greedy as before, but the special forms of collective greed which make for war may be controlled.

Let extra-national progress be recognised publicly as the highest and as the one indisputable good, in which all may share, and let it be known that in comparison, national pride is small and trivial, and so palpable a truth may not impossibly spread among the leaders of men. Fashion will do the rest. Those who contribute to extra-national advancement are certainly in all ages few, but though separated in time and space, *they* truly have ingredients in common. The bond which unites them is a thousand times more real than that which unifies a modern nation. If their collective consciousness could be awakened, as that of each separate nation has been, it might constitute a definite force for the direction of public opinion. Truth and beauty, science and art, wisdom and loveliness, these are extra-national possessions. They are the only aims which in the long run are worth pursuing. They are the treasure whose glory cannot pass away.

Reformers know that the only means of redeeming the criminal or the vicious lie in distraction, in changing the current of their thoughts. I am not without hope that if the strength and gifts of those who would form an Extra-National League were earnestly put forth in all their manifold forms, the thoughts of men might be turned into a new current and a public opinion might be aroused that would not merely in silence look on war as an evil, but might fix a stigma of shame on those who promote courses leading to war.

It is curious to reflect that the mediaeval Church did at times exert an extra-national authority somewhat analogous—not indeed for the repression of war, but for the enforcement of high standards of conduct. When Gregory VII made the Emperor Henry stand, under sentence of excommunication, shivering in his shirt in the yard at Canossa, I am perfectly aware that he was then exerting an irresistible force, Superstition, to which we cannot now appeal. But though we are no longer able to invoke hell-fire, there are pains more concrete that men really fear.

To stand well with their fellows, and especially with those whom they regard as their superiors is a motive that appeals even to stupid and selfish people. If the fomenter of national ambitions, instead of being praised as he now is, were made to feel that he is a mischief-maker, one of the worst enemies of mankind, that in the judgment of the wise he is vulgar, an object of contempt, a hindrance to the progress of the world, we might see an advance towards the abolition of war more rapid than pessimists imagine.

If someone declares, as he may with truth, that though sixty or seventy years ago, the extra-national ideals of art and science had taken a place in public esteem higher than that they now hold, they were soon, nevertheless, displaced by the ideals of trade and nationality, and that this sequence will recur, I would reply that I have a better hope, and for this reason. Science has introduced two new conditions. First, by its own operation, it has caused the earth to shrink so that a unity of public opinion, once created, can now be maintained which in the time of our grandfathers would have been impossible. But the second and prodigious new condition is the growth of science itself, then a doubtful wonder, now risen to majesty and volume all-enveloping, humbling the vanity of nations. Size tells at last. The magnitude and cosmic scope of natural knowledge stand revealed; and gazing into that deep perspective men will behold their pride in true proportion.

COMMON-SENSE IN RACIAL PROBLEMS

The Galton Lecture

Eugenics Review, 1919

It shews, I think, remarkable catholicity that the Eugenics
Education Society should have invited me to deliver the Galton
Lecture, inasmuch as though engaged in studies cognate with
your own and of a kind which furnish some of the basic materials
upon which the eugenist builds, I have never seen my way to
take a definite part in its activities nor even to become a member
of your body. In introduction I should like to explain the position which in common with several genetical colleagues both
here and in the United States I have thought it best to maintain
in this respect. Whoever is occupied with the practical investigation of genetic physiology can scarcely be out of sympathy
with your objects. Witnessing, as such a man does every day of
his life, the consequences of the working of the laws of heredity,
the knowledge that the destinies of mankind are governed by
the same laws is to him an all-pervading truth. Of this fact he
needs no reminder. The course of heredity varies in detail with
the organism and the characteristic under investigation, but
the nature of the control which heredity exerts is the same in all
living things. Every creature that has life arises by the division
of a pre-existing cell, and the nature of the offspring will be
determined by that of the parent until men gather grapes of
thorns or figs of thistles. When the breeder watches the descent
of qualities and of defects down the lines of his stock, to the
layman it may seem that these things are merely parables, but
the physiologist knows better. *De te fabula narratur.* Nevertheless the pursuit of truth is one thing and its application is another.
Few have combined these objects with success. At least in the
earlier stages of inquiry, to be committed even in general terms
to anything savouring of a policy will not strengthen such
authority as a worker may be able to claim in his own
province.

The terms of your membership are very wide and commit to
nothing beyond a desire to educate the world in a knowledge
of the truth, a wholly admirable purpose, but corporations almost invariably come to hold, or to be credited with holding,
corporate views and corporate principles, which are seldom in
practice compatible with perfect freedom. The eugenist and the

geneticist will, I am convinced, work most effectively without organic connection, and though we have much in common we should not be brigaded together. Genetics are not primarily concerned with the betterment of the human race or other applications, but with a problem of pure physiology, and I am a little afraid that the distinctness of our aims may be obscured. Alliances between pure and applied science are as dangerous as those of spiders, in which the fertilising partner is apt to be absorbed.

The truth which the eugenist is urging upon a reluctant and unheeding world is in essence this: that the physiological fact of the diversity of mankind is of prime importance in every consideration of human affairs: that all measures for the regulation of public or private conduct which ignore this fundamental fact are entered upon in defiance of common-sense, and that the consequences of such defiance are stupendous, and farreaching to a degree that can as yet be only dimly estimated. Those who have comprehended and realised this manifest truth plead further that since the diversity of type is certainly transmitted to posterity according to fixed and ascertainable rules, it behoves the human race to make the phenomena of heredity and racial physiology the objects of zealous study. The question at issue is whether the facts of physiology are to be ignored or to be accepted as the common ground from which conduct is to be directed. At the present time for the statesmen in whose hands the destinies of the world still remain, the facts of nature do not exist. Men are not animals propagated according to physiological systems, fixed like those of Chemistry or Astronomy, but voters, and how voters are propagated, with what consequence to themselves or to the succeeding generations, it is superfluous to ask or consider.

I cannot better bring out the distinction between what may be called on the one hand the rational or natural, and on the other the conventional or political view of our problem than by reference to what passed on the occasion of a recent Galton Lecture. The Dean of St Paul's delivered an address full of stimulus and penetration, indicating many indubitable consequences which recent legislation must certainly entail upon the composition of our population, results altogether outside the purview of those from whose action they ensue. Sir Auckland Geddes, in proposing the vote of thanks, after sufficiently indicating his own mode of thought by asking us to look with complacency on the danger of over-population—that overwhelming menace to the peace of the world and to the stability of civilisation—proceeded to affirm that "in politics, in the affairs with which Governments have to deal, it is not accurate knowledge that matters: it is emotion",[1] concluding with an exhortation that we should let ourselves go on the great wave of emotion sweeping the nation towards the millennium which the Ministry of Reconstruction, unhampered by accurate knowledge, was then preparing for us. As I listened to that speech with its presumptuous repudiation of knowledge in favour of sentiment, my latent sympathies with the movement which Galton inaugurated were kindled into activity, and I am proud indeed to deliver a lecture in which his name is commemorated.

The charge most often brought against the eugenic doctrine is that it aims at perpetuating a rash and subversive interference with habits and manners in pursuit of some cold and calculated purpose. I suppose that that is what some people mean by eugenics. Foolish legislation passed or promoted in certain American States gives colour to such opinions. I have heard also of busybodies who, in the name of eugenics, have made some irresolute young people gratuitously miserable. That crude interpretation is, so far as I see, based neither on scientific fact nor on common-sense. Everyone who has studied these problems at all would advise the State to put such control on the feeble-minded members of the population as to prevent their propagation. They are examples of a peculiar physiological condition, not very difficult to recognise, and when they interbreed, as at present they frequently do, they have no normal children but infallibly add to the asylum and institute population. As to the propriety and I may add the humanity of exercising control over these persons we are all agreed, but I know no warrant for direct legislative interference beyond that obvious and altogether special case.

The sterilisation of habitual criminals has been mooted in America. We require to know far more as to the genetics and aetiology of criminality before such a question can even be profitably discussed. Criminals are often feeble-minded, but as regards those who are not, the fact that a man is for the purposes of Society classed as a criminal tells me little as to his value, still less as to the possible value of his offspring. It is a fault inherent in criminal jurisprudence based on non-biological data that the law must needs take the nature of the offences rather than that of the offenders as the basis of classification. A change in the right direction has begun, but the problem is

[1] Reported in *Eugenics Review*, XI, 1919, 19.

difficult and progress will be very slow. Pending the institution of a proper classification it must happen that we all know, or know of persons convicted, perhaps even habitually, whom the world could ill spare. Therefore I hesitate to proscribe the criminal. Proscription, we may remember, is a weapon with a very nasty recoil. Might not some with equal cogency proscribe army contractors and their accomplices the newspaper patriots? The crimes of the prison population are petty offences by comparison, and the significance we attach to them is a survival of other days. Felonies may be great events locally, but they do not induce catastrophes. The proclivities of the war-makers are infinitely more dangerous than those of the aberrant beings whom from time to time the law may dub as criminals. Consistent and portentous selfishness, combined with dulness of imagination are probably just as transmissible as want of self-control, though destitute of the amiable qualities not rarely associated with the genetic composition of persons of unstable mind.

Eugenics is represented as a cold and ascetic faith. It is expected that

> Priests in black gowns will be walking their rounds
> And binding with briars our joys and desires.

I must grant that the doctrine is easily capable of such perversion. Galton himself, in a well-known passage which I cannot read without a shudder, speaks of the Bohemian habits "ingrained in the nature of the men who inhabited most parts of the earth now overspread by the Anglo-Saxon and other civilised races". He then declares that the Bohemian element in our own race is destined to perish, and "the sooner it goes the happier for mankind". I heard almost the same opinion in Germany before the war, and the speaker did not confine himself to general terms but specified the Latin races as the element which he said his countrymen regarded as destined to perish. In that and some other utterances of Galton's we are reminded that, like most men of great intellectual activity, his mind contained many dissimilar ingredients. With great skill and discernment in literature he combined a lurking contempt for the other arts which perhaps prompted these unsympathetic remarks about Bohemianism. With extraordinary elevation of outlook he was not without a respect for material success, much as his grandfather, though a quaker, made a fortune as a manufacturer of small arms. In the eugenic paradise I hope and believe that there will be room for the man who works by fits and starts, though Galton does say that he is a futile person who can no longer earn his living and ought to be abolished. The pressure of the world on the families of unbusinesslike Bohemians, artists, musicians, authors, discoverers and inventors, is severe enough in all conscience. In well-ordered communities their support should be a first charge on the State. They are literally the salt of the earth, without whom the savour of life would be flat and wearisome indeed. There is no more painful reading than the annual Civil List, which in the name of England allots to genius in distress sums which a Government official or a tradesman would despise.

Broadcloth, Bank balances and the other appurtenances of the bay-tree type of righteousness are not really essentials of the eugenic ideal. My notion of it is the exact contrary. Genetic discovery will put a new power into the hand of man. Will he use it to diminish his scope? Rather I should expect that a recognition of the wholesome teaching of biology would favour a wise and pagan sense of facts, teaching us to see things as they are. That knowledge must surely make for width and generosity, not for narrowness or restraint. Eugenic education should work not like the puritan campaign of Prohibition in America by abolishing one of the most precious and beneficent of pleasures, but rather by the obviation of suffering. With such an example before us we may well dread any development which invests authority with new powers. Prohibition, if it is maintained, will rank among the great disasters which from time to time have checked human progress. It is a reversal of civilisation, a forgoing of the fruits of one of the great discoveries by which man has learned to control nature and make his lot on earth tolerable. To abolish wine because men get drunk is like abolishing steel because men fight with it. Those who perpetrated that act of tyranny will not stop there. Neither tobacco, nor art, nor literature is safe. All this, to be sure, has happened before, but mercifully nothing came of it. Thomas Bowdler, F.R.S., was so convinced of the value of his expurgated Shakespeare and his "purified" Gibbon that he thought no other version would subsequently be published. Bowdler's opinion did not prevail, but in his day propaganda was not invented, and the art of raising waves of emotion for political purposes was imperfectly understood. The success of Prohibition is enough to make us regret the decay of monasticism; for in the Middle Ages uncomfortable people of that kind naturally gravitated into the monkeries, thereafter troubling the laity less.

But though each of us has his personal predilections we can

only make rough estimates of the worth of the several types and of their value to the world. Quantitative reckonings are still very far off, and meanwhile we must remain content with academic aspirations, praying only that in that day humanity may not be measured by the scale which would be appropriate to a Charity Organisation Society or a Board of Guardians, who I am told are able to distinguish the deserving from the undeserving poor.

To those who fear that the prevalence of eugenic ideas may have some such consequence I would remark that though this or some other country may not improbably submit itself to the government of a censorious clique, it will not be by the consent of those who are familiar with biological fact. The limits of responsibility may be clear to lawyers, but to us biologists they are very hazy indeed, and parochial views of man and his destiny do not commonly flourish in a biological atmosphere.

Two entirely different aspects of eugenic policy are to be distinguished. The one is personal, the other public. From the point of view of a young man or woman contemplating marriage it may be disquieting that one or other of the parties may have doubtful elements in their family history. From the point of view of Society it does not follow that the contribution of such a marriage, even if the trouble recurs, must be detrimental. When people discuss this question they usually have in view one of the two commonest family stigmata, tuberculosis or insanity. As regards tuberculosis there is, as yet, no clear proof that special susceptibility is hereditary in the ordinary sense, still less have we any evidence as to the genetic scheme by which it may be transmitted. A moment's reflection will shew how difficult it must be to learn anything with certainty on these points. We are concerned with an infective disease, to which the whole race is probably susceptible more or less. Its incidence and course are greatly influenced by climate, occupation and other conditions. In family life continual opportunities of infection occur. To trace the descent of special degrees of susceptibility in regard to such a widespread disease is almost impossible by present methods of investigation. On the eugenic bearing of tuberculosis therefore science has as yet nothing to say, and common-sense not much.

As regards insanity, the forms of mental disease are manifold and still most imperfectly distinguished. Apart from the fact that the pathology of these diseases is obscure, their development may be favoured or hindered by circumstances, and the age of onset is liable to great fluctuations. Hence the compilation of reliable charts of descent is impracticable. We cannot usually tell whether the normal or the abnormal state is the dominant, to use the technical term, and whether the defect can be transmitted by the healthy collaterals or not. Thus no system of descent can be predicated with much confidence, nor can genetic analysis be hopefully attempted in the present state of knowledge. We must remember, however, that even a fact so obvious as the syphilitic origin of general paralysis was only recently established. In former days general paralytics would certainly have been scored as insane, to the utter confusion of the pedigrees. The whole problem may be greatly elucidated by fresh pathological advances.

The existence of family defects may naturally cause anxiety to the persons concerned. They may, however, easily worry unduly on that account. Encouragement may be derived from many considerations, mainly arithmetical or actuarial, which I cannot now develop. But I would especially emphasize a doubt whether from the point of view of society, which is that in which we are here concerned, families which have suffered from definite stigmata may not contribute at least their proper share to the success and delight of mankind. We should hesitate to assert that either special susceptibility to tuberculosis, or any form of mental instability is associated with genius either directly or collaterally, but the frequency of such association has often been noticed, and I cannot deny that it is sufficient to suggest the reality of some positive connection. At least I imagine that by the exercise of continuous eugenic caution the world might have lost Beethoven and Keats, perhaps even Francis Bacon, and that a system might find advocates under which the poet Hayley would be passed and his friends Blake and Cowper rejected.

In so far as eugenics has yet attracted public notice it is chiefly to such particular problems that attention has been given. They are nevertheless beside the main point. No responsible person is proposing to subject human society to eugenic discipline. But those who would abhor any such proposal have to realise that a biological discrimination is no new thing, but on the contrary an inevitable consequence of almost any considerable legislative change. To make the world appreciate this simple truth I conceive to be the chief function of education in eugenics. Galton was the first to perceive this, or at least to demonstrate it with effect, and to shew that whether we like it or not, the social condition of mankind, both physical and intellectual, is the direct product mainly of the working of heredity. He saw that nations, unconsciously by their own acts of policy, favouring one class or discouraging another, change the genetic composition of succeeding generations. Though no such object be in contemplation, they decide the nature of their posterity, just as John Ellman decided the type of sheep with which the pastures of the South Downs should be stocked. The action of the farmer is deliberate, and as the sheep breeds more quickly than man and the matings are controlled, the result is seen much sooner. Also whereas the sheep farmer usually wants only one homogeneous type for his district, human society requires a mixture of a great number of distinct types. There is no other essential difference between the two procedures. We are now, and from the dawn of social government man has always been, interfering and tampering with the composition of his race. All that is new is that we have begun to see something of the ways in which the process works, and to think of it in biological terms. In his various writings Galton gives simple illustrations of this theme. He pointed out how the celibacy of the clergy kept down the numbers of the intellectual strains, since the Church was the only sphere in which such persons could find a congenial career. Again he indicated how in France the strict catholic families, and especially the Bretons, threaten to replace other components of the population for the reason that they obey the Church's ban on the practice of restricting the number of children by methods until lately almost universally adopted by the prudent remainder. These and other similar examples of the direct racial consequences of customs or laws instituted with altogether different objects will be familiar to you all. But the outstanding lesson taught by eugenic education is that the changes which must inevitably follow on *every* considerable interference in the distribution of wealth and of opportunity are ultimately racial. Neither custom nor law can be changed materially without introducing a discriminating influence on the prospects of the several varieties of which society is composed, and the more fully the extreme congenital diversity of the several types is realised, the more will the magnitude and extent of these racial effects become apparent.

Fun is made of the lament of the landed gentry in the last century that the country was going to the dogs, but we see now that they were perfectly right. They were, of course, thinking of their own class. They did go to the dogs and went quickly. Their very names are already partially extinct after a survival of several centuries. Their place is taken by financiers and tradespeople; and could those old gentlemen return to earth, they would see what a country looks like that retains only travesties of the things they valued. I am not concerned with the question whether they and their seed were worth perpetuating or not. I daresay they were a dull lot and their smart successors may be the better men—or women. Many would agree that a Sunday party at the Chequers or Glen was a more satisfactory and recuperative experience than a whole season's shooting with Squire Western. But I merely call attention to the fact that the county families are gone, largely by the operation of genetic process. Conditions supervened in which they could not maintain families and which were favourable to other breeds or "genotypes" as Johannsen calls them. Up and down the earth no doubt the strain exists. We saw many specimens among the Australian troops, descendants of those expropriated families, but those that remain here are, for the most part, merged in the general population, and their ancient homes know them no more.

Dean Inge in his Galton Lecture shewed how recent legislation must similarly work for the extinction of the intellectual middle classes. The value of money was suddenly halved as a consequence of measures which the Government improvised in its conduct of the war. The savings of the middle classes, which contributed to their stability, were correspondingly reduced. Being thrifty, as well as intelligent, these classes, though endangered, would before long have reinstated themselves, but further legislation has virtually prohibited thrift, rendering any savings on the modest scale attainable by persons not engaged in trade, altogether futile. The intellectual middle class, numerically by the nature of the case a small body, will therefore be obliterated, giving place to the thriftless majority, and to a class that a breeder might designate as culls.

Concurrently another influence is operating which cannot be without effect on racial composition. By the institution of abundant scholarships and other machinery for detecting and encouraging the abler children, practically every boy or girl in the elementary schools who shews marked mental aptitude is offered an opportunity of continuing education. Many are taking this opportunity. Some will reach the higher grades of industry, others will take their place among the professional classes. A process of sifting or gleaning is thus going on which must gradually remove, or as some would say raise, the more intelligent elements out of the industrial classes, thereby sensibly lowering the mental capacity of those classes. This consequence, so obvious to any one accustomed to think in genetical terms, was not contemplated by the promoters of the education movement, who commonly regard mankind as a homogeneous plastic substance which can be modelled to taste, unaware that they are in reality disentangling strands of permanently heterogeneous material. This process of disentangling will make for social peace in so far as it tends to allay feelings of discontent based on a real injustice, but it can scarcely postpone the doom of the intellectual class. Indeed their end may thereby be hastened, for the ranks of their destroyers, no longer containing dissentients who might in some degree understand their claims, may then be closed.

We are approaching a phase already reached in the United States, in which society consists of a small number of unstable and transient families possessing fabulous wealth—and the others whose incomes are more or less at the same level, having little or no property to bequeath. Here again it is beside my purpose to inquire whether this distribution makes for collective happiness or the contrary; nor do I know any means by which that question could be answered. But so long as we are divided into nations I have no doubt what the loss of the intellectual class will mean to the prospects of a state striving against equal competitors, or what in the long run it will mean to the development of mankind. That development was arrested for about a millennium by the domination of the Church. It may be suspended indefinitely by the edict of the proletariat. We may have made the world safe for democracy, but we have made it unsafe for anything else. If posterity takes any interest in history they will observe that the unusual feature of the Victorian epoch was not the exceptional distinction of the notables which it produced, for perhaps that was more evident to their contemporaries than it will ever be again, but the truly extraordinary circumstance that at that time intellectual distinction was held in public estimation as a thing of great worth. How and by whose example the mass which is congenitally incapable of appreciating art, literature or science was for a brief interval cowed into doing homage to an unknown god is most difficult to explain, but so it was. The unnatural phenomenon passed quickly by, but it left its trace on the fortunes of the intellectual class, consolidating their position for the moment though inducing a false sense of security and reconciling them to concessions which they can never recall. Released from that momentary enchantment of fashion the people have resumed their proper habits, gratifying their natural tastes in more congenial and less exacting ways. In so far as this is a return to candour and simplicity there is little to regret; but to those who have witnessed the rapid transition from a period when learning, the arts, and even pure science stood high in general reverence, to the present time when science is tolerated as a source of material advantage, when chaos is acclaimed as art, and learning supplanted by schools of commerce, the rarity not merely of intellectual producers but of intellectual consumers will need no further demonstration. The whole number of such persons in this country scarcely reaches to thousands, scattered sporadically among a population wholly different in tastes and capacity. Their existence is precarious indeed.

Genius it may be said can be trusted to force itself along the destined path, but even genius lives on the hope that some day recognition may come. Hitherto that recognition has been found in the intellectual middle class. But the amenities of life, leisure, reading, travel, social intercourse, are essential preparations for intellectual appreciation of any but the most meagre order, and the practical certainty that a man's contemporaries will not even have leisure to look at his work, let alone cultivation sufficient to appreciate it, may numb the bravest heart. Will they make books when there are none to understand? "What", as Bishop Stubbs lamented, when shewing a friend his library, "what is the use of a library now to me, a man who hasn't time to take a Seidlitz powder?"

The Dean of St Paul's, with invincible optimism, after pointing out such racial consequences as I have sketched with a weaker hand, concluded by predicting that after these events have passed by, especially after the great decline in population which I agree is to be expected, we shall emerge into a serener atmosphere. I wish I could see so far. Pending the exhaustion of the coal I anticipate a continuance of the new dark age. But without seeking to penetrate the remoter future, we can perceive that interference with racial composition is proceeding under our very eyes, though by the popular imagination any such action is supposed to be possible only in the dreams of a visionary.

I have spoken of the landed gentry as having become almost extinct in this country, and of the probability that the intellectual classes will also disappear. Let us consider by what steps these processes of obliteration are brought about, and what exactly becomes of racial elements thus submerged. It is here that the popular view differs from the genetical. I may illustrate the process by reference to what we know of other breeds that have been lost in consequence of change in the vagaries of fashion. Bull dogs and bull terriers are at the present time two very distinct breeds, each breeding approximately true. By reference to early illustrations and descriptions their history can be made out with fair accuracy. If some fancier nowadays wanted dogs of the pattern that he saw represented in an engraving, say of the end of the eighteenth century, probably no one could supply him. The factorial elements of which those older types consisted no longer exist in combination with each other. The modern bull dog and bull terrier between them have some of these characteristics. The Boston terrier might supply others, and among other varieties it might be possible to find missing ingredients such as special colours, or shapes of head or ear. A skilful breeder could, by combining these ingredients, reconstruct a given pattern, provided that no element has wholly disappeared. Observe that the types of bull dogs disappeared through specialisation. If, however, the modern breeds were to become unfashionable, these types would disappear as many have done in the past, because no one would trouble to breed them. Such offspring as existing dogs might leave would be merged in the general mongrel population. Conceivably our present types could be reconstructed by selection from that mass, but it would be a difficult task. This illustration corresponds in some respects fairly closely with what we see in human strains. For 150 years ago, few breeds of dogs had been approximately fixed. There were no club standards and each breeder followed his own ideas. An assemblage of the various dogs in those days would have been exceedingly like an assemblage of say Englishmen at the present time. Some would be, technically, thoroughbred in various respects, others would be wholly mongrel. We have not in our mixed community strains pure for any but the grosser distinctions, but there are plenty of incipient strains. The landed gentry were such an incipient strain. The clerical, educational and especially the theatrical world provide other examples. Readers of *Hereditary Genius* will be aware of more. So long as things go well with them the homogeneity of the strain increases, as that of the professional classes certainly has done in the past 100 years; but if, by Act of Parliament or change of fashion, their special product is made comparatively worthless, not only is their prospect of rearing families diminished immediately, but their caste becomes merged in the general population, and their children will be of a less special type. The statesman, knowing nothing of genetical principles, if he considers the problem at all, perhaps supposes that a displaced type takes to some other occupation. The probable success of these transformations may be estimated by those who will imagine a Clydesdale entering for the Derby, or Mr Henry Chaplin exchanging parts with his namesake, Charlie.

Until we can make the analyses of the descent of human faculty which have now been proved for so many animals and plants, there will of course be sceptics who take refuge in the defence that human pedigrees follow other rules or no rules at all. Of course there will be intricacies to unravel in regard to the mental faculties, sex-limitation and other difficulties of a technical kind, but if man's generations were annual, in the twenty years that have elapsed since the study of genetics was properly inaugurated, we should already have established an outline of the system of his inheritance. Of the facts hitherto ascertained there is only one which is in any way singular—the comparative frequency of new dominants. That is rather exceptional, though not suggestive of fundamental difference, and it may very well be an illusion due to the attention with which the variations of man have been studied.

We may be perfectly satisfied that the ordinary rules or rules very like them hold in regard to man, and that in obedience to these rules the intellectual class may disappear, merged in the unsegregated mass of the dominant population, for most of whom any but the most rudimentary effort of the mind is a physical impossibility.

Racial discrimination is, then, in constant progress. Might we not gain by recognising this fact and allowing it prominence in political philosophy? It is improbable that all states will always be ruled without regard to knowledge. I have said that we have no accurate method of estimating the value of the several types, nor can we declare the proportions in which each should be represented in a community, but that is no reason for treating populations as though we had still to discover that they are made up of living organisms, obeying certain principles in their propagation. Mathematics has long been recognised as a study helpful to the political economist, and the two sciences have often been pursued in common. I should have thought there was a closer affinity between economics and biology, and I am convinced that biological principles in so far as they relate to the nature of variety, the permanence of type, and the laws of hereditary transmission should not be without influence on economic thought. These phenomena lie at the root of political science.

In so far as economic philosophy is a constructive and applied science it is endeavouring to ascertain by what distribution of wealth and political power the well-being of a nation may be best promoted. Yet in the attempt to solve that intricate problem a knowledge of the biological structure of the community is usually treated as superfluous. The first essentials for any serious investigation of these questions are analytical data as to the distribution of faculty and similar particulars, which our present returns to the Registrar-General do not aim at supplying. To decide from these returns how the birth-rate is distributed among the various grades, even among the various ages, of the parents, is not readily possible. The sudden rise of the birth-rate has been made the occasion of rejoicing on the part of the unthinking. If we could see a parade of the parents who have made themselves responsible for this excess, I wonder if we should take so much pride in their performance, and whether we should not simply see in this output of spawn one more manifestation of the recklessness engendered by a period of spurious prosperity.

The Registrar's returns might be made to give that information, but they do not; because, though governments are manipulating living units, social physiology is no concern of theirs. "It is not accurate knowledge that matters", as the late Minister of Reconstruction told you.

Man is from the naturalist's point of view a domesticated and most variable animal, and to make a domesticated species, recently derived from various stocks, into a breed of equal and similar individuals is only possible in the peculiar and highly special circumstances which the fancier can provide. We are a heterogeneous group of dissimilar beings, and it is time that the greatness of this dissimilarity were brought home to all civilised communities. No one perhaps at this time would venture to assert that men are born equal, but few realise *how* unequal they are. By an ingenious calculation Galton found that in many types of competition the difference between the performance of the winner and that of the second man is commonly three times that of the distance separating the second man from the next best, and he concluded that if there are only to be two prizes, the first prize ought to be three times the value of the second prize. That is one expression of the heterogeneity of a mixed population. Has anyone considered the implications of this natural heterogeneity in political economy? Who dare tell voters that? They must all have prizes. Soon, moreover, the prizes are to be all equal, an episode which must lead rapidly to the grand finale in which there are no prizes at all.

If it be too much to expect that those who aspire to form

judgments in the problems of social politics should be trained in biological science, perhaps something might be done by applying those methods of pictorial instruction adopted in elementary schools. Just as the children learn the dissimilarities and attributes of the various animals, the giraffe, the elephant, the bat, and the toad, from wall-pictures, so in the schools of economics, political clubs, and other centres of debate there might be hung graphic representations exhibiting the prodigious diversities of our population. The physical differences would be easy to illustrate in this way, but there must also be graphic representations of the far greater intellectual diversities. Everyone knows that people differ, but the towering majesty of great minds, the commanding range and variety of their knowledge, the intensity of their penetration, who shall measure these powers? There is no rod by which the genius of Shakespeare might be translated into feet and inches, but his vocabulary has been counted and perhaps the club orator might be induced to count his own. We cannot sound the depth of Newton, but to learn that the performance of a good Senior Wrangler might be judged by skilled examiners to be twice as good as that of a decent second Wrangler, and thirty-two times as good as that of a low junior optime would help to give the rest of us some idea where we come in.

I wish Galton had designed a set of such charts, portraying these kings over all the children of pride. I cannot imagine anything better fitted to teach humility, the only foundation of contentment.

It is not inequalities in the distribution of wealth, or of opportunity, or of political power which are to be deplored, but the totally distinct evil that those inequalities do not follow with sufficient accuracy the congenital inequality of faculty. The inequality of faculty is a natural and physiological fact. It can only be obliterated by measures which would treat genius and capacity as the enemies of mankind. The effort of the social reformer should instead be turned to the means by which reward may be more closely correlated with performance. Under a public opinion which, like our own, rates and rewards commercial success more highly than any other class of achievement, this counsel may sound fatuous enough, but let us at least know the point at which trouble begins.

If we cannot grade the various kinds of excellence in a comparative order of merit there might be general agreement that certain classes of attainment, of which a thoroughly comprehensive schedule could be prepared, should entitle the

immediate posterity to some favour. Prizes or allowances for
unexceptionable marriages have, I believe, been contemplated,
a course which, frankly, I feel would be intolerable. A proper
place for recognition is the assessment of the death duties. It is
there that considerations of heredity should be allowed their
weight. In estimating the proper incidence of the death duties
one of the questions to be considered is what fraction of the
estate represents the probable deterioration of the offspring
from the position reached by the parent, and thus on eugenic
grounds should be forfeited to the State.[1] Genetic analysis has
not nearly reached the stage of successful prediction as to the
distribution of faculty among the immediate offspring, but the
broad principle that even under a system of random mating the
performances of ascendants is an indication of the probable
quality of the offspring would not be questioned by any geneticist. The death duties might be graded in accordance with
this expectation. For the present, strict analyses being impossible, cruder methods of estimation must be applied, but
I suppose to those who could shew sixty-four quarterings of
scheduled excellences this form of taxation might be remitted
altogether. With the advance of genetic science I would venture
to predict that the same privilege might be extended to those
whose first-class quarterings of certain kinds (in the right
places) amounted to eight or even four, but that is at present a
personal and perhaps heretical opinion. In actual practice the
performances of the children would be often ascertainable,
and might be brought to account in the calculation of the
assessment.

The problem of the Upper House might be treated on similar
principles. What is wrong is not that the House of Lords is
constituted on a hereditary basis, but that without further
proved qualification the privilege of membership is extended
in perpetuity. Lancashire reckons three generations from clogs
to clogs, and perhaps three generations from commoner to commoner might represent the actuarial expectation under a system
of almost random mating, which, by the way, is not prevalent
in the peerage. But it would be not impossible here also to
compile a schedule of various attainments which, on either side
of the parentage, might justify a renewal of the prerogative for
a further life or lives.

To give quantitative expression to these values might be
difficult but it would not be unattainable. I imagine to Galton

[1] The French law which forbids the alienation from the family of more
than a minor part of an estate has much to commend it.

himself it would have been fairly easy. His genius shewed itself especially in the ease with which he found numerical measures for differences which to the ordinary mind seemed qualitative and incapable of anything like exact estimation. In his definition of the "eminent" man as being the foremost in 4000 and the "illustrious" as being the best in a million, an estimate which he tested and justified by various independent methods, he shewed the lines on which such estimations should be attempted.

There are plenty of ways in which common-sense might be applied to racial problems. That we shall see any such application in our own country seems to me in a high degree unlikely. But we are not the only nation in the world, and a competitor may not improbably learn from us, appropriate our discoveries, and enter into our labours. The consequences of bringing biological knowledge to bear on the composition of society must be enormous, rapidly accomplishing aims of a magnitude that statesmen perhaps have never conceived. Our own concern in these developments will probably be that of spectators. Equality of political power has been bestowed on the lowest elements of our population. This is nearing the final stage of democratic decay, in which the lowest not only have the power but exercise it, a sequel which the next generation may witness. I am not aware that any community so heterogeneous as our own has ever made this experiment. Our immediate posterity will learn something of the consequences of un-applied biology. The force of the intellectual and professional class is assuredly prodigious and there is a bare chance that they may exert it in some co-ordinated form, against which the rest could of course offer no effective resistance. Recent history does not encourage us to expect that any such thing will come to pass. The truth has been recognised too late.

NOTE. Some years later Will was invited to lecture in Cambridge on *Eugenics*. Mr Michael Pease kindly allows me to quote from his reply:

28. 1. 1925.

...The fact is I never feel Eugenics is my job. On and off I have definitely tried to keep clear of it. To real Genetics it is a serious—increasingly serious—nuisance diverting attention to subordinate and ephemeral issues, and giving a doubtful flavour to good materials. Three times I have come out as an Eugenist, yielding to a cheap temptation, and on each occasion I have wholly missed even that humble mark; I don't mean to try again. My Galton lecture, which I thought would be a famous clap-trap, had the unique distinction of being the only Galton lecture to which no single newspaper would make allusion, much less report. I infer it got home on to somebody's nerves all right. I have tried to republish these papers with others more or less cognate, but publishers know their public and refuse with contumely. My eugenic career I regard as closed, and serve me right for dabbling in taboo waters. The kind of thing I say on such occasions is what no reformer wants to hear, and the Eugenic ravens are croaking for Reform....

EVOLUTIONARY FAITH AND MODERN DOUBTS

Address to American Association for the Advancement of
Science. Toronto, 1922

Science, LV, 1412. Jan. 1922

I visit Canada for the first time in delightful circumstances.
After a period of dangerous isolation, intercourse between the
centres of scientific development is once more beginning, and
I am grateful to the American Association for this splendid
opportunity of renewing friendship with my western colleagues
in Genetics, and of coming into even a temporary partnership
in the great enterprise which they have carried through with
such extraordinary success.

In all that relates to the theme which I am about to consider
we have been passing through a period of amazing activity and
fruitful research. Coming here after a week in close communion
with the wonders of Columbia University, I may seem behind
the times in asking you to devote an hour to the old topic of
evolution. But though that subject is no longer in the forefront
of debate, I believe it is never very far from the threshold of our
minds, and it was with pleasure that I found it appearing in
conspicuous places in several parts of the programme of this
meeting.

Standing before the American Association, it is not unfit that
I should begin with a personal reminiscence. In 1883 I first
came to the United States to study the development of Balanoglossus at the Johns Hopkins summer laboratory, then at
Hampton, Va. This creature had lately been found there in an
easily accessible place. With a magnanimity, that on looking
back I realise was superb, Professor W. K. Brooks had given
me permission to investigate it, thereby handing over to a young
stranger one of the prizes which in this age of more highly
developed patriotism, most teachers would keep for themselves
and their own students. At that time one morphological
laboratory was in purpose and aim very much like another.
Morphology was studied because it was the material believed
to be most favourable for the elucidation of the problems of
evolution, and we all thought that in embryology the quint-
essence of morphological truth was most palpably presented.
Therefore every aspiring zoologist was an embryologist, and
the one topic of professional conversation was evolution. It

had been so in our Cambridge school, and it was so at Hampton.

I wonder if there is a single place where the academic problems of morphology which we discussed with such avidity can now arouse a moment's concern. There were of course men who saw a little further, notably Brooks himself. He was at that time writing a book on heredity, and, to me at least, the notion on which he used to expatiate, that there was a special physiology of heredity capable of independent study, came as a new idea. But no organised attack on that problem was begun, nor had anyone an inkling of how to set about it. So we went on talking about evolution. That is barely forty years ago; to-day we feel silence to be the safer course.

Systematists still discuss the limits of specific distinction in a spirit, which I fear is often rather scholastic than progressive, but in the other centres of biological research a score of concrete and immediate problems have replaced evolution.

Discussion of evolution came to an end primarily because it was obvious that no progress was being made. Morphology having been explored in its minutest corners, we turned elsewhere. Variation and heredity, the two components of the evolutionary path, were next tried. The geneticist is the successor of the morphologist. We became geneticists in the conviction that there at least must evolutionary wisdom be found. We got on fast. So soon as a critical study of variation was undertaken, evidence came in as to the way in which varieties do actually arise in descent. The unacceptable doctrine of the secular transformation of masses by the accumulation of impalpable changes became not only unlikely but gratuitous. An examination in the field of the interrelations of pairs of well characterised but closely allied "species" next proved, almost wherever such an inquiry could be instituted, that neither could both have been gradually evolved by natural selection from a common intermediate progenitor, nor either from the other by such a process. Scarcely ever where such pairs co-exist in nature, or occupy conterminous areas do we find an intermediate normal population as the theory demands. The ignorance of common facts bearing on this part of the inquiry which prevailed among evolutionists, was, as one looks back, astonishing and inexplicable. It had been decreed that when varieties of a species co-exist in nature, they must be connected by all intergradations, and it was an article of faith of almost equal validity that the intermediate form must be statistically the majority, and the extremes comparatively rare. The plant breeder might declare that he had varieties of Primula or some other plant lately constituted, uniform in every varietal character, breeding strictly true in those respects, or the entomologist might state that a polymorphic species of a beetle or of a moth fell obviously into definite types, but the evolutionary philosopher knew better. To him such statements merely shewed that the reporter was a bad observer, and not improbably a destroyer of inconvenient material. Systematists had sound information but no one consulted them on such matters or cared to hear what they might have to say. The evolutionist of the 'eighties was perfectly certain that species were a figment of the systematist's mind, not worthy of enlightened attention.

Then came the Mendelian clue. We saw the varieties arising. Segregation maintained their identity. The discontinuity of variation was recognised in abundance. Plenty of the Mendelian combinations would in nature pass the scrutiny of even an exacting systematist and be given "specific rank". In the light of such facts the origin of species was no doubt a similar phenomenon. All was clear ahead. But soon, though knowledge advanced at a great rate, and though whole ranges of phenomena which had seemed capricious and disorderly fell rapidly into a co-ordinated system, less and less was heard about evolution in genetical circles, and now the topic is dropped. When students of other sciences ask us what is now currently believed about the origin of species we have no clear answer to give. Faith has given place to agnosticism for reasons which on such an occasion as this we may profitably consider.

Where precisely has the difficulty arisen? Though the reasons for our reticence are many and present themselves in various forms, they are in essence one; that as we have come to know more of living things and their properties, we have become more and more impressed with the inapplicability of the evidence to these questions of origin. There is no apparatus which can be brought to bear on them which promises any immediate solution.

In the period I am thinking of it was in the characteristics and behaviour of animals and plants in their more familiar phases, namely, the zygotic phases, that attention centred. Genetical research has revealed the world of gametes from which the zygotes, the products of fertilisation, are constructed. What has been there witnessed is of such extraordinary novelty and so entirely unexpected that in presence of the new discoveries we would fain desist from speculation for a while. We see long courses of analysis to be travelled through, and for some time to come that will be a sufficient occupation. The evolutionary systems of the eighteenth and nineteenth centuries were attempts to elucidate the order seen prevailing in this world of zygotes and to explain it in simpler terms of cause and effect: we now perceive that that order rests on and is determined by another equally significant and equally in need of "explanation". But if we for the present drop evolutionary speculation it is in no spirit of despair. What has been learned about the gametes and their natural history constitutes progress upon which we shall never have to go back. The analysis has gone deeper than the most sanguine could have hoped.

We have turned still another bend in the track and behind the gametes we see the chromosomes. For the doubts—which I trust may be pardoned in one who had never seen the marvels of cytology, save as through a glass darkly—can not, as regards the main thesis of the *Drosophila* workers, be any longer maintained. The arguments of Morgan and his colleagues, and especially the demonstrations of Bridges, must allay all scepticism as to the direct association of particular chromosomes with particular features of the zygote. The transferable characters borne by the gametes have been successfully referred to the visible details of nuclear configuration.

The traces of order in variation and heredity which so lately seemed paradoxical curiosities have led step by step to this beautiful discovery. I come at this Christmas season to lay my respectful homage before the stars that have arisen in the west. What wonder if we hold our breath? When we knew nothing of all this the words came freely. How easy it all used to look! What glorious assumptions went without rebuke. Regardless of the obvious consideration that "modification by descent" must be a chemical process, and that of the principles governing that chemistry science had neither hint, nor surmise, nor even an empirical observation of its working, professed men of science offered very confidently positive opinions on these nebulous topics which would now scarcely pass muster in a newspaper or a sermon. It is a wholesome sign of return to sense that these debates have been suspended.

Biological science has returned to its rightful place, investigation of the structure and properties of the concrete and visible world. We cannot see how the differentiation into species came about. Variation of many kinds, often considerable, we daily witness, but no origin of species. Distinguishing what is known from what may be believed we have absolute certainty that new forms of life, new orders and new species have arisen on the earth. That is proved by the palaeontological record. In a spirit of paradox even this has been questioned. It has been asked how do you *know* for instance that there were no mammals in palaeozoic times? May there not have been mammals some-where on the earth though no vestige of them has come down to us? We may feel confident there were no mammals then, but are we sure? In very ancient rocks most of the great orders of animals are represented. The absence of the others might by no great stress of imagination be ascribed to accidental circum-stances.

Happily however there is one example of which we can be sure. There were no Angiosperms—that is to say "higher plants" with protected seeds—in the carboniferous epoch. Of that age we have abundant remains of a world-wide and rich flora. The Angiosperms are cosmopolitan. By their means of dispersal they must immediately have become so. Their remains are very readily preserved. If they had been in existence on the earth in carboniferous times they must have been present with the carboniferous plants, and must have been preserved with them. Hence we may be sure that they did appear on the earth since those times. We are not certain, using certain in the strict sense, that the Angiosperms are the lineal descend-ants of the carboniferous plants, but it is very much easier to believe that they are than that they are not.

Where is the difficulty? If the Angiosperms came from the carboniferous flora why may we not believe the old comfortable theory in the old way? Well, so we may if by belief we mean faith, the substance, ὑπόστασις, the foundation of things hoped for, the evidence of things not seen. In dim outline evolution is evident enough. From the facts it is a conclusion which in-evitably follows. But that particular and essential bit of the theory of evolution which is concerned with the origin and nature of *species* remains utterly mysterious. We no longer feel, as we used to do, that the process of variation now contem-poraneously occurring is the beginning of a work which needs merely the element of time for its completion; for even time can not complete that which has not yet begun. The conclusion in which we were brought up, that species are a product of a sum-mation of variations, ignored the chief attribute of species that the product of their crosses is frequently sterile in greater or less degree. Huxley, very early in the debate, pointed out this grave defect in the evidence, but before breeding researches had been made on a large scale no one felt the objection to be serious. Extended work might be trusted to supply the deficiency. It has not done so, and the significance of the negative evidence can no longer be denied.

When Darwin discussed the problem of inter-specific sterility in the *Origin of Species* this aspect of the matter seems to have escaped him. He is at great pains to prove that inter-specific crosses are *not always* sterile, and he shews that crosses between forms which pass for distinct species may produce hybrids which range from complete fertility to complete sterility. The fertile hybrids he claims in support of his argument. If species arose from a common origin, clearly they should not always give sterile hybrids. So Darwin is concerned to prove that such hybrids are by no means always sterile, which to us is a commonplace of everyday experience. If species have a common origin, where did they pick up the ingredients which produce this sexual incompatibility? Almost certainly it is a variation in which something has been added. We have come to see that variations can very commonly—I do not say always—be distinguished as positive and negative. The validity of this distinction has been doubted, especially by the *Drosophila* workers. Nevertheless in application to a very large range of characters, I am satisfied that the distinction holds, and that in analysis it is a useful aid. Now we have no difficulty in finding evidence of variation by loss—examples abound; but variations by addition are rarities, even if there are any which must be so accounted. The variations to which inter-specific sterility is due are obviously variations in which something is apparently added to the stock of ingredients. It is one of the common experiences of the breeder that when a hybrid is partially sterile, and from it any fertile offspring can be obtained, the sterility, once lost, disappears. This has been the history of many, perhaps most, of our cultivated plants of hybrid origin.

The production of an indubitably sterile hybrid from completely fertile parents which have arisen under critical observation from a single common origin is the event for which we wait. Until this event is witnessed, our knowledge of evolution is incomplete in a vital respect. From time to time a record of such an observation is published, but none has yet survived criticism. Meanwhile, though our faith in evolution stands unshaken, we have no acceptable account of the origin of "species".

Curiously enough, it is at the same point that the validity of the claim of natural selection as the main directing force was most questionable. The survival of the fittest was a plausible account of evolution in broad outline, but failed in application to specific difference. The Darwinian philosophy convinced us that every species must "make good" in nature if it is to survive, but no one could tell how the differences—often very sharply fixed —which we recognise as specific, do in fact enable the species to make good. The claims of natural selection as the chief factor in the determination of species have consequently been discredited.

I pass to another part of the problem, where again, though extraordinary progress in knowledge has been made, a new and formidable difficulty has been encountered. Of variations we know a great deal more than we did. Almost all that we have seen are variations in which we recognise that elements have been lost. In addressing the British Association in 1914 I dwelt on evidence of this class. The developments of the last seven years, which are memorable as having provided in regard to one animal, the fly *Drosophila*, the most comprehensive mass of genetic observation yet collected, serve rather to emphasize than to weaken the considerations to which I then referred. Even in *Drosophila*, where hundreds of genetically distinct factors have been identified, very few new dominants, that is to say positive additions, have been seen, and I am assured that none of them are of a class which could be expected to be viable under natural conditions.

If we try to trace back the origin of our domesticated animals and plants, we can scarcely ever point to a single wild species as the probable progenitor. Almost every naturalist who has dealt with these questions in recent years has had recourse to theories of multiple origin, because our modern races have positive characteristics which we cannot find in any existing species, and which combinations of the existing species seem unable to provide. To produce our domesticated races it seems that ingredients must have been added. To invoke the hypothetical existence of lost species provides a poor escape from this difficulty, and we are left with the conviction that some part of the chain of reasoning is missing. The weight of this objection will be most felt by those who have most experience in practical breeding. I cannot, for instance, imagine a round seed being found on a wrinkled variety of pea except by crossing. Such seeds, which look round, sometimes appear, but this is a superficial appearance, and either these seeds are seen to have the starch of wrinkled seeds or can be proved to be the produce of stray pollen. Nor can I imagine a fern-leaved Primula producing a palm-leaf, or a star-shaped flower producing the old type of *sinensis* flower. And so on through long series of forms which we have watched for twenty years.

Analysis has revealed hosts of transferable characters. Their combinations suffice to supply in abundance series of types which might pass for new species, and certainly would be so classed if they were met with in nature. Yet critically tested, we find that they are not distinct species and we have no reason to suppose that any accumulations of characters of the same order would culminate in the production of distinct species. Specific difference therefore must be regarded as probably attaching to the base upon which these transferables are implanted, of which we know absolutely nothing at all. Nothing that we have witnessed in the contemporary world can colourably be interpreted as providing the sort of evidence required.

Twenty years ago, de Vries made what looked like a promising attempt to supply this so far as *Oenothera* was concerned. In the light of modern experiments, especially those of Renner, the interest attaching to the polymorphism of *Oenothera* has greatly developed, but in application to that phenomenon the theory of mutation falls. We see novel forms appearing, but they are no new species of *Oenothera*, nor are the parents which produce them pure or homozygous forms. Renner's identification of the several complexes allocated to the male and female sides of the several types is a wonderful and significant piece of analysis introducing us to new genetical conceptions. The Oenotheras illustrate in the most striking fashion how crude and inadequate are the suppositions which we entertained before the world of gametes was revealed. The appearance of the plant tells us little or nothing of these things. In Mendelism, we learnt to appreciate the implication of the fact that the organism is a double structure, containing ingredients derived from the mother and from the father respectively. We have now to admit the further conception that between the male and female sides of the same plant these ingredients may be quite differently apportioned, and that the genetical composition of each may be so distinct that the systematist might without extravagance recognise them as distinct specifically. If then our plant may by appropriate treatment be made to give off two distinct forms, why is not that phenomenon a true instance of Darwin's origin of species? In Darwin's time it must have been acclaimed as exactly supplying all and more than he ever hoped to see. We know that that is not the true interpretation. For that which comes out is no new creation.

Only those who are keeping up with these new developments can fully appreciate their past significance or anticipate the next step. That is the province of the geneticist. Nevertheless,

I am convinced that biology would greatly gain by some co-operation among workers in the several branches. I had expected that Genetics would provide at once common ground for the systematist and the laboratory worker. This hope has been disappointed. Each still keeps apart. Systematic literature grows precisely as if the genetical discoveries had never been made and the geneticists more and more withdraw each into his special "claim"—a most lamentable result. Both are to blame. If we cannot persuade the systematists to come to us, at least we can go to them. They too have built up a vast edifice of knowledge which they are willing to share with us, and which we greatly need. They too have never lost that longing for the truth about evolution which to men of my date is the salt of biology, the impulse which made us biologists. It is from them that the raw materials for our researches are to be drawn, which alone can give catholicity and breadth to our studies. We and the systematists have to devise a common language.

Both we and the systematists have everything to gain by a closer alliance. Of course we must specialise, but I suggest to educationists that in biology at least specialisation begins too early. In England certainly harm is done by a system of examinations discouraging to that taste for field natural history and collecting, spontaneous in so many young people. How it may be on this side, I cannot say, but with us attainments of that kind are seldom rewarded, and are too often despised as trivial in comparison with the stereotyped biology which can be learnt from text-books. Nevertheless, given the aptitude, a very wide acquaintance with nature and the diversity of living things may be acquired before the age at which more intensive study must be begun, the best preparation for research in any of the branches of biology.

The separation between the laboratory men and the systematists already imperils the work, I might almost say the sanity, of both. The systematists will feel the ground fall from beneath their feet, when they learn and realise what genetics has accomplished, and we, close students of specially chosen examples, may find our eyes dazzled and blinded when we look up from our work-tables to contemplate the brilliant vision of the natural world in its boundless complexity.

I have put before you very frankly the considerations which have made us agnostic as to the actual mode and processes of evolution. When such confessions are made the enemies of science see their chance. If we cannot declare here and now how species arose, they will obligingly offer us the solutions with which obscurantism is satisfied. Let us then proclaim in precise and unmistakable language that our faith in evolution is unshaken. Every available line of argument converges on this inevitable conclusion. The obscurantist has nothing to suggest which is worth a moment's attention. The difficulties which weigh upon the professional biologist need not trouble the layman. Our doubts are not as to the reality or truth of evolution, but as to the origin of *species*, a technical, almost domestic, problem. Any day that mystery may be solved. The discoveries of the last twenty-five years enable us for the first time to discuss these questions intelligently and on a basis of fact. That synthesis will follow on an analysis, we do not and cannot doubt.

PROGRESS IN BIOLOGY

An Address delivered March 12, 1924, on the occasion
of the Centenary of Birkbeck College, London

Nature, May 1924

An address on progress in Biology during the last hundred
years has one element of simplicity; since, with scarcely a tinge
of exaggeration, the whole subject from its inception may be
held included. Though the materials studied by biologists are
those which have been the objects of man's curiosity from the
earliest times, yet the biological way of looking at them was
new, and biology was a term deliberately selected to proclaim
the consciousness of a new hope. Treviranus—Gottfried Rein-
hold, 1776–1837, of Bremen, elder brother of the well-known
botanist Ludolf Christian (1779–1864)—was the first to use the
word *Biology* (1802). He complained that the current treatments
both of zoology and botany were lacking in penetration and in
comprehensiveness, and that their practitioners were too often
giving a divided attention, with an eye ever wandering towards
medicine and other applications. The catalogues of plants and
animals, and barren descriptions compiled in the name of those
sciences, are a beginning, not an end. They are the materials
which the science of living things is to absorb and co-ordinate.
The problem before the biologist is, What is Life? and the collections have value in so far as they contribute to a solution of
that problem.

Treviranus says all this, and says it very well, with a lofty but
well-controlled imagination. If some one objects that he is
offering old things in a new form he will not deny that, but he
claims that to see them in the new form is no trifling help.
Surely he has proved right. The new word connoted a new
thought. Though it was not given to him to see further into the
mystery than his contemporaries, nor indeed so far as many,
notably Bichat and Lamarck, he was looking in the right direction, and he polarised the attention of many more.

Sir William Lawrence introduced the word into English in
those lively *Lectures on Physiology* (ed. 1818) which, in spite—or
perhaps in consequence—of much detraction, went through
many editions and were very widely read. He too was aware
that the term *Biology* implied a certain ambition.[1]

[1] See especially pp. 52 and 58–60, ed. 1823.

We see other signs that about a century ago the study of life underwent a renaissance. Before biology was dreamed of there were plenty of good naturalists, men devoted and exact, full of curiosity, gathering truth where they could, in the right spirit of science, but it is at this critical period that we first hear those graver notes of more resolute endeavour which thenceforth constantly recur. The gain in depth and precision was of course promoted by the development of microscopy, but the aspirations of the first biologists to obtain a fuller understanding of the nature of life had been excited, and their minds were prepared before that visible revelation came in unquestionable form. I like at least to think that the questions were asked before the instrument-makers came on the field. Sage Sidrophel, who

> made an Instrument to Know
> If the Moon shine at Full or no,

has been a great begetter of modern researches, but less fortuitous discoveries are the more honourable and command a warmer admiration.

However that may have been, as a matter of history the first great advance was the recognition of the cell; and nowadays, when we talk of biology, we mean the study of life in terms of cells—their structure, properties, and behaviour. The microscope led soon, though not immediately, to the cell theory, and cellular biology began. It is not my purpose to repeat that familiar story. I ask you to observe, however, that, little as we yet know of life, it is not a century since we began to be certain about even the proximate and easily accessible phenomena which characterise life. The rudiments of chemistry, to be sure, were still only dimly apprehended, and it may be thought that, in the absence of chemical knowledge, sound ideas about life could scarcely be expected. Perhaps; but surely the essential feature which—apart from psychical attributes—differentiates living creatures from all other systems whatever, might have been distinguished by ordinary observation, at any time. Living organisms are systems which have the power of continual and spontaneous division. In that they are unique. *How* they divide could not even be imagined, much less investigated, without the microscope; but nothing in the history of discovery is more curious than this fact, that until well into the nineteenth century, men should not have known familiarly that living things come into existence solely by a process of orderly division. The conception of the cell as a unit was necessary to give anything like accuracy to this knowledge, but it was not essential, and may not improbably be replaced hereafter.

Scarcely less remarkable is the fact that upon this process of division attention has only quite recently been concentrated. The other attributes of life, chemical and physical, have proved both more amenable to experimental study and more prolific in fine discoveries immediately applicable to medical practice; with the result that to them—at least in our English use—the once comprehensive term physiology has become exclusively attached. An overburdened study must discard something, but it is scarcely fortunate that the process of genetical division, the central phenomenon of physiology, should in its several manifestations—heredity, variation, segregation, differentiation, and the regulation of form—be unfamiliar ground to academic physiologists.

I have referred to the moment from which our survey begins as a time of renaissance. The student of to-day sometimes scarcely realises what happened then. A hundred years ago you might find an undescribed—even an unnamed—animal any day on the British coasts. Next to nothing was known of the development of any creature whatsoever, animal or plant. The student's microscope, of course, did not exist, and indeed was only coming into general use in my own student days. Neither my revered master, the professor of zoology of that time in Cambridge—a man of great learning and distinguished gifts—nor his colleague, the professor of botany, could use a microscope to any purpose; and I doubt if either of them would have seen much in a section of anything. No doubt we were a bit behind the rest of the world just then. They were ahead of us in France, and very greatly ahead in Germany. Reputation, if not fortune, might occasionally be made by a little judicious dumping of foreign products. That state of things happily soon passed away, and an international standard of scientific public opinion in biology became universal.

Though in 1824 little was known of what we now call morphology, the unicellular organisms had not attracted serious attention at all. Such creatures, even some bacteria, had been seen, but no one had felt much concern about them. If you care to attach history to dates, let me give you two easy ones, worth remembering. 1838 is the year in which Schleiden gave formal expression to the cell theory, and being also the date of Ehrenberg's *Infusionsthierchen* with the first good pictures of bacteria and many infusoria, is doubly memorable. Ten years later we come to 1848, a date which in significance to mankind may rank with 753 B.C., or, indeed, any of the greatest anniversaries.

In written history there are only a few cardinal points—the invention of printing, the First Folio Shakespeare, Newton's *Principia*, the discovery of oxygen, with some dozen more. Of these most certainly one came in 1848, when Pasteur first demonstrated to Biot his observations on the asymmetry of tartrates. From that one small bit of clean experimenting, interpreted by penetrative genius of the highest order, has grown modern surgery, a rational investigation of disease, bacteriology, much of biochemistry, hygiene, with all the consequences that those developments have had on vital statistics, the structure of populations, and contributions to almost all the arts and the multifarious activities of civilised man. 1848 was the year of the revolutions. The public was not thinking much about bacteria or biology. Mobs were in the streets and the fermentations then most regarded were metaphorical. But the world as we see it to-day is rather the product of biological discovery than of democratic institutions, and there are moments when your modern Esau would wisely barter a lot of his political rights for some honest antiseptic.

The great corpus of knowledge grows by solid increments, definite, predicable discoveries of fact. Rarely is any piece of interpretation an event of equal consequence. In that small group of fertile theories, by common consent we class the *Origin of Species*. For the public of 1859 Charles Darwin's book seemed, and for most of the laity to-day it still seems, like some meteor of the heavens, to have been a revelation unheralded. That of course we know it was not. Apart from earlier adumbrations, some vague, some clear, the doctrine of evolution first took categorical form in the hands of Lamarck (1809). For fifty years the new ideas were strenuously debated, especially in France. The discussion, nevertheless, had reached England, though our contributions were largely on the non-placet side. Chambers (1844) was a timid evolutionist; Herbert Spencer, though no naturalist, a bold one (from 1852). Lyell in 1832 had argued against "evolution", adopting the word. Sir W. Lawrence had (1818) collected many illustrations of variability, but maintains that none transgress the limits of specific difference, and he took a firm stand against the Lamarckian teaching of the transmission of acquired characters, which he declared was contrary to experience—the first, I believe, actively to denounce that illusion.

The very considerable work of Godron, *De l'Espèce*, appeared in book-form in 1859,[1] the year of the *Origin*. Godron was a

[1] Following previous publication in *Mém. Acad. Stanislas*, Nancy, 1848–49.

most competent botanist. His collection of evidence as to variation was the fullest then compiled. Most of it was afterwards incorporated in Darwin's *Animals and Plants under Domestication*, but the book is still of value. Godron examines "la théorie de l'évolution successive de l'espèce", and rejects it. Darwin, using very similar evidence—though greater in quantity and more varied in character—with the introduction of the one wholly new consideration, natural selection, succeeded in making the doctrine of transformation acceptable. Natural selection was an undeniable *vera causa*, whereas few had felt quite comfortable about Lamarck's appeal to the effects of conditions. The presentation, moreover, was set forth in language so suave and cogent that the reader gladly resigns himself into the hands of a master. With the advances in knowledge, the additions from embryology, from palaeontology, geographical distribution, and many other convergent lines, the truth of the doctrine of evolution became in broad outline finally established.

This much being admitted by all, we may inquire why had so many good naturalists, whose information was sound, resisted so firmly? With the exception of natural selection, every essential element in Darwin's case had been present to their minds before. To suppose that all these writers had suffered from theological obsessions is absurd. Lawrence, for example, flouted authority with great enjoyment. They were genuinely convinced, some of them probably against their will, that the mutability of species was contrary to observation. Remember that up to Linnaeus few, if any, troubled to consider whether species were or were not mutable; but after species had been declared to be immutable, those who proposed to maintain the doctrine of mutability might be expected to prove their case.

What weighed, then, with Godron and careful men of his type was that the variability they observed did not result in the production of new species, and that in particular, as they insisted, the new forms derived from a single common origin, when interbred, do not produce offspring of impaired fertility as so many genuine species do. This critical link in the evidence is equally absent from Darwin's case. He satisfies the mind about so many other difficulties that this one is allowed to pass, and the reader, learning that many putative species give hybrids fully fertile and that between these and the totally sterile hybrids all gradations can be seen, feels that this objection must have been removed, though as a matter of fact it remains.

The progress of the last twenty years has brought us to a position from which we can at last begin to discuss these problems

26-2

fruitfully. No general principles governing the incidence of interspecific sterility have been ascertained. We there find ourselves in a tangle of empirical and as yet unrelated phenomena, specific like those of chemistry. Expectation founded on our estimates of divergence is constantly at fault. The oxen, the *Canidae*, the finches, the ducks, the pheasants: who can say right off which crosses in these orders give hybrids fertile in both sexes, or in one sex only, or in neither? Which species cannot be crossed at all? Like chess openings, these things, no doubt, are governed by principle, but the principle is not obvious. For the right answers we depend largely on memory. They have been ascertained by accumulated experience and are not easily found "over the board".

We are clear that the forms of life are the products of evolution, but we are equally sure that specific distinctions are not culminating terms reached by the accumulation of small differences. The variations by which they have arisen are not yet known to us, but we are satisfied that the particular account of their origin which is the one Darwin chiefly favoured is incorrect.

Of the origin of specific distinctions we have, as I have said, no acceptable account. Appeals to adaptative value are here beset by the gravest improbability. The Darwinian principle that the fixity of a character is a measure of its value to the possessor is not tenable. The sharpest and most permanent specific differences are constantly to be found as characteristics which no one by the utmost exercise of ingenuity has been able to represent as other than trivial in adaptative significance. Whatever stress we are disposed to lay on natural selection, it does not assist us here.

We must frankly admit that modern discoveries have given little aid with the problem of the origin of adaptation. Darwin in 1844 regarded all adaptation as the consequence of natural selection. His letters of that date speak of Lamarck with contempt. With the lapse of time, nevertheless, we find him frequently and increasingly appealing to the transmitted effects of the conditions of life, and between the two he sometimes does not distinguish so clearly as we would wish. His most urgent task was to make evolution an acceptable principle, and one argument failing he would invoke the other, until in the edition of 1876 certain passages read uncommonly like Lamarck obscured. Seizing upon one which is, to say the least of it, ambiguous, the irreverent Samuel Butler makes the flippant comment: "This comes of tinkering. We do not know whether we are on our head or our heels. We catch ourselves repeating

'important', 'unimportant', 'unimportant', 'important', like
the King when addressing the jury in *Alice in Wonderland*".[1]
All this matters little now. In response to Weismann's challenge that critical proof of the transmission of adaptive re-
sponses to environmental influences should be produced, none
satisfactory has been forthcoming. In one respect only, I think,
we have to recognise positive evidence which Weismann did
not perhaps sufficiently anticipate. The germ cells have in certain experiments been injured by special and violent treatment
to which the parents have been exposed, with permanent consequences to the posterity. But there is no question here of
adaptation, and such evidence does not make the origin of the
adaptive mechanisms more easy to understand.

We have reached contemporary developments. The study of
variation, and indeed of several branches of what we now call
Genetics, especially cross-breeding, had been pursued with
vigour in the 'sixties and 'seventies, but had totally lapsed. Renewal of those inquiries led at once to an advance. We saw
that the received ideas as to the magnitude of variations, and
especially as to the interrelations of the domesticated breeds,
were largely erroneous. As in regard to the incidence of sterility
in interspecific crosses, so in regard to variation, we found ourselves among an intricate mass of empirical observations,
obeying none of the principles which the orthodoxy of the time
presupposed. The incidence of variation was utterly capricious,
and was determined neither by utility, nor by the antiquity of
the feature, nor by the conditions of life, nor by any other
ascertainable circumstance.

Most of the genetical work of the early time had been perfunctory and unsystematic. Godron, Naudin, Verlot, Carrière,
Morren, and many more, had all seen interesting things, but
they had not looked close enough. A single man, Mendel, had
worked in a different fashion. Again, by one small bit of clean
experimenting, a fact of a new class had been discovered. The
evidence of this new witness shewed us whole ranges of phenomena in their right perspective and proportions. We had at
once a rationale which disposes of such outstanding mysteries
as reversion and the determination of sex. Only those who remember the utter darkness before the Mendelian dawn can
appreciate what has happened. Stories which then seemed mere
fantasies, are now common sense. When I was collecting examples of variation in 1890, I remember well reading the
fanciers' tales about dun tumbler pigeons being almost always

[1] *Luck or Cunning*, 1887, pp. 185–186.

hens, and about the "curious effects of crossing" with cinnamon canaries, but I would never have dared to repeat them, any more than Darwin ventured to quote Girou de Buzareingues (1828) to the effect that in cattle the milking-character was mainly transmitted by the bull—a proposition with which the researches of Pearl and others have now made us familiar.

Though Mendelian analysis has done all this, and very much more of which I will presently speak, it has not given us the origin of species. It has finally closed off a wrong road. I notice that certain writers who conceive themselves to be doing a service to Darwinism, take thereupon occasion to say that they expected as much, and that from the first they had disliked the whole thing. I would remind them that the class of evidence to which we were appealing was precisely that to which Darwin and every other previous evolutionist had appealed. Mendelian analysis led to the discovery of the transferable characters, not merely in sporadic instances but as a group, and the study of their behaviour enabled us to avoid endless misinterpretations into which our predecessors had consistently fallen. If we now have to recognise that the transferable characters do not culminate in specific distinctions, the acknowledgment will not come from us alone. The old belief of systematists that real species differ from each other in some way not attainable by summation of varietal characters is no longer contestable, and we know now upon what to concentrate. It is no occasion for dismay. We have not to go back very far. We do not understand specific differences, nor can we account for the adaptative mechanisms. Was it to be expected that we should? Biology is scarcely a century old, and its intensive study is of yesterday. There is plenty of time ahead.

The identification of the transferable characters and their linkages has led to a further discovery of the greatest—I might almost say, of romantic—brilliancy, which must have consequences as yet inestimable. Morgan and his colleagues have, as is well known, proved that some, probably all, of this group of characters are determined by elements transmitted in or attached to the chromosomes. It may be, as Bridges has indicated in regard to sex, that the visible distinctions are produced not so much by the presence or absence of a bit of special chromosome material, as of an interaction between the several chromosomes as a whole, and much depends on that issue; but however that may be, henceforth the study of evolution is in the hands of the cytologists acting in conjunction with the experimental breeder. As to what the rest of the cell is doing, apart from the chromosomes, we know little. We think that in plants the presence or absence of chloroplasts may be a matter of extranuclear transmission. Perhaps the true specific characters belong to the cytoplasm, but these are only idle speculations.

While all this has been going on we learn of advances developing from a totally different quarter—palaeontology. Those whose work has lain in other fields can form only a dim and tentative understanding of these new lines of discovery. We look eagerly to the palaeontologists for a full exposition. We have heard that they, especially the group of investigators connected with the American Museum, have collected wonderful series, in numbers hitherto never attainable, ranging through many geological epochs, demonstrating a continuity of succession between very dissimilar forms of life. For an introduction to this subject I am greatly indebted to Professor D. M. S. Watson. In connection with these observations we hear frequent use of the term orthogenesis, a word introduced by Eimer to express the notion that evolution proceeds along definitely directed lines.[1] Eimer was both a vigorous opponent of natural selection and a confirmed Lamarckian. His idea had been enunciated at various times by others, but in spite of a superficial attractiveness such short cuts have seemed too facile, and to be avoided in the absence of irresistible evidence that they are right. Nevertheless, what we have learned of variation, especially of the incidence of parallel variations, has taught us that many varietal forms owe their origin to a process of unpacking a definite pre-existing complex, with the consequence that, given the series of varieties to which one species is liable, successful predictions may sometimes be made as to the terms which will be found in allied series. This is not what is meant by orthogenesis, but the phenomena have features in common.

These symptoms of order in variation have prepared our minds, and there may well be a sense in which orthogenesis will be found to denote a valid principle. Granting that a gradual and secular evolution in one direction is demonstrated, much turns on the evidence that can be produced as to the other variations by which these changes have been accompanied. We anxiously await such details. Especially are we curious as to the nature of the characters concerned. Are they such as in our contemporary experiments we have found to be transferable, and thus likely to be subject to clean segregation? Secondly—a question much more difficult to answer—is it possible that, though undoubted as indications of the course of an

---

[1] *Artbildung u. Verwandtschaft bei den Schmetterlingen*, II, 1895, p. 3.

actual evolution, the most positive indeed which can be imagined, they should be interpreted as evidence of the origin of species in that stricter sense to which genetics has introduced us?

A sound analytical classification of the several kinds of characters in respect of their modes of variation is greatly to be desired. We have determined the transferable characters as one group, and we no longer confound them with the essential elements conferring specificity. Segregation is of course often seen in species crosses, but as to the behaviour of these critical elements we know as yet very little.

Of a third group we may presently learn from the palaeontologists. Independent of all these substantive characters we shall distinguish what I have called the Meristic group, as a fairly homogeneous class of phenomena recognisable without much difficulty though still not precisely defined.

That is a place to which I always look for one of the great discoveries about the nature of life. The phenomena of Meristic repetitions, especially in their most obvious manifestations as seen in simple patterns, would appear to be amenable to analysis. Who can look at the stripes on a zebra's hide—to take one of a thousand such illustrations—and not see them as a series of waves? Further, who can compare the hide of *Equus zebra* with that of *E. grevyii* and not see that in *E. grevyii* the wave-length of the same vibrations is approximately halved? It is in the analysis of pattern that mathematical treatment might properly be applied to biology. If some physicist would examine our patterns and, treating the problem as one of ordinary mechanics, set himself to consider how the forces must be disposed to produce those patterns, I am not without hope that he might find a clue to the nature of the forces themselves.

The future of biology lies not in generalisation, but in closer and closer analysis. It is the lack of analytical penetration that we so miss in the nineteenth-century evolutionists. Phenomena the most diverse are confounded together and discussed under some common name, for example, variation. Their aim is always to unify, never to distinguish. Never are we reminded that every appeal must ultimately be to the mechanics of cell-division. That is the one true and logical unification. The cell, as Cuvier said of the living organism long ago, is a vortex of chemical and molecular change. Matter is continually passing through this system. We press for an answer to the question, How does our vortex spontaneously divide? The study of these vortices is biology, and the place at which we must look for our answer is cell-division.

ADDRESS TO THE SALT SCHOOLS, SALTAIRE, SHIPLEY

7 December 1915. (Unpublished)

You are a body interested in education, and in asking me, a biologist, to deliver this Presidential Address, I suppose it is your wish that I should speak of the problems of education as they look in the light of biological knowledge.

Like so many other things education in the active has aspects different from those it has in the passive. My own direct acquaintance with the subject is mostly of the passive kind. I underwent the treatment in its most drastic form. I have given a little of it, though not much: but having lived most of my life in Cambridge I have been in constant association with teachers and the taught. I have been continually in what is called an educational atmosphere—so I have watched an enormous number of cases. To speak still in the language of metaphor the majority of those cases have not recovered. I don't mean that they have actually perished or sunk into permanent mental disablement—though even that is true of a considerable number—but from the treatment provided for them at vast expense they have got very little, and I am sure they would have done about as well had they never undergone the process.

A minority no doubt have been trained in the rudiments of some profession, such as law or engineering. They have been fitted for an occupation—but of the truths that enable man to see himself in his true position in time and space, the knowledge that gives him a sense of balance and proportion in his progress through life, the majority acquire no knowledge at all.

The knowledge I mean is natural knowledge: knowledge of the structure and properties of the materials of which the earth and our bodies are made, the outline, that is to say, of chemistry and physics; the main facts, too, regarding the nature of animals and plants, the properties of living things, the relation of man to the other animals: something, too, of the history of man —I do not mean political history, the date of the Reform Bill or of Waterloo, the genealogy of the house of Hanover, or any of those particulars which, however interesting, are mere trivial ripples on the tide of events—but the true broad history of man, the history of civilisation, the rise and disappearance of races, their succession on the face of the earth, what their attributes have been and are, knowledge that will give an understanding of the composition of a mixed population like our own, its capabilities and its limitations.

Surely it is the first business of education thus to introduce the young to the ground plan of the natural world. Not merely for those who govern, for those who teach and generally for the classes in command, but for us who have to live is this true. Put in that form few would dare to dispute so self-evident a proposition. Yet our whole scheme of government, and for the most part our system of public education, is constructed as if this knowledge of the facts and properties of nature, though doubtless a technical requisite for scientific men and for certain industries, were a superfluity with which the population at large has no particular concern.

I feel that I can speak with a certain freedom of these things to those who are interested in the welfare of the Salt Schools, since from the prospectus I see that these fundamentals of education, the two great realities of life, science and art, are very conspicuous elements in it. That is as it should be and perhaps it is as much as public opinion yet sanctions. Would it were so throughout our secondary schools. But we must now go further. We demand not merely the universal teaching of science as a part of secondary education, but that it should in essence form the basis of such education, and thus bring about a revolution in the attitude of the nation towards natural knowledge.

The work of the last century has wholly changed man's outlook on the world. That accession of knowledge which we call science has made the good of life more easy to attain and the evil more easy to avoid. Man's power over nature has increased immeasurably, and we know that we are still only at the beginning of the exploration of natural truth. After ages of slowest progress science has begun to grow fast. In such a year as this we may wonder sometimes whether that progress has not permanently stopped, but we know in our hearts that in the quiet places men's minds are still working, boring into the unknown. The continuity is imperilled. It may perhaps be broken. Almost a generation of the splendid youth of Europe will have been swept away—destroyed in sheer waste: but the body of knowledge remains, and from it the stream of progress will surely again break forth.

When I visit my old haunts in Cambridge everything seems changed. The students are gone. The colleges, even the class-rooms, are full of soldiers. But one thing is not changed: the great Library, the centre of the place, that visible, warm, engendering lap of Alma Mater, continues as it was, ready to kindle zeal for learning and inquiry in fresh generations of her sons.

Whether invented first in Germany or in Holland, the printing press gave the true charter of man's mental liberty, and, short of cosmic cataclysm, the work that it has done will endure.

Natural knowledge is the basis of all power, the one source of rational conduct; it is the light which shews man in his true natural perspective, that makes him at home on this planet, and steadies him among events. You may ask, why do I use *natural knowledge*, the old seventeenth-century term, in preference to our modern word *science*. I do so for two reasons. First because science has come to bear a narrow sense. We use it especially for the work done in laboratories and for technical knowledge which has to be acquired by special means. Now of course we must have laboratories and technical appliances in our schools. We must have them far more abundantly than hitherto, but I do not urge that all should be compelled to use them. For many I conceive that would be mere waste of time and trouble. It would simply repeat the old blundering narrowness of our grammar schools in a new field.

But I say *natural knowledge* should be the basis of education because I mean that that basis should be something wider and more inclusive than is commonly denoted by the word science. I mean *science*; but in that broader sense which includes all knowledge that has natural reality and significance. It is in the spirit rather than in the subject-matter of education that I conceive our fault has chiefly lain. There are books of Homer and not a few passages in the Bible, which, illustrated and expounded by a competent teacher, may serve to convey essential parts of science as fruitfully as many an hour's work in a laboratory; though taught in the way that satisfies our educationists, lessons on the same subjects may be destitute of any reality or significance whatever.

I look back on my school education as a time of scarcely relieved weariness, mental starvation and despair. There came at last a moment when I was turned into a chemical laboratory and for the first time found there was such a thing as real knowledge which had a meaning and was not a mere exercise in pedantry. Our staple was of course Latin and Greek, of which I made nothing. Some emotional pleasure came towards the end of my school course from the Greek tragedies, but otherwise those years were almost blank. Now what I and thousands of other boys like me discover in after life is that by those very same materials, perhaps more than by any others, we might have been "waked to ecstasy" and to the joy of development.

We were classed as not caring for literature and "bad at languages". But we learn in after life that as a class we derelicts care for literature as much as most people, and can pick up languages quickly enough when we feel the need of them. No one has any difficulty in learning a thing he really wants to learn. We had no reason for wanting to learn the Classics. Latin and Greek indeed were not taught as literature nor even as languages. They were taught as lessons, mere pedantic lessons, something laboriously contrived in the class-room for the class-room without value or meaning in or relation to the real life of the world. They tell me sometimes that all this is changed, and I am happy to think that at the hands of a few exceptional teachers real live teaching is being given. But I have boys of my own at school now and I know from what I see and hear that the ways of the schoolmaster remain substantially what they were.

And why is this? For the only reason that causes anything to endure: because the people like it so. Through causes perhaps untraceable it has become the fashion of our race to demand an ingredient of humbug in whatever is offered them. We have not "that terrible habit of looking at things frankly" which I lately heard a distinguished professor of art refer to as a characteristic of the German mind. We are afraid of natural knowledge. Its nakedness shocks our sense of propriety. Our system of manners and of morals is contrived to make us affect to be something different from what we are.

> You *don't* like what you only like too much;
> You *do* like what if given you at your word
> You find abundantly detestable.

This appetite for humbug pervades not only education but even the arts of daily life. In some great works I once saw a pathetic sight. A man, obviously very skilful and intelligent, was employed for his whole time engraving ferns on the sides of fish-knives. He was, I am sure, doing this not because he wanted to do it, nor because anyone thinks a picture of ferns appropriate or in any way an ornament to a fish-knife, but simply because some cunning tradesman had succeeded in fixing that hideous affectation on a servile public; and so for many a year we shall continue to have ferns engraved on our best fish-knives.

We live in a world of make-believe and pretence, and we are

at least consistent in that from their earliest years we fill our children with pretence knowledge.

In parenthesis I may remark that it is possible to present even laboratory science so pedantically that the pupils may come to loathe it as much as grammar. I have myself seen both chemistry and zoology so taught; but it is not very easy to do this with those subjects, for the pupil generally sees for himself the reality of the subject though veiled by the pedantry of a dull teacher. The pupil besides does not meet these things only in the class-room and has other means of finding for himself that they are parts of the living growing body of knowledge.

It is a question whether that almost hypochondriacal interest in the stupid school games which now wastes the intelligence of our young people may not be due to the fact that we give them so little else that they care to think about. Of course all healthy young people like to play games, and quite right, too. It is the continual thinking about games that is so morbid and devastating. The amount learnt at school is so small that many parents declare quite openly that they send their sons to school more for the games and the discipline of companionship than for education in the old sense. No one doubts the enormous value of that discipline, but we should supply it with less destruction of the mind if we followed the penetrating suggestion of Sir Joseph Thomson and sent the children to school for the holidays and kept them at home to learn.

Consider the waste and extravagance of our present system. Almost all parents who can afford it send their sons away to school. We do this in pure vanity and affectation. Of all mad institutions in England, this, to our friends abroad, seems perhaps the maddest. Those unfamiliar with England can scarcely believe that for several days at three seasons of the year the London railway stations are blocked with streams of cabs distributing thousands of unhappy children at random to all parts of the country. But their homes have to be run just the same in their absence. After beginning school boys get almost nothing from the home-life. Without the incalculable waste that this displacement involves, the children of Germany or Holland for example, living in their own homes, get a better education at a mere fraction of the cost.

If we will not amend our ways for any higher reason we must change them because they are not safe. As Professor Frankland lately said in the course of a deputation to the Board of Trade: "Profound neglect of science in general, and of chemical science in particular, is something specifically British, and it would be difficult to find any other civilised country in which a comparable state of things exists ". Such a charge is a grave one, but anyone who reads the newspapers or who observes the utterances of our public men knows that it is true.

Public references to scientific subjects generally contain gross blunders, and are not rarely flippant or jocose. In these respects things are worse than they were. Those in authority in this country not only are ignorant of the subject-matter of science, but they do not even realise what it is about or what it is going to do.

Everyone is aware that science has created great industries and the public has suddenly discovered that science is the source of all power in war: but it has yet to be realised that every part of the life of a modern state is now amenable to natural knowledge. The lawyer-politician regards even the population of these islands as a simple aggregate of similar units. Among them may be distributed equal food, equal education, or equal political power, and the ideal of the more philanthropic of statesmen is that under this treatment the result will be a healthy and uniform population, like a field of wheat in which each plant has equal food and equal space to grow in, and develops its growth in equal degree. But no modern population in the least resembles a field of wheat. There are isolated communities closely intermarrying, like those of some Yorkshire dales, in which a small number of families have followed the same occupation for many generations, have shed their aberrant elements, and are now fairly uniform in type, but these in the modern world are altogether exceptional. By most careful selection a high-class wheat has been bred true to a type and each plant is born sensibly identical and with equal powers. Nothing like this is true of the English population, even of a county or a town. These populations are mixtures of many strains, and their members have many distinguishing features and aptitudes both bodily and mental. From our modern knowledge of heredity we are sure that, could we study the family descent of these characteristics, as we can those of our cultivated animals and plants, we should soon prove that they are transmitted according to definite systems. Each of us is born with a definite physical and mental composition which we can by no means change. We are all men just as dogs are all dogs; but both in their forms and instincts and in their powers of hereditary transmission, men of a mixed race like ours differ among themselves as much as the sheep-dog differs from the greyhound or the bull dog.

Occasionally we notice connections between the mental and the physical characteristics. Just as yellow mice commonly shew a tendency to lay on excessive fat, so a certain sort of gingery hair-colour in men is not rarely associated with a hot temper. I was lately sitting with a well-known anthropologist in a place in London much frequented by the higher official classes. We were talking of the shapes of heads, especially the round and the long type. Our neighbours were round-headed men and I asked my friend to point out examples of long heads. He said with a smile, "I don't think we should find many *here*", and as it happened we were not able to find a single one. It is not to be suggested that a man with a long skull, as such, is of inferior ability, but few come to distinction in the official world. The varieties of brains, however, are more numerous and important than the varieties of skulls. Most people are accustomed to suppose that other people's mental processes are like their own, but if they take the trouble to watch and to inquire, they will find that this is not so. To take an example: many, probably most of us, have what is called a visual or picture memory. When we try to call to mind an incident we have a mental picture of it. We can see a passage lately read, in its proper place on the page of a book—right or left—top or bottom. But people are not uncommon who have no such faculty, or at least cannot consciously evoke it. We who think in pictures cannot imagine how the non-visualisers think at all: but plenty of them have very good brains, though I cannot believe they get as much out of life as we do. Similarly the horses that run in the American trotting races are of two kinds, trotters proper, and pacers. Pacing is a gait very different from trotting, being more like what we call ambling. Most horses fall naturally into the one gait or the other. By various systems of training the gaits can be modified to some extent, but the distinction is physiological and fundamental, which appears at once from the pedigrees. For I learnt on inquiry that horses bred from two pacer-parents are always pacers, and not a single case could be produced of a natural trotter being born to such parentage. The offspring of trotters may or may not be pacers, but the offspring of pacers are pacers only, just as the offspring of two chestnuts are chestnuts only, and not bays or browns. Pacing is what is called technically a "Mendelian recessive" and breeds true.

A population, then, like our own is a medley of many kinds of dissimilar individuals, with most various faculties and tastes. The reality is something entirely different from the fiction which lawyers or theologians assume as the basis of their polity. Natural knowledge makes short work of these fantasies. Now what does such an institution as public education really do in application to this composite population? First and most obviously it provides opportunity. Opportunity of mental growth and development. What comes of the opportunity depends on the living material, and for its deficiencies nothing can atone. When I said in opening that most cases submitted to the treatment of education get little good from it, I only said what most teachers in their hearts know with sorrow to be true. The man's powers are greater than those of the child, but so they would be in the absence of education. Generally the clever child becomes a clever man, and stupid children grow up stupid, though, owing to the idiosyncrasies of development rather than to differences in education, even that is not always true.

Reading such stories as that of George Stephenson, a man who, with no opportunities, developed powers which have altered the course of civilisation, some have been tempted to imagine that we need only provide the schooling that Stephenson had to create for himself and Stephensons would abound. That is a dream which will never come to pass. One has only to study the faces in a crowd and ask oneself respecting each in turn whether the owner could conceivably have been fitted to fill an important place in the world, and in nearly every case the suggestion is felt to be absurd. That at least is true in the south of England, though I admit that in the north intelligent faces are not quite so rare. What public education can do, what such magnificent institutions as that which I have the honour to address in fact do, is to give the rare Stephensons a better chance. To a slight degree they cultivate the population at large, but their effective work for civilisation is the lifting of exceptional men. And for the mass, the undistinguished mass, what can be done? I would reply—tell them the truth. Give them natural knowledge. Let them learn what they and the world really are.

When the multiplicity of mental types is understood it will be realised that education should be as varied as it is possible to make it. The courses should offer something of everything. I have no patience with that thoughtless catchword of educationists that whatever is learnt should be thorough, and that smatterings are useless. On the contrary the minds of young people want that "fine confused feeding" that the Scotchman gets in his haggis. One mind is fired by machinery, another by poetry, another by chemistry, by tales of daring, by art in its various forms, some, I am told, by algebra, and so on. A child is like some new kind of caterpillar. One must try it with many different kinds of food to see what it can eat. As a practical hint I should like every house that contains any children of intelligence to have a good simple Encyclopaedia. I don't mean a collection of treatises, but one like Chambers's, with clear short articles on all kinds of topics and all notable persons. Nothing will so quickly help a boy to find out what he likes, and it is only the thing that he likes that he will ever do really well.

The first great thing that education does is to provide opportunity. I wonder if it sometimes occurs to you what the chief consequence of universal education will be to the future of the race. We are accustomed to regard public education as the most democratic of institutions. In reality it will be the means of creating the most powerful and permanent of aristocracies. For year by year we are now sorting the population. We are carefully, and as I hold, rightly, lifting from the mass of the population everyone who has talent in any degree. We enable such persons to rise into the upper layers of society and thus build up an aristocracy of intelligence and power. By the workings of heredity such an aristocracy acquires a high degree of permanence. The old aristocracy has largely gone under, not because it had not great qualities, but because those qualities were not of a kind that count for much in the modern world. People ignorant of biology suppose that the mixture of types in each of our social layers is something natural to society, but in reality it is a product of the industrial upheaval which in its turn is the consequence of the discovery and application of science.

In Lancashire they have a saying that it is "three generations from clogs to clogs". If that is actually their experience, I can only suppose that in Lancashire they choose their wives with insufficient care. In all seriousness an approximate and fairly stable separation of the classes based on their relative powers is by no means unattainable. Not only must that segregation be greatly promoted by popular education, but by every kind of reform which tends to make success proportional to merit.

You may remind me that I am not speaking in Plato's *Republic* but in the dregs of a wretched actuality. This, it may be said, is no time for the contemplation of distant truth. Rather I would urge that in such contemplation lies the best hope of a return to the sanity of nations. The tragedy that is overwhelming Europe has resulted from the uncontrolled growth of national sentiment. After a long quiescence these feelings, stirred by the writers and speakers of many countries, have once more burst into flame, threatening the destruction of civilisation. This spirit of nationality, whether it masquerade under its more glorious name of patriotism, or comes forth without disguise in its true shape as selfish pride, is but a poor thing in the light of natural knowledge. To be strongly moved by the spirit of nationality a man should have real ground for thinking that his own nation is best, and that the world would really gain by the supremacy of descendants of his type. But in fact, that emotion moves most deeply those who know least of other nations and often very little of their own. The chained dog who has never left his yard is the readiest to fly at strangers. They are strangers and that is enough. So it is with the nations. Knowledge, more and more knowledge of natural fact, is the only cure for these ephemeral emotions, dissipating them by turning thought to other things, as the child is cured of a vicious habit when its attention is turned to something real to do.

Patriotism is glorious because it stirs the hearts of great men to noble deeds. We glow when we read their stories; but noble deeds are not the prerogative of any cause. What men have done for their country in the field, other heroes have dared in silence for the world. Think of the history of X-ray work, now in safe daily use as one of the greatest aids to surgery. In perfecting the system four pioneers endured prolonged suffering and mutilation: Blackett, Cox, Harnack, and Hall Edwards, well knowing the risk they ran. Think also of the men who in the laboratory have taken plague, leprosy, yellow fever, and the rest, in the same spirit of devotion. By their sufferings we and all nations have been redeemed.

Consider the meaning of nationality at the present time. What is a nation in this age of rapid travel, of continual immigration and intermarriage? Lawyers, who know and care nothing about nature, can answer the question at once by turning up their books. That the answer they give has no natural meaning and is different for every nation does not matter to them in the least, and I notice as consistent with the legal spirit that our government of lawyers should parade a scheme of division of Europe by nationality as a solution of our present troubles. Is it his birth-place that receives the allegiance of a man?

The test of blood is as useless; for we are almost all the mixed product of many breeds. There is only one designation that is not a fiction of the law. We are all men, born into a splendid and terrible world in which for a while our lot is to enjoy and to suffer. The one reasonable aim of man is that life shall be as happy as it can be made with as much as possible of joy and as little as possible of pain. There is only one way of attaining that aim: the pursuit of natural knowledge.

We are all citizens of one little planet. We are as it were a ship's company marooned on an unknown and mysterious island. There is no time to quarrel about our origins. We have food to find and shelter to prepare. Of what that island can provide for our comfort we know still very little. Let us in peace explore the place. It is full of wonderful things and for aught we know we may yet find even the elixir of life.

EVOLUTION AND EDUCATION

From *The New Educator's Library*
(Sir Isaac Pitman, Ltd., 1922). *Written* 1915

The recognition of Evolution as the mode by which the human race came into existence has reacted in various ways on conceptions of education. In recent years, the study of the nature of variation and heredity (known as Genetics), the phenomena by which we must suppose Evolution to proceed, has made rapid progress. The knowledge thus acquired limits in several ways our expectations as to the results which education can attain. That education can modify the composition and development of such a people as our own is not in doubt; but even the preliminary acquaintance with what may be called racial physiology (recently acquired) has greatly promoted an understanding both of the possibilities of modification and of the way in which these changes are actually effected by the institution of public education. The conclusions to which genetic science points run counter to many notions long popularly entertained. It was, for example, assumed both by physiologists and by laymen that the effects of cultivation or training in the case of both animals and plants were, in greater or less degree, transmitted to the offspring, and that in the course of generations these effects would accumulate. This theory was prominently developed by Lamarck, and was adopted, with few exceptions (e.g. Sir W. Lawrence), by all writers on these subjects, notably by Charles Darwin. Weismann was the first to induce the world seriously to examine the foundations of this doctrine. He shewed not only that the little evidence favourable to such a belief was, in reality, worthless; but also that the physiological mechanism of heredity, in so far as it can be observed, was such that the occurrence of any transmission is in a high degree improbable. The results of the modern accurate study of heredity are entirely in harmony with this negative conclusion. There is now scarcely any doubt that the germ-cells of which the offspring are composed possess from the beginning ingredients determining their powers and attributes; and that, with rare and doubtful exceptions, it is not in the power of the parent, by use, disuse, or otherwise, to increase or diminish this total. It is not impossible that injury to the germ-cells may be effected by starvation of the parent, by excessive doses of drugs (such as alcohol), and similar violent treatment, though there is little definite evidence that even in this limited degree the destiny of the offspring can be changed; but that the development of a faculty in the parent by education or practice causes an increase in that faculty in the offspring is recognised by most students of the subject to be altogether unproven and probably impossible.

It is true that in the last decade some have again revived the view brilliantly expounded by Samuel Butler (*Life and Habit*, 1878), and also by Hering, that living things may, through their generations, have a continuous accumulation of "unconscious memory". Just as learning to read or to play a musical instrument requires close attention and extreme effort in the early stages—though afterwards these acts may be performed almost without conscious attention at all—so, it is argued, may even the ordinary reflex actions, such as respiration or digestion, have been acquired as a summation of effort originally conscious. Such a fascinating proposition, if well supported, would have enormous consequences, and man's outlook on the world would be profoundly modified. It is, however, maintained chiefly as providing a complete account of the origin of adaptations with which no other current theory of evolution has successfully dealt, rather than by appeal to direct evidential proofs. Living things do continually display purposeful faculties which seem as if they *must* result from the inheritance of parental experience. A bird builds the nest peculiar and appropriate to its species. Conceivably, however, the bird remembers the nest in which it was reared, and copies that when its own time comes. But insects do similar things, though the parents died when the offspring were eggs. Parasitic ichneumons, for instance, find the larvae on which their young are to feed, though hidden deep in a tree-trunk. Somehow they perceive the hidden larva, and lay their eggs in such a way that the young will reach it. Nature abounds with such examples. We can say that the "instinct" of the ichneumon is fired or let off by the perception—probably scent—of its food-larva, just as the first drop of drink may excite the craving for alcohol in the youth who inherits that vice; but that is a mere description of the phenomenon and no account of its causation. Nevertheless it must be recognised as a fact that the purposeful acts of animals are, in many cases, first made in response to external stimuli. Sometimes they are, no doubt, rightly interpreted as directly imitative of the similar acts made by the parent in the presence of the young, but unimpeachable examples of actual teaching given by the parent are rare. It has, for instance, been often reported, on fairly good authority, that diving birds teach their young to dive. The significance of the evidence must, in these cases, obviously be largely a matter of interpretation. Fundamental instincts are evidently called into play by trifling circumstances. Chickens are sometimes said to peck up food as soon as they are hatched. That is not true; but, as every one who has used incubators knows, at about twenty-four hours after hatching, chicks make vigorous but, at first, ill-directed strokes at any small coloured or shining objects. They peck in this way, especially at each other's claws. Some hours later they acquire precision, and can seize bits of food with certainty. The instinct appears to be excited largely, if not entirely, through the sense of sight. Imitation greatly aids. A single chick may not learn to feed itself for several days, but there is no difficulty when several are together. On the other hand, the disposition to run to the hen's "cluck", which is manifested very early in chicks hatched under a hen, is not developed in incubated chicks of similar age, who evidently attach no meaning to the sound. An apparatus is present ready to act if the appropriate stimulus is given at the right time; but for want of that stimulus, it remains inoperative. It is tempting to suppose that the apparatus, the readiness to make the right response to various stimuli, is a manifestation of "unconscious memory"; but since, as we have said, there is no good reason to suppose that even the simplest experiences of the parent are at all transmitted to a succeeding generation, the suggestion of continuous memory as applicable to education can only be defended on grounds which to the biologist are mystical and unconvincing.

The racial changes which may admittedly follow on the institution of popular education are seen by biologists to be produced in a very different way. One of the chief facts demonstrated by modern research is the heterogeneity of the individuals of which most species of animals and plants are composed. Conspicuously is this true of man, and of the mixed races in a very high degree. Applied for even a few generations to composite populations, universal education can effect remarkable changes by re-arrangement of the constituent members. Opportunity is given for the more intel-lectual individuals in the various classes of the community to improve their position. A natural selection of the intelligent from the several social layers is thus given an increased scope. By changes in public opinion, which in an educated community tends to discredit the less intelligent, the process is accelerated. As a result of this sorting process, a considerable reconstitution of the layers or classes may be effected. The intelligence of the community then seems to have been raised, and undoubtedly it may shew a higher average of mental efficiency; but the alteration accrues by change in the distribution of opportunity and the selective process, not by any physiological transmission of the cumulative effects of education. Discussion often arises whether various non-European races submitted to our education will be found capable of assimilating themselves to our mental standards. The question is raised both in reference to peoples like the Chinese, immemorially civilised, and to races of low intellectual type, such as the negro or the American Indians. Reformers and philanthropists are disposed to treat these two classes of cases as similar, and to argue that in the course of generations any race exposed to education can develop along the lines which we have followed. To the biologist, it is clear that no answer of general application can be given. The problem is special to each race, and is simply a question whether the race does or does not contain individuals capable of responding to the treatment. For, while it must be supposed that a vast country like China, among the divers races of its inhabitants, many of whom have for ages shewn intellectual capacity of a very high order, may almost certainly contain human material possessing all kinds of attributes, nothing in the history of the negro races indicates even the sporadic existence of such material among them. Unless, however, the aptitude is already present, there is no likelihood that it can be introduced except by cross-breeding, and the possibility of a change in intellectual type is not essentially distinct from that of a change in colour. It must be understood that we are here considering the *capacity* of races to respond to our education— not the question whether, given the capacity, a particular race is likely to do so. Reference was made to the possibility of a change of type being effected by crossing; and, in considering any practical example, a large and very uncertain allowance must be made for the consequences of such events. As we have learnt from Mendelian studies in heredity, the results of crossing are by no means so simple as was formerly supposed. A feature or attribute may, indeed, be introduced from a foreign source and eventually become widely disseminated among a population, though the general appearance and characteristics may remain sensibly unchanged in other respects.

Judgments in regard to the intellectual evolution of races are further obscured by the vast changes produced solely by the exercise of the faculty of *imitation*. The part that fashion and imitation have played in the history of civilisation is still imperfectly understood and by no means fully appreciated, and it is not inconceivable that these phenomena have been significant in determining the course of Evolution. (The reader may be referred to the remarkable essays of G. Tarde: *Les Lois de l'Imitation* [ed. 6], 1911.) The large physiological departures from the normal—variations in the biological sense—are rare. The bulk of the population remains of an older type: yet by the irresistible instinct towards imitation, the race, as a whole, may, to use Tarde's happy expression, be "polarised" under the influence of a few dominant minds, so as to present a semblance of uniformity which masks their real composition. Even among European nations which pass for educated, only a small part of the population really assimilates education in any considerable degree. With the majority, the process is carried but a small way, little permanent effect being produced, and signs are not wanting that the failure is due to congenital want of aptitude. Remembering this fact, that among contemporary peoples the type which can in any sense be termed intellectual is always rare, it is evident that a large ostensible change may be induced in a population by the presence among it of a comparative minority who can respond fully and readily to education. A spurious transmutation of the people as a whole is completed by imitation in its manifold forms. With the prosecution of some far more rigorous analysis than can yet be applied to human populations, it may be possible to trace with some accuracy the principles here indicated, but at present we can only recognise their operation.

THE PLACE OF SCIENCE IN EDUCATION

From *Cambridge Essays on Education*. (Cambridge
University Press, 1917)

That secondary education in England fails to do what it
might is scarcely in dispute. The magnitude of the failure will
be appreciated by those who know what other countries accomplish at a fraction of the cost. Beyond the admission that
something is seriously wrong there is little agreement. We are
told that the curriculum is too exclusively classical, that the
classes are too large, the teaching too dull, the boys too much
away from home, the examination-system too oppressive, ath-
letics overdone. All these things are probably true. Each cause
contributes in its degree to the lamentable result. Yet, as it
seems to me, we may remove them all without making any
great improvement. All the circumstances may be varied, but
that intellectual apathy which has become so marked a characteristic of English life, especially of English public and social
life, may not improbably continue. Why nations pass into these
morbid phases no one can tell. The spirit of the age, that
"polarisation of society" as Tarde[1] used to call it, in a definite
direction, is brought about by no cause that can be named as
yet. It will remain beyond volitional control at least until we
get some real insight into social physiology. That the attitude
or pose of the average Englishman towards education, knowledge and learning is largely a phenomenon of infectious imitation we know. But even if we could name the original source
of the mischief, the person—for in all likelihood there was such
an one—whom English society in its folly unconsciously selected
as a model, the knowledge would advance us little. The psychology of imitation is still impenetrable and likely to remain
so. The simple interpretation of our troubles as a form of sloth—
a travelling along lines of least resistance—can scarcely be
maintained. For first there have been times when learning and
science were the fashion. Whether society benefited directly
therefrom may, in passing, be doubted, but certainly learning
did. Secondly there are plenty of men who under the pressure
of fashion devote much effort to the improvement of their form
in fatuous sports which otherwise applied would go a considerable way in the improvement of their minds and in widening
their range of interests.

[1] *Les Lois de l'Imitation*, 1911, p. 87.

Of late things have become worse. In the middle of the nineteenth century a perfunctory and superficial acquaintance with recent scientific discovery was not unusual among the upper classes, and the scientific world was occasionally visited even by the august. These slender connections have long withered away. This decline in the public estimation of science and scientific men has coincided with a great increase both in the number of scientific students and in the provision for teaching science. It has occurred also in the period during which something of the full splendour and power of science has begun to be revealed. Great regions of knowledge have been penetrated by the human mind. The powers of man over nature have been multiplied a hundredfold. The fate of nations hangs literally on the issue of contemporary experiments in the laboratory; but those who govern the Empire are quite content to know nothing of all this. Intercommunication between government departments and scientific advisers has of course much developed. That, even in this country, was inevitable. Otherwise the Empire might have collapsed long since. Experts in the sciences are from time to time invited to confer with heads of Departments and even Cabinet Ministers, explaining to them, as best they may, the rudiments of their respective studies, but such occasional night-school talks to the great are an inadequate recognition of the position of science in a modern State. Science is not a material to be bought round the corner by the dram, but the one permanent and indispensable light in which every action and every policy must be judged.

To scientific men this is so evident that they are unable to imagine what the world looks like to other people. They cannot realise that by a majority of even the educated classes the phenomena of nature and the affairs of mankind are still seen through the old screens of mystery and superstition. The man of science regards nature as in great and ever increasing measure a soluble problem. For the layman such inquiries are either indifferent and somewhat absurd, or, if they attract his attention at all, are interesting only as possible sources of profit. I suspect that the distinction between these two classes of mind is not to any great degree a product of education.

It is contemporary commonplace that if science were more prominent in our educational system everybody would learn it and things would come all right. That interest in science would be extended is probable. There is in the population a residuum of which we will speak later, who would profit by the opportunity; but that the congenitally unscientific, the section from which the heads of government temporal and spiritual, the lawyers, administrators, politicians, the classes upon whose minds the public life of this country almost wholly depends, would by exhibition of scientific diet at any period of life, however early, be essentially altered seems in a high degree unlikely. Of the converse case we have long experience, and I would ask those who entertain such sanguine expectations, whether the results of administering literature to scientific boys give much encouragement to their views. This consideration brings us to the one hard, physiological fact that should form the foundation of all educational schemes: the congenital diversity of the individual types. Education has too long been regarded as a kind of cookery: put in such and such ingredients in given proportions and a definite product will emerge. But living things have not the uniformity which this theory of education assumes. Our population is a medley of many kinds which will continue heterogeneous, to whatever system of education they are submitted, just as various types of animals maintain their several characteristics though nourished on identical food, or as you may see various sorts of apples remaining perfectly distinct though grafted on the same stock. Their diversity is congenital.

According to the proposal of the reformers the natural sciences should be universally taught and be given "capital importance" in the examinations for the government services, but, cordially as we may approve the suggestion, we ought to consider what exactly its adoption is likely to effect. The intention of the proposal is doubtless that our public servants, especially the highest of them, shall, while preserving the great qualities they now possess, add also a knowledge of science and especially scientific habits of mind. Such is the "ample proposition that hope makes". Does experience of men accord with it at all? Education, whether we like it or not, is a selective agency. I doubt whether the change proposed will sensibly alter the characters of the group upon whom our choice at present falls. Rather, if forced upon an unwilling community, must it act by substituting another group. The most probable result would not be that the type of men who now fill great positions would become scientific, but rather that their places would be taken by men of an altogether distinct mental type. At the present time these two types of men meet but little. They scarcely know each other. Their differences are profound, affecting thoughts, ways of looking at things, and mental interests of every kind. If either could for a moment see the world with the vision of the other he would be amazed, but to do so he would need at least to be born again, and probably, as Samuel Butler remarked, of different parents. No doubt the abler men of either type could learn with more or less effort or unreadiness the subject-matter and principles of each other's business, but anyone who has watched the habits of the two classes will perceive that for them in any real sense to exchange interests, or that either should adopt the scheme of proportion which the other assigns to the events of nature and of life, presupposes a metamorphosis well-nigh miraculous.

The Bishop of London speaking lately on behalf of the National Mission said that nature helped him to believe in God, and as evidence for his belief referred to the fact that we are not "blown off" this earth as it rushes through space, declaring that this catastrophe had been averted because "Some one" had wrapped seventy miles of atmosphere round our planet.[1] Does anyone think that the bishop's slip was in fact due to want of scientific teaching at Marlborough? His chances of knowing about Sir Isaac Newton, etc., etc., have been as good as those of many familiar with the accepted version. I would rather suppose that such sublunary problems had not interested him in the least, and that he no more cared how we happen to stick on the earth's surface than St Paul cared how a grain of wheat or any other seed germinates beneath it, when he similarly was betrayed into an unfortunate illustration.

So too on the famous occasion—always cited in these debates —when a Home Secretary defended the Government for having permitted the importation of fats into Germany on the ground that the discovery that glycerine could be made from fat was a recent advance in chemistry, he was not shewing the defects of a literary education so much as a want of interest in the problems of nature, and the subject-matter of science at large. It is to be presumed indeed that neither fats, nor glycerine, nor the dependent problem how living bodies are related to the world they inhabit, had ever before seemed to him interesting. Nor can we suppose they would, even if chemistry were substituted for Greek in Responsions.

The difficulty in obtaining full recognition for science lies deeper than this. It is a part of public opinion or taste which may well survive changes in the educational system. Blunders about science like those illustrated above are soon excused. Few think much the worse of the perpetrators, whereas a corresponding obliviousness to language, history, literature, and

[1] Reported in *Evening Standard*, 11 September 1916.

indeed to learning other than their own which we of the scientific fraternity have agreed to condone in our members is incompatible with public life of a high order. Both classes have their disabilities. That of the scientific side is well expressed in an incident which befell the late Professor Hales. Examining in the Little-Go *viva voce*, he asked a candidate, with reference to some line in a Greek play, what passage in Shakespeare it recalled to him, and received the answer "Please, sir, I am a mathematical man". Some, no doubt, would rather ignore gravitation. When, for example, one hears, as I did not long since, several scientific students own in perfect sincerity that they could not recall anything about Ananias and Sapphira, and another, more enlightened, say that he was sure Ananias was a name for a liar though he could not tell why, one is driven to admit that ignorance of this special but not uncommon kind does imply more than inability to remember an old legend. We may be reluctant to confess the fact, but though most scientific men have some recreation, often even artistic in nature, we have with rare exceptions withdrawn from the world in which letters, history and the arts have immediate value, and simple allusions to these topics find us wanting. Of the two kinds of disability which is the more grave? Truly gross ignorance of science darkens more of a man's mental horizon, and in its possible bearing on the destinies of a race is far more dangerous than even total blindness to the course of human history and endeavour; and yet it is difficult to question the popular verdict that to know nothing of gravitation though ridiculous is venial, while to know nothing of Ananias is an offence which can never be forgiven.

That is the real difficulty. The people of this country have definitely preferred the unscientific type, holding the other virtually in contempt. Their choice may be right or wrong, but that it is reversible seems unlikely. Such revolutions in public opinion are rare events. Democracy moreover inevitably worships and is swayed by the spoken word. As inevitably, the range and purposes of science daily more and more transcend the comprehension—even the educated comprehension—of the vulgar, who will of course elevate the nimble and versatile, speaking a familiar language, above dull and inarticulate natural philosophers.

In these discussions there is a disposition to forget how very largely natural science is already included in the educational curriculum both at schools and universities. Schools subsidised by the Board of Education are obliged to provide science-teaching. The public schools have equipment, in some cases a superb equipment, for teaching at least physics and chemistry. At the newer universities there are great and vigorous schools of science. Of the old universities Cambridge stands out as a chief centre of scientific activity. In several branches of science Cambridge is without question pre-eminent. The endowments both of the university and the colleges are freely used for the advancement of the sciences. Not only in these material ways are scientific studies in no sense neglected, but the position of the sciences is recognised and even envied by those who follow other kinds of learning. The scientific schools of Cambridge form perhaps the dominant force among the resident body of the University, and except by virtue of some great increase in the endowments, it would be impossible to extend further the scientific side of Cambridge and still maintain other forms of intellectual activity in such proportion as to preserve that healthy co-ordination which is the life of a great University.

At Oxford the case is no doubt very different. The measure in which the sciences are esteemed appears only too plainly in the small proportion of Fellowships filled by men of science. Progress has nevertheless begun. At the remarkable Conference called in May 1916, to protest against the neglect of science, it was noticeable that the speakers were, in overwhelming majority, Oxford men.[1]

Among the educational institutions of England there is no general neglect to provide teaching of natural science and much of the language used in reference to the problem of reform is not really in accord with fact. Probably no boy able to afford a good secondary school, certainly none able to proceed to a university, is debarred from scientific teaching merely because it does not "form an integral part" of the curriculum. This alone suffices to prove that the real cause of the deplorable neglect of science is to be sought elsewhere. The fundamental difficulty is that which has been already indicated, that public taste and judgment deliberately prefers the type known as literary, or as it might with more propriety be designated, "vocal". In the schools there is no lack of science teaching, but the small percentage of boys whose minds develop early and whose general capacity for learning and aptitude for affairs mark them out as leaders, rarely have much instinct for science, and avoid such teaching, finding it irksome and unsatisfying.

[1] Two Cambridge men spoke, one being Lord Rayleigh, the Chairman, and ten Oxford men, besides one originally Cambridge, for several years an Oxford professor.

These it is, who going afterwards to the universities, in preponderating numbers to Oxford, make for themselves a congenial atmosphere, disturbed only by faint ripples of that vast intellectual renaissance in which the new shape of civilisation is forming. With self-complacency unshaken, they assume in due course charge of Church and State, the Press, and in general the leadership of the country. As lawyers and journalists they do our talking for us, let who will do the thinking. Observe that their strength lies in the possession of a special gift, which under the conditions of democratic government has a prodigious opportunity. Uncomfortable as the reflection may be, it is not to be denied that the countries in which science has already attained the greatest influence and recognition in public affairs are Germany and Japan, where the opinions of the ignorant are not invited. But facts must be recognised, and our government is likely to remain in the hands of those who have the gift of speech. A general substitution of scientific men for the "vocal" could scarcely be achieved, even if the change were desirable. The utmost limit of success which the conditions admit is some inoculation of scientific interest and ideas upon the susceptible members of the classes already preferred. That a large proportion of those persons are in the biological sense resistant to all such influences must be expected. Granting however that a section, perhaps even the majority, of our βέλ-τιστοι may prove unamenable to the influences of science, no one can doubt that under the present system of education a proportion of not unintelligent boys in practice have little option. From earliest youth classics are offered to them as almost the sole vehicle of education. They do sufficiently well in classics, as they probably would on any other curriculum, to justify themselves and their advisers in thinking that they have made a good beginning to which it is safer to stick. The system has a huge momentum, and so, holding to the "great wheel" that goes up the hill, they let it draw them after. In their protest against the monotony of the courses provided for young boys the reformers are right. The trouble is not that science is not taught in the schools, but that in schools of the highest type, with certain exceptions, the young boys are not offered it.

Realising the determinism which modern biological knowledge has compelled us to accept, we suspect that the power of education to modify the destinies of individuals is relatively small. Abrogating larger hopes we recognise education in its two scientific aspects, as a selective agency, but equally as a provision of opportunity. In view therefore of the congenital diversity of the individual types, that provision should be as diverse and manifold as possible, and the very first essential in an adequate scheme of education is that to the minds of the young something of everything should be offered, some part of all the kinds of intellectual sustenance in which the minds of men have grown and rejoiced. That should be the ideal. Nothing of varied stimulus or attraction that can be offered should be withheld. So only will the young mind discover its aptitudes and powers. This ideal education should bring all into contact with *beauty* as seen first in literature, ancient and modern, with the great models of art and the patterns of nobility of thought and of conduct; and no less should it shew to all the *truth* of the natural world, the changeless systems of the universe, as revealed in astronomy or in chemistry, something too of the truth about life, what we animals really are, what our place and what our powers, a truth ungarbled whether by prudery or mysticism.

But presented with this ideal the schoolmaster will reply that something of everything means nothing *thorough*. I know the objection and what it commonly stands for. It is the cloak and pretext for that accursed pedantry and cant which turns every sort of teaching to a blight. Thoroughness is the excuse for giving boys grammar and accidence in the name of Greek: diagrams, formulae and numerical examples in the name of science. Stripped of disguise this love of thoroughness is nothing but an indolent resolve to make things easy for the teacher, and, worse still, for the examiner. Live teaching is hard work. It demands continual freshness and a mind alert. The dullest man can hear irregular verbs, and with the book he knows whether they are said right or wrong, but to take a text and shew what the passage means to the world, to reconstruct the scene and the conditions in which it was written, to shew the origins and the fruits of ideas or of discoveries, demand qualities of a very different order. The plea for thoroughness may no doubt be offered in perfect sincerity. There are plenty of men, especially among those who desire the office of a pedagogue, whose field of vision is constricted to a slit. If they were painters their work would be, in the slang of the day, "tight". One small group of facts they see hard and sharp, without atmosphere or value. Their own knowledge having no capacity for extension, no width or relationship to the world at large, they cannot imagine that breadth in itself may be a merit. Adepts in a petty erudition without vital antecedents or consequences, they would willingly see the world shrivel to the dimensions of their own landscape.

Anticipating here the applause of the reforming party, to avoid misapprehension let it be expressly observed that pedantry of this sort is in no sense the special prerogative of teachers of classics. We meet it everywhere. Among teachers of science the type abounds, and from the papers set in any Natural Sciences Tripos, not to speak of scholarship examinations of every kind, it would be possible to extract question after question that ought never to have been set, referring to things that need never have been taught, and knowledge that no one but a pedant would dream of carrying in his head for a week.

The splendid purpose which science serves is the inculcation of principle and balance, not facts. There is something horrible and terrifying in the doctrine so often preached, reiterated of course by speaker after speaker at the "Neglect of Science" meeting, that science is to be preferred because of its utility. If the choice were really between dead classics and dead science, or if science is to be vivified by an infusion of commercial, utilitarian spirit, then a thousand times rather let us keep to the classics as the staple of education. They at least have no "use". At least they hold the keys to the glorious places, to the fulness of literature and to the thoughtful speech of all kindred nations, nor are they demeaned with sordid, shop-keeper utility. This was plainly in the mind of the Poet Laureate, who speaking at the meeting I have referred to, said well that "a merely utilitarian science can never win the spiritual respect of mankind". The main objection that the humanists make to the introduction of natural science as a necessary subject of education, is, he declared, that science is not spiritual, that it does not work in the sphere of ideas. He went on very properly to shew how perverse is such a representation of science, but, alas, in further recommendation of science as a safe subject of instruction he added that the antagonism of science to religion is ended, and that the contest had been a passing phase. Reading this we may wonder whether we are in fairness entitled to Dr Bridges's approval. "Tastes sweet the water with such specks of earth?" Since he spoke of the "unscientific attitude" of Professor Huxley as a thing of the past, candour obliges us to insist emphatically that the struggle continues and must perpetually be renewed. Huxley was opposing the teaching of science to that of revelation. In these days the ground has shifted, and supernatural teachings make preferably their defence by an appeal to intuition and other obscure phenomena which can be trusted to defy investigation. Against all such apocryphal glosses of evidential truth science protests with equal vehemence, and

B   28

were Huxley here he would treat Bergson and his allies with the same scorn and contumely that he meted out to the Bishop of Oxford on the notorious occasion to which Dr Bridges made reference. As well might we decorate our writings with Plantin title-pages, shewing the author embraced by angels and inspiring muses, as recommend ourselves in these disguises.

Agnosticism is the very life and mainspring of science. Not merely as to the supernatural but as to the natural world must science believe nothing save under compulsion. Little of value has a man got from science who has not learned to be slow of faith. Those early lessons in the study of the natural world will be the best which most frankly declare our ignorance, exciting the mind to attack the unknown by shewing how soon the frontier of knowledge is reached. "We don't know" should be ever in the mouth of the teacher, followed sometimes by "we may find out yet". Not merely to the investigator but to the pupil the interest of science is strongest in the growing edges of knowledge. The student should be transported thither with the briefest possible delay. Details of those parts of science which by present means of investigation are worked out and reduced to general expressions are dull and lifeless. Many and many a boy has been repelled, gathering from what he hears in class that science is a catalogue of names and facts interminable.

In childhood he may have felt curiosity about nature and the common impulse to watch and collect, but when he begins scientific lessons he discovers too often that they relate not even to the kind of fact which nature is for him, or to the subjects of his early curiosity and wonder, but to things that have no obvious interest at all, measurements of mechanical forces, reaction-formulae, and similar materials.

All these, it is true, man has gradually accumulated with infinite labour; upon them, and of such materials has the great fabric of science been reared: but to insist that the approaches to science shall be open only to those who will surmount these gratuitous obstacles is mere perversity. Men's minds do not work in that way. How many would discover the grandeur of a Gothic building if they were prevented from seeing one until they could work out stresses and strains, date mouldings, and even perhaps cut templates? Most of us, to be sure, enjoy the cathedrals more when we acquire some such knowledge, and those who are to be architects must acquire it, but we can scarcely be astonished if beginners turn away in disgust from science presented on those terms.

It is from considerations of this kind that I am led to believe that for most boys the easiest and most attractive introduction to science is from the biological side. Admittedly chemistry is the more fundamental study, and some rudimentary chemical notions must be imparted very early, but if the framework subject-matter be animals and plants, very sensible progress in realising what science means and aims at doing will have been made before the things of daily life are left behind. These first formal lessons in science should continue and extend the boy's own attempts to find out how the world is made.

I shall be charged with running counter both to common sense and to authority in expressing parenthetically the further conviction that, in biology at least, laboratory work is now largely overdone. Whether this is so at schools I cannot tell, but at the universities whole mornings and afternoons spent in making elaborate preparations, drawings and series of sections, are frequently wasted. These courses were devised with the highest motives. Students were to "find out everything for themselves". Generally they are doing nothing of the kind. It may have been so once, but with text-books perfected and teaching stereotyped, the more industrious are slavishly verifying what has been verified repeatedly, or at best acquiring manipulative skill. The rest are doing nothing whatever. They would be better employed taking a walk, devilling for some investigator, browsing in museums or libraries, or even arguing with each other. Certainly a few lessons in the use of indexes and books of reference would be far more valuable. Students of every grade must of course do some laboratory work, and all should see as much material as possible. My protest is solely against those long, torpid hours compulsorily given to labour which will lead to nothing of novelty, and serves only to teach what can be got readily in other ways. There are a few whose souls crave such employment. By all means let them follow it.

But whatever is good for maturer students, biology for schoolboys should be of a less academic cast.

The natural history of animals and plants has the obvious merit that it prolongs the inborn curiosity of youth, that its subject-matter is universally at hand, accessible in holidays and in the absence of teachers or laboratories, and best of all that through biological study the significance of science appears immediately, disclosing the true story of man's relation to the world. From natural history the transition to the other sciences, especially to chemistry and physics, is easy and again natural. In the study of life many of the fundamental conceptions of those sciences are met with on the threshold, and boys whose

aptitudes are rather of the physical order will at once feel the
impulse to follow nature from that aspect. Biology is the more
inclusive study. A man may be a good chemist and miss the
broad meaning of science altogether, being sometimes indeed
more devoid of such comprehension than many a philosopher
fresh from Classical Greats.

In appealing for a progress from the general to the particular
I am not blind to the dangers. Biology for the young readily
degenerates into a mawkish "nature-study", or all-for-the-best
claptrap about adaptation, but a sure remedy is the strong tonic
of agnosticism, teaching one of the best lessons science has to
offer, the resolute rejection of authority.

Some take comfort in the hope that all subjects may be taught
as branches of science, but the fact that must permanently postpone arrival at this educational Utopia is that a great proportion of teachers are not and can never be made scientific.
Nothing proceeding from such persons will by the working of
any schedule, regulation, or even Order of the Board be ever
made to bear any colourable resemblance to science. Moreover
as has already been indicated, there are plenty of pupils also
who will flourish and probably reach their highest development
taught by unscientific men, pupils whose minds would be steril-
ised or starved by that very nourishment which to our thinking
is the more generous. Were we a homogeneous population one
diet for all might be justifiable, but as things are, we should
offer the greatest possible variety.

From Rousseau onwards educationists, deriving their views,
I suppose, from some metaphysical or theological conception
of human equality, speak continually of the "mind of the child"
as if the young of our species conformed to a single type. If the
general spread of biological knowledge serves merely to expose
that foolish assumption there would be progress to record. Dr
Blakeslee,[1] a well-known American biologist, lately gave a good
illustration of this. In a paper on education he shewed photo-
graphs of two varieties of maize. The ripe fruits of both are
colourless if their sheaths be unbroken. The one, if exposed to
the light before ripening, by rupture of its sheath, turns red. The
second, otherwise indistinguishable, acquires no red colour
though uncovered to the full sun. If these maizes were two
boys, not improbably the one would be caned for failing to
respond to treatment so efficacious in the case of the other.
When we hear that such a man has developed too exclusively
one side of his nature, with what propriety do we assume that

[1] *Journ. of Heredity*, VIII, 1917, p. 53.

he had any other side to develop? Or when we say that such-and-such a course of study tends to make boys too exclusively literary, or scientific, or what not, do we not really mean that it provides too exclusively for those whose aptitudes are of these respective kinds? Living in the midst of a mongrel population we note the divers powers of our fellows and we thoughtlessly imagine that if something different had happened to us, we can't say what, we should have been able to rival them. A little honest examination of our powers shews how vain are such suppositions. The right course is to make some provision for all sorts, since unscientific teaching and unscientific persons will remain with us always.

Teaching of this universal and undifferentiated sort, provided for all in common, should be continued up to the age at which pupils begin to shew their tastes and aptitudes, in general about 16, after which stage such latitude of choice should be given as the resources of the school can provide.

Of what should the undifferentiated teaching consist? Coming from a cultivated home a boy of ten may be expected to have learned the rudiments of Latin, and at least one modern language, preferably French, *colloquially*, arithmetic, outlines of geography, tales from Plutarch and from other histories. Going to a preparatory school he will read easy Latin texts *with translations* and notes; French books, geography including the elements of astronomy, beginning also algebra and geometry. At 12 dropping French except perhaps a reading once a week, he will begin Greek, by means of easy passages again with the translations beside him, continuing the rest as before. Transferred at 14½ to a public school he will go on with Latin, starting Latin prose, Greek texts, again read fast with translations. He will now have his first formal introduction to science in the guise of biology, leading up to lessons and demonstrations in chemistry and physics. At about 16½ he may drop classics *or mathematics* according as his tastes have declared themselves, adding modern languages instead, continuing science in all cases, greater or less in amount according to his proclivities.

Boys with special mathematical ability will of course need special treatment. Moreover provision of German for all has avowedly not been made. For all it is desirable and for many indispensable. But as the number who read it for pleasure, never very large, seems likely to diminish, German may perhaps be reserved as a tool, the use of which must be acquired when necessary.

Such a scheme, I submit, makes no impossible demand on

the time-table, allowing indeed many spare hours for accessory subjects such as readings in English or history. Note the
main features of this programme. The time for things worth
learning is found by dropping *grammar* as a subject of special
study. There are to be no lessons in grammar or accidence as
such, nor of course any verse compositions except for older boys
specialising in classics. *Mathematics* also is treated as a subject
which need not be carried beyond the rudiments unless mathematical or physical ability is shewn. For other boys it leads
literally nowhere, being a road impassable.

All the languages are to be taught as we learn them in later
life, when the desire or necessity arises, by means of easy passages with the translation at our side. Our present practice not
only fails to teach languages but it succeeds in teaching how
not to learn a language. Who thinks of beginning Russian by
studying the "aspects" of the verbs, or by committing to memory the 28 paradigms which German grammarians have devised on the analogy of Latin declensions? Auxiliary verbs are
the pedagogue's delight, but who begins Spanish by trying to
discriminate between *tener* and *haber*, or *ser* and *estar*, or who
learns tables of exceptions to improve his French? These things
come by use or not at all.

If languages are treated not as lessons but as vehicles of
speech, and if the authors are read so that we may find out
what they say and how they say it, and at such a pace that we
follow the train of thought or the story, all who have any sense
of language at all can attend and with pleasure too. What
chance has a boy of enjoying an author when he knows him
only as a task to be droned through, 30 lines at a time? Small
blame to the pupil who never discovers that the great authors
were men of like passions with ourselves, that the Homeric
songs were made to be shouted at feasts to heroes full of drink
and glory, that Herodotus is telling of wonders that his friends,
and we too, want to hear, that in the tragedies we hear the voice
of Sophocles dictating, choked with emotion and tears; that
even Roman historians wrote because they had something to
tell, and Caesar, dull proser that he is, composed the *Com-
mentaries* not to provide us with style or grammatical curiosities, but as a record of extraordinary events. To get into touch
with any author he must be read at a good pace, and by reading
of that kind there is plenty of time for a boy before he reaches
17 to make acquaintance with much of the best literature both
of Greek and Latin.

Education must be brought up to date; but if in accomplishing that, we lose Greek, it will have been sacrificed to obstinate formalism and pedagogic tradition. The defence of classics as a basis of education is generally misrepresented by opponents. The unique value of the classics is not in any begetting of literary style. We are thinking of readers not of writers. Much of the best literature is the work of unlettered men, as they never tire of telling us, but it is for the enjoyment and understanding of books and of the world that continuity with the past should be maintained. John Bunyan wrote sterling prose, knowing no language but his own. But how much could he read? What judgments could he form? We want also to keep classics and especially Greek as the bountiful source of material and of colour, decoration for the jejune lives of common men. If classics cease to be generally taught and become the appanage of a few scholars, the gulf between the literary and the scientific will be made still wider.[1] Milton will need more explanatory notes than O. Henry. Who will trouble about us scientific students then? we shall be marked off from the beginning, and in the world of laboratories Hector, Antigone, and Pericles will soon share the fate of poor Ananias and Sapphira.

I come now to the gravest part of the whole question. We plead for the preservation of literature, especially classical literature, as the staple of education in the name of beauty and understanding: but no less do we demand science in the name of truth and advancement. Given that our demand succeeds, what consequences may we expect? Nothing immediate, as I fear. In opening the discussion it was argued that even if scientific knowledge be widely diffused, any great change in the composition of the ruling classes is scarcely attainable under present conditions of social organisation. Even if science stand equal with classics in examinations for the services the general tenor of the public mind will in all likelihood be undisturbed. Yet it is for such a revolution that science really calls, and come it will in any community dominated by natural knowledge. Science saves us from blunders about glycerine, shews how to economise fuel and to make artificial nitrates, but these, though they decide national destinies, are merely the sheaf of the wave-offering: the harvest is behind. For natural knowledge is destined to give man not only a direct control of the material world but new interpretations of higher problems. Though we in England make a stand upon the ancient way, peoples elsewhere

[1] Few realise how deep it is already. A contemporary man of science once complained to me that he found it hard to keep up, having, as he put it, "never read the life of Apollo".

will move on. Those who have grasped the meaning of science, especially biological science, are feeling after new rules of conduct. The old criteria based on ignorance have little worth. "Rights", whether of persons or of nations, may be abstractions well-founded in law or philosophy, but the modern world sooner or later will annul them.

The general ignorance of science has lasted so long that we have virtually two codes of right and duty, that founded on natural truth and that emanating from tradition, which almost alone finds public expression in this country. Whether we look at the cruelty which passes for justice in our criminal courts, at the prolongation of suffering which custom demands as a part of medical ethics, at this very question of education, or indeed at any problem of social life, we see ahead and know that science proclaims wiser and gentler creeds. When in the wider sphere of national policy we read the declared ideals of statesmen, we turn away with a shrug. They bid us exalt national sentiment as a purifying and redeeming influence, and in the next breath proclaim that the sole way to avert the ruin now menacing the world is to guarantee to all nations freedom to develop, "unhindered, unthreatened, unafraid". So, forsooth, are we to end war. Nature laughs at such dreams. The life of one is the death of another. Where are the teeming populations of the West Indies, where the civilisations of Mexico or of Peru, where are the blackfellows of Australia? Since means of subsistence are limited, the fancy that one group can increase or develop save at the expense of another is an illusion, instantly dissipated by appeal to biological fact, nor would a biologist-statesman look for permanent stability in a multiplication of competing communities, some vigorous, others worthless, but all growing in population. Rather must a people familiar with science see how small and ephemeral a thing is the pride of nations, knowing that both the peace of the world and the progress of civilisation are to be sought not by the hardening of national boundaries but in the substitution of cosmopolitan for national aspiration.

CLASSICAL AND MODERN EDUCATION

Nature, September 1921

Following that general misgiving as to our national system of education which, long felt by thoughtful men, found loud and continual expression during the war, Mr Asquith, then Prime Minister, appointed (1916) Committees to consider the position of natural science and of modern languages respectively. After these Committees had reported, a third Committee was set up (1919) to investigate the position of classics in our educational system. The Report of this Committee,[1] recently issued, is a comprehensive document, full of interesting materials, readable and scholarly, as from the character of the Committee might be expected. The history of classical teaching in the several parts of the United Kingdom, its rise and recent decline, are set out in detail, with an abundance of information never before collected. As to the main inference, no mistake is possible. The classical element in British education is disappearing, and will probably soon be gone altogether.

In the Public Schools few boys are learning Greek, and even Latin, though still generally taught in middle and lower forms, tends more and more to be dropped higher up. None of the new Provided Schools has yet been able to develop a classical tradition and few of them teach Greek.... The danger with which we are faced is not that too many pupils will learn Latin and Greek, but that the greater part of the educated men and women of the nation will necessarily grow up in ignorance of the foundations on which European society is built.

The course of events has been exactly that which the defenders of compulsory classics at Oxford and Cambridge foresaw as the consequence of any weakening of front. Classics were maintained in education solely by the authority of the two old universities. Fearing the financial consequences of competition, they reduced the minimum demanded until it became ridiculous, the inevitable result being that Greek had to be dropped, with Latin soon to follow. The reformers were, of course, mostly persons who set no great store by classical education, but they were aided by many representatives of the humanities, who believed, or were persuaded, that the inherent

[1] *Classics and Education.* Report of the Committee appointed by the Prime Minister to inquire into the Position of Classics in the Educational System of the United Kingdom. Pp. 308. (London: H.M. Stationery Office.) 2s. net.

value of classical training was so obvious that it would hold its
own without protection. They forgot that, on their abdication,
the decision would pass into the control of those who knew the
classics only as a symbol of exclusion, with the Board of Education naturally well disposed towards any movement which
could be represented as popular.

Probably emanating from that group of the reformers, there
are passages in the Report which maintain an undertone of
hope. Wonders have been achieved by a few resolute and devoted scholars in some of the most modern universities. This is
"of good augury", and "with the enthusiasm born of free choice
of subject" there may yet be a revival. Numerous recommendations on points of detail are suggested to this end. The regulations, especially those relating to "Advanced Courses" in
secondary schools, and the examination schedules should not
be weighted unduly against the classics. In every large district
there should be at least one school where Greek teaching can be
had and provision is made for boys and girls with literary tastes,
and generally the Committee pleads that in every branch of
educational administration classical education should be re-
spected as a thing of great worth.

The value of the classics has never been better set forth. This
part of the Report is admirable good sense, and approaches to
eloquence as nearly as a Report to a Prime Minister can do.
In the classics a man "obtains access to literature, both in prose
and poetry, which in the judgment of many is absolutely the
noblest in the world; but if that claim be not admitted, it is at
least unique, inimitable, and irreplaceable. We have here a
spiritual value not easily reckoned...". Not merely are the
works of antiquity "classic in the sense that they belong to the
highest class of human achievement", but they have the peculiar merit of introducing the student to a world which is not our
own, though presenting problems closely akin to ours, thus
promoting a certain power of understanding and of judgment
in fundamentals. The student has "attained this access to
beauty and this power of understanding by means of a peculiar
course of training which requires the exercise of many different
powers of the mind, and forms a remarkable combination of
memory-training, imagination, aesthetic appreciation, and
scientific method. For better or worse, the study of the classics
is quite a different thing from the learning of languages pure
and simple". Even the merely verbal exercises start the "invaluable habit of thinking out the real meaning of words and
phrases before attempting to translate them". The exposure of

the inadequacy of translations is especially convincing. "Few people would seriously maintain that we can get 'all we want' out of an English translation of Victor Hugo or Goethe, or a French translation of Shakespeare or Burke."

The attitude of the Committee towards grammar is symptomatic of a welcome change. Hitherto the classical teacher has refused to put grammar anywhere but first. He would surrender nothing. By this pedantry thousands have been repelled. No one doubts that grammatical exercises are a fine educational instrument, but in comparison with the rest that classical education can do, grammar is such a small, poor thing. Had the scholastic world repented of this error when the warning came, the classics might have survived as the staple of at least a complete education. The remarks of the present Committee on methods of teaching are all that could be wished. It is advised that "great stress should be laid on the subject-matter and the historical background of the texts read, though not to the prejudice of exact training in the language". We may be thankful for this concession to common sense.

Probably it comes too late. Time was when modern languages, and especially science, were admitted grudgingly, and could be treated with scant respect. There is a sadly humbled tone now, and the classical apologist comes delicately before his judges. All he begs for is an equal chance; for instance, that in the "first examination" the requirements in other subjects should not be so exacting as to discourage the candidates from offering at least Latin as well as one modern foreign language, and that natural science should not be made compulsory.

It is true that the classics offer an access to beauty and give a power of understanding which nothing supplies so readily and so well, but the members of the Committee are sanguine men if they expect their recommendations to be adopted. The mind of the country is set on other things. "The civilisation of the modern Western world is grounded upon the ancient civilisation of the Mediterranean coast", as they rightly say in their exordium. The understanding of those who know nothing of these origins is hopelessly imperfect and starved. To this theme the Committee often recurs. The members have had the good fortune to meet with much evidence indicating a growing appreciation of the value of the classics, which is epitomized in striking passages of the Report. Mr Mansbridge is quoted to the effect that a widespread demand for classical teaching may be expected amongst working people, who are greatly interested in the civilisation of Greece, "in spite", as he adds, "of deep-rooted prejudice against a nation which had such a sharp division of the classes "—a naïve and significant illustration of the instructive value of classical experience. The social reformer may learn something by contemplating the peculiar and, as he holds, reprehensible system of Athens, a spectacle from which he naturally shrinks, as the feminist might from inspecting the dreadful example of the termite queens, or the hen hornbill plastered into the nest by her husband.

The Committee is under the impression that the scientific world especially concurs in its opinion. It is very doubtful whether the representatives of labour or of science who testified before them are truly representative of the mass with whom authority now rests. As the Committee remarks in another place, discussing the policy of the local education authorities, "it is unlikely that any body of ratepayers would consent to special financial provision for the encouragement of classics". It is: most unlikely. Let them raise the question in any place of common resort, or even, say, at a laboratory tea, and they will carry away no illusions about growing appreciation of the classics. They will find themselves in a world which cares not a jot that "all our modern forms of poetry, history, and philosophy" originated with the Greeks, and has only a scant curiosity as to whether Western civilisation is grounded on that of the Mediterranean or of some other coast. So complete is the break already that the younger students scarcely know that classical education can be seriously defended, and regard any tenderness for it as mere perversity and affectation. We are probably witnessing that rare and portentous event, a break in the continuity of civilisation. In the Press, in the arts, and, most singular of all, in learning of various kinds, the same phenomenon appears. The modern room of a picture gallery tells the same story as the pages of a scientific periodical. The new generation means to go a lot more easily than the last. Precision of language and finish are out of fashion and superfluous; as they would say, they have no use for them. A narrower range suffices. Any deep background of knowledge is only a source of perplexity. Richness and abundance are uncongenial to modern pragmatism. A simpler diet, consisting largely of ready-prepared and familiar substitutes, such as home provides, is preferable. All this is very curious and most interesting to observe. The world may become more contented, but it is likely to be duller, and that simultaneously with these changes a revival of interest in the classics can occur seems highly improbable.

The Report betrays a consciousness that the subject dealt
with is one wider than the nominal problem of classical education, and the members of the Committee know that they are
in reality pronouncing on a great social question. They are
haunted by timidity and obsessed with the democratic nostrum
of equal rights and opportunities; but, though fighting for their
lives, they dare not make a firm stand. They should have declared boldly that learning, classical and natural, though com-
prising many parts, is one indivisible whole. Never was it so
urgently necessary that the unity of the intellectual world should
be maintained and strengthened. The natural and permanent
division of society is between them and the rest. Instead of
seeing in science a competitor, the classical advocates should
have welcomed natural knowledge as an indispensable and
essential part of complete education. Spontaneous curiosity is,
as they truly say, the only safe foundation for the continuous
life of any study; but curiosity is a function of active minds,
which alone are entitled to the privilege of direction. Freedom
of choice is a counsel of perfection; a mere vanity unless the
choosers have themselves wisdom, and the knowledge by which
choice must be guided. In default, the decision must be made
by informed authority, and must be enforced by compulsion.

If the continuity of civilisation is to be preserved, there can
be no question of abandoning the classics, but in the name of
truth and advancement no less must science be presented to all
who pretend to complete education. They must acquire a
"widespread knowledge, however elementary, of the ancient
world", and an equally widespread knowledge of the elements
of natural truth. The rudiments of classics, of natural science,
and of a modern language can be easily mastered by any boy
of ability before he is seventeen. The feebler will no doubt drop
behind. They will find their place below.

CLASSICAL EDUCATION AND SCIENCE MEN

Précis of evidence offered to the Prime Minister's
Committee on Classics, June 1920

(Unpublished)

My views on the relations between scientific and classical education were lately given in an Essay which appeared in the *Cambridge Essays on Education*, 1917. I regret I have no spare copies.

Education is often represented as a process by which boys originally homogeneous are converted into specialised types. I submit that all schemes of education should be planned in accordance with the physiological fact that living material is, as regards aptitudes, naturally heterogeneous. To provide adequately for those who have these various aptitudes—which may often be latent until puberty—the teaching should be as varied as possible, including elements of every kind of knowledge, to be presented in their most attractive forms, without pedantry, mysticism or prudery.

Classical teaching should, in my opinion, be maintained for all who can afford a complete education. I have continually supported compulsory classics at Cambridge, being convinced that without this requirement by the Universities they will cease to be a staple of education even in the Public Schools. Whittled as it was to nothing the Greek test became ridiculous and has now disappeared, with Latin soon to follow.

The classical teachers are themselves very greatly to blame for the contempt in which their subject is usually held by scientific and practical men. They have steadily refused to put grammar anywhere but first. It is possible to know a language enough for many purposes both of use and enjoyment with very slender equipment in grammar. A normal mind learns a language first, and the grammar afterwards, if at all. That grammar has value as an instrument of education is not in dispute; but by emphasising the importance of grammar an artificial obstacle is created. Few pupils who have no special taste for literature ever reach the point of seeing that anything worth having lies beyond this obstacle which they fail to surmount.

It may be maintained that the hard grind of grammar serves a purpose in keeping out those whose minds are not formed for education. There is more truth in this contention than most of us would care to admit. To repel deliberately is a dangerous course; it should be remembered that the unlettered specialist, though sometimes a bore or even a barbarian, is a valuable and very powerful person. He has his place in the learned world, and it would have been better to have made a friend of him from the beginning.

Both classics and science should be taught to all; not in pursuit of the fancy that unscientific minds can thereby be made to think scientifically; nor in the hope that the gift of speech can be bestowed on the cramped and inarticulate; but because classics in the broad sense and science are the two complementary halves of the intellectual achievement of mankind. Many boys can assimilate both, and all who are worth educating will need one or the other. Till boys are about 17 no one, save in rarest cases, knows what they will be or upon what they will thrive: therefore they should be given both and in abundance. In a well-ordered school they should both be equally staples of education, till the age of about 16½. If grammar as such is dropped there will be plenty of room in the time-table for classics and science and for some collateral subjects also. The pedant's notion that teaching in a given subject is useless unless it goes so far as to be accounted "thorough" is fallacious and contrary to fact. I have no fear of "smatterings". (What would Shakespeare have been without smatterings?) The great thing is to be introduced to many kinds of knowledge early.

Of course every educated young man must have learned at least two modern languages. But I do not think he will get from them what is to be had from classics. They come too near common life. They are too obviously useful. From classics he may learn a catholic respect for things outside his own little world, and something of the intellectual history of man and his variations. From contemporary literature he will not acquire that sense of size and proportion that good classical teaching is likely to give him.

But especially I would urge that to recognise education as sufficient which does not include classics is to break the continuity of European civilisation, and to forgo the fulness of our own inheritance. Defenders of classical education sometimes plead that it gives literary style and precision. In my judgment they weaken their case. Plenty of scholars write indifferently and we all know masters of English who had no scholarship. The right defence of classics lies in their value to us not as writers but as readers. When once the break with the past is completed, as it probably will be, the literature which has grown in the main stream of our civilisation will be un-intelligible. For many of us it is so already.

If classical teaching becomes the prerogative of the literary caste, not only must continuity with the past be broken, but any coherence among the contemporary intellectual community will be most difficult to preserve—a consequence even more serious. United, the representatives of learning of every kind might exercise a prodigious authority and force, the only effectual defence against the ignorant politician and tradesman. The best hope of creating such unity is to provide a common education, literary and scientific, in which all have at one time shared. That either group should feel the methods and objects of the other to be alien to them is deplorable. Learning of every kind and zeal for its development should be a common bond among thoughtful men of every type. The real division is between them and the rest.

The thoughtful and the learned *are* a separate caste. They need not dissipate their influence by again sub-dividing themselves. Communication among their several kinds is in any case not easy, but it would be facilitated if in a common education each had learned at least the essential rudiments of the other's business.

24 *June* 1920

REVIEWS

EVOLUTION FOR AMATEURS

The Evolution Theory. By Dr August Weismann. Translated with the author's co-operation by Professor J. Arthur Thomson and Margaret R. Thomson. London: Arnold. 2 vols. 1904

The Speaker, 24 June 1905

To Professor Weismann the gratitude of naturalists is ever due for two excellent services. He it was who first taught us to distinguish the "soma", or body, from the germ, thus ridding evolutionary science of the distracting belief that the experience of the organism is transmitted to its offspring. Formerly "use and disuse" were good enough answers to any troublesome conundrum of adaptation. Weismann's demand for evidence that in a single case such effects were transmitted brought this vague reasoning to an end. The inheritance of acquired characters was then seen to be an assumption needing independent proof, and, when proof was called for, there was no reply that a critical mind could accept as valid. How much laborious argumentation collapsed when this keystone was withdrawn we need not now recall; but those who are now constructing a sound science of heredity on the basis of physiological fact know that it was by Weismann's thorough demolition that their ground was cleared.

It was, moreover, through his ingenious speculations as to the mechanism of heredity that efforts were concentrated on a determination of the exact processes by which germ cells are formed. Whatever be the interpretation, the visible facts are now known, a direct consequence of Weismann's stimulus and initiative, which will bear fruit hereafter.

But even with this record well in mind, it is impossible to pass a lenient judgment on *The Evolution Theory*. It should have appeared thirty years ago. Then Natural Selection was a new idea. Variation and heredity were unexplored territories. The struggle for existence was an imposing force, and the temptation to suppose it the sole factor in evolution was great. Such a view had many attractions. It economised hypotheses, provided a complete system, and saved further trouble. Chaos at one end, order at the other; promiscuous *ad libitum* variation in every direction; Natural Selection guiding and eliminating, there was the solution ready made. Professor Weismann from the first was among those who cherished this simple creed, claiming in set terms that the principle of selection was "omnipotent". Wherever order was to be found in the living world there was the working of selection manifest. Even trivial features of form or habit that are constant or regular owed their constancy and their regularity to purposeful adaptation under selection's moulding hand. They are permanent and regular because selected, and selected because they contribute to the welfare of their possessors. All further hypotheses are superfluous. Such is the great theme expounded in the present treatise, wherein Professor Weismann sets before us his matured views.

But, as often happens, the endeavour to avoid assumptions has introduced some very grave ones, more dangerous because unavowed. For in pursuit of his idea the selectionist did not ask himself for proof that the beginning of life was chaos, nor did he pause to consider how living things could exist for a day or an hour in that primeval state of misfit from which Natural Selection was to deliver them. Natural Selection was to create adaptations; but could an unadapted being *live*? Professor Weismann feels no such difficulty. He says explicitly that since the simplest creatures now living shew traces of order in their structure, they must therefore be the products of aeons of selection. But what would the most primitive Amoeba of them all look like before its parts in order stood? Does not life imply chemical processes of high complexity? And how are those processes to be conceived apart from order when the structure and properties of the simplest and stablest substance are instinct with order?

And if we grant that organisms started without adaptation— as consonant with the magnificence of the unknown and because we must begin somewhere—at the next step we meet another assumption no less serious. Natural Selection being the all-sufficing "cause", variation must be random and fluctuate fortuitously. But this is a proposition we can test by observation and experiment. Variation, all agree, is going on still. Why not look and see if it is at random? Unfortunately for Professor Weismann's philosophic scheme, this is now being done. The answer thus far is very positive, and in flat contradiction to the selectionist's assumption. For to the births of time is added one which to selectionists seems more than common monstrous— the discovery that new forms arising by sudden variations may have just those characteristics of definiteness which Professor

Weismann would claim as the very hall-mark of continued selection; though, by the nature of the case, these new varieties have not undergone any such process. If thirty years ago it could be conjectured in ignorance that variation might be chaotic, many know better to-day, and therefore we say this book comes too late.

The Evolution Theory can in truth be taken as a serious contribution only by those who are content to treat the discoveries of recent years as non-existent. On the rare occasions when he refers to the ascertained facts of variation, Professor Weismann betrays a not unnatural apprehension. Professor de Vries, a pioneer of the new methods, is told, for example, that he "obviously overrates the value of his facts", and he is duly admonished for neglect of "what lies before him—the other aspect of the transmutation of species, to which the attention of most observers since Darwin and Wallace has been almost exclusively devoted—I mean the origin of adaptations". Might not de Vries reply that it is just because most observers can see no more than lies before them that we have been halting so long? If he values his facts rather highly they at least are facts, and they make a pleasant change from guess-work.

The newly discovered principles of heredity to which Mendel's work has led us are scarcely more welcome to Professor Weismann; for at a stroke they destroy most of the speculations as to the mechanism of heredity which in the pre-experimental era ran an illustrious course. So he hastens to assure his readers that plenty of exceptions to these principles are known, and that consideration of the whole matter may be deferred.

Indeed, with the gospel of selection to be preached, everything else must be deferred. Order means fitness: whatever is is useful. That is the text constantly reiterated through 800 pages. Our author does not shrink from the crudest expression of his faith. "We venture to maintain that everything in the world of organisms that has permanence and significance depends upon adaptation, and has arisen through a sifting of the variations which presented themselves, that is, through selection....From the first beginnings of life, up to its highest point only what is purposeful has arisen, because the living units at every grade are continually being sifted according to their utility, and the ceaseless struggle for existence is continually producing and favouring the fittest." As we read we think of that other philosopher who, observing how the nose fits the spectacles and the feet the shoes, deduced with the same confidence and more brevity that all was for the best. The passage quoted above might well have run on *par conséquent ceux qui ont avancé que tout est bien, ont dit une sottise: il fallait dire qui tout est au mieux.*

And as the Westphalian sage could turn in some very tight places, so can he of Freiburg. Let us pass from the general to the particular. Among things designed, as it might appear, to confute the selectionist, are the facts of regeneration. Many insects, for example, can grow perfect legs to replace those lost by injury, having often elaborate apparatus for this sole end, never otherwise brought into action. We are told categorically in italics that *every essential part of a species is not merely regulated by natural selection, but originally produced by it.* How, then, did Natural Selection produce this mechanism? First reciting that the loss of limbs helps the insect to escape its enemies, Professor Weismann adds a more novel conjecture. Bordage, he tells us, observed that stick-insects often die by getting caught by the leg in moulting. "Of 100 Phasmids nine died in this way, twenty-two got free with the loss of one or more legs, and only sixty-nine survived the moult without any loss at all." So, while creating this mechanism Selection had an eye solely to the maimed, choosing among them those which, shewing symptoms of regeneration, left offspring maimed again in their turn, and so on for countless ages. May we ask what the unmutilated were doing all this time? Loss of limb might have been supposed a heavy handicap. The marvel of the phenomenon grows when one learns against what desperate odds it was brought off. Why go this long way round? It is like travelling from the City to Waterloo *via* Kamtchatka. Natural Selection would have saved a lot of time and other things by merely making the cast skin fit a bit easier. That it would soon have done if the uninjured had prevailed and the maimed been eaten as usual. And this is that parsimonious force which in another mood put down the wings of the island birds to save tissue! Lastly, in fairness to Nature, we may perhaps hint that the moult difficulty is supposititious, and that if the 100 unfortunate Phasmids had not been in confinement their skins would have come off well enough.

We miss from this treatise those frank passages in which Darwin was wont to own himself at a loss. The author's bearing as he picks his way among the facts is not reassuring. It is a favourite assertion of selectionists that in domesticated forms the selected parts are more differentiated—as by hypothesis they should be—than those which man has neglected.

Unexceptionable illustrations of this "principle" are however, somewhat rare. Amongst others we are asked to observe that in the gooseberry, of which a hundred varieties can be distinguished by the fruits, the flowers of the several varieties are alike. But how about the stems, spines, leaves, manner of growth, all characters that have not been the subject of special choice? If Professor Weismann will gather his own gooseberries this summer, or prune them in December, he will learn to know the sorts as easily by the plant-characters as by the fruits. For those who prefer book-knowledge the beautiful plates in Thory's monograph may suffice. Professor Weismann is preferring the exception to the rule. What potato-grower needs to dig the tubers to name the variety? In apples, plums, pears, chrysanthemums, pelargoniums, violets, and hosts more plants, the vegetative parts are just as characteristic as the parts man requires and selects. The truth is, as a gardener once said to his scientific master, "What you say, sir, is all very well for anyone to read in a book, but it won't do at the potting-bench".

A little knowledge of domesticated breeds is a very dangerous thing. All evolutionists now agree that the origin of the domesticated forms is a guide to that of the wild. But Professor Weismann inverts this logical sequence. Having made up his mind that wild forms arose by continuous selection directed to each character, he lightly applies this proposition to the domesticated. Without any reserve, he declares, "Thus it is not by crossing of different breeds, but by a patient accumulating of insignificant variations through many generations that the desired transformations are brought about. That is the magic wand by which the expert breeder produces his different breeds, we might almost say, as the sculptor moulds and remoulds his clay model according to his fancy". Now, this may do for the scientist public, to whom the book is addressed, but the fancier can only smile at such statements. Professor Weismann makes the assertion with special reference to pigeons, of all unlikely things. Let him take an opportunity of joining the pigeon or poultry fancy, and he will soon undeceive himself. He will discover that within his own lifetime dozens of new breeds have been created by crossing, and he may even get fairly complete histories of the actual ways in which the various unit-characters were combined to make them.

Again and again we long to cross-examine Professor Weismann's evidence. He has seen one side only of the facts. Wider knowledge, or a moment's reflection as to the possible answers to his contentions would have led to the suppression of many passages. He is at his best perhaps in his criticism of Nägeli's once famous transcendental thesis that evolutionary progress was due to a mysterious "phyletic force", which impelled living things along the path of progress. With reason on his side Weismann shews this speculation to be a substitution of a greater mystery for a less. It was, indeed, a mere confession of faith. Not one scrap of tangible evidence was ever produced pointing to the existence of any such "force". We should, however, admire the exposure more if we could feel that it was called for, or that these warnings were really needed. "Phyletic force" makes no doubt a good *repoussoir* to omnipotent Selection, and that perhaps is why we hear so much of it from Professor Weismann.

We wonder, too, whether our author himself is not entering that very maze wherein Nägeli was lost. "Use and disuse" in the old familiar senses being barred by common agreement, the need of some substitute presses sore on the pure-selectionist. This we are now offered in the shape of *Germinal Selection*, depending on "the struggle of the parts of the germ-plasm"—an extension of Roux's *Kampf der Teile*. But the criteria which decide the fate of the germinal parts are their *prospective* contributions to the developed organism. Natural Selection, being here in her economical mood, is to eliminate the parts which, if allowed to develop, would not have "pulled their weight". We thus meet a proleptic Natural Selection, dealing in "futures"—as transcendental a conception as any Nägelian could desire.

The loyal evolutionist, who knows the facts, cannot but regret this book. It is avowedly written for amateurs. That they will find it so readable makes matters worse. In the uncritical it will create a complacent satisfaction that the truth does not warrant, and to the critical almost every page gives occasion for derision. The professional will not find it of much service. The facts are mostly such as have been used before, and references to authorities are rarely given. The book is a piece of special pleading.

It should be the last of its kind for some years. Evolution has passed out of the speculative stage. Experimental methods are fast revolutionising the notions of the founders of the science, to whom from Lamarck to Weismann be all honour according to their several degrees, and until this experimental work is advanced somewhat further, the popular treatise should be held up. It must be either out of date or premature.

The preface tells pathetically that among men of science only two have given thorough-going adherence to the author's views, and we cannot expect *The Evolution Theory* to add to the number. Of the two one is Professor J. Arthur Thomson, the present translator. He has done his work remarkably well, and scarcely a sentence has a foreign twang. There are, however, symptoms that, like the author himself, he has experience rather of the laboratory than the field. *Polyommatus phlaeas*, for instance, has an English name, which is not "the little red-gold fire-butterfly", but the Small Copper. Without the scientific name, too, we should not have recognised an old friend as the "Bannermoth".

HEREDITY IN THE PHYSIOLOGY OF NATIONS

The Principles of Heredity. By G. ARCHDALL REID, M.B., F.R.S.E.
London: Chapman and Hall

The Speaker, 14 October 1905

What will happen when civilised society thoroughly grasps
what heredity means? There are signs, of which Mr Archdall
Reid's book is only one, that that time may not be very far off.
It is a mere accident that recognition of the plain facts has been
delayed so long. Were the physiology of inheritance slightly
less complex, its paramount importance would long ago have
been evident to all, and man would have perceived that this is
the point at which he can really shape his own destiny. The
steady application of a breeding law would accomplish more
in three generations than all the criminal and sanitary enactments that the centuries have devised. Mr Galton has been
proclaiming this truth to a sceptical world for forty years.
Nature, to use his antithesis, is much; nurture incomparably
little. Circumstances have lately combined to bring these
matters into prominence. Physical deterioration, the alarming
increase in the relative numbers of the insane, the utility of
teaching the minds of starving children, the relation of the
State to the unemployed, and all questions of grave national
anxiety—they are problems of national physiology, and as
physiological problems they are at last beginning to be studied.

Political economists have hitherto incurred no reproach if
their doctrine were not based on physiological evidence. That
is not their department; and though illustrative references to
such topics are considered becoming in their writings, neither
economist nor politician has been expected to go to physiology for his fundamental facts. Man, for the purposes of all
social enactments, has been a purely ideal conception. Society
is assumed to consist of homogeneous units, all similar, and
endowed with similar faculties and rights—that of procreating
their kind being permitted to, and even to some extent en-
joined on, all. This resolute determination to treat humanity
as physiologically an undifferentiated mass ranks, indeed,
among the particular glories of civilisation. No one supposes
that view to reflect the facts of nature. On the contrary, we
recognise in the feudal system a nearer approach to the natural
plan. The "ladder of lords", reaching from the *minuti homines*

below to the king on his throne and so up to God above, might
stand for a frank if unconscious recognition of the struggle for
existence as a scheme of social order. But at length the disturbance which biologists call variation supervened—historians
have other names for it—and new types of fitness grew up from
below. The system became unstable; the "ladder" eventually
collapsed, and now we are all equal again. This phase, obviously transitional, must in its turn come to an end. That it
should have lasted so long is remarkable, and with great probability the naturalist may surmise that the continuance of the
equality system has only been made possible by the fact that in
the same period the supply of necessities on the whole increased
in a higher ratio than the demands of population.

But with that intensification of the social strain, of which a
few of the symptoms were named above, there is a disposition to
look more closely into the elements of these problems, and to
reflective minds it is evident that society is here embarking on
an inquiry of the deepest significance. As to the main results of
that inquiry, there can be no doubt whatever. Many are aware
that of late years our knowledge of heredity has greatly
advanced. The general principles which Mr Galton's genius
detected are already giving place to precise and specific laws,
ascertained by experimental tests. It is easy to foresee that the
complications which have obscured the working of heredity
must before long be unravelled by research, and at least the
outlines of the process will be clear. Sooner or later the attention of the world at large will be attracted, and the appearance
of Mr Archdall Reid's *Principles of Heredity* has brought that
moment perceptibly nearer. The main thesis of the book has
even formed the subject of a holiday symposium in the *Morning
Post*. Such prominence may be due to the bearing of heredity
on deterioration and thence indirectly on universal conscription
—the favourite panacea of that journal—but none the less this
publicity bears witness to the magnitude of the issue, and it
must hasten the crisis which will assuredly be reached when the
facts of heredity are familiarly realised. To that realisation
Mr Archdall Reid contributes in no small degree, but he would
have done still more had he given the reader a really comprehensive survey of those facts. The book treats too much of things
in general. It is disappointing that a volume, apparently a
treatise on heredity, should not even contain a description of
the various types of inheritance already identified, or of the
methods of research employed in these studies. The work consequently must be regarded rather as a valuable addition to the

popular literature of evolution, with which the author has a wide acquaintance, than as a contribution to science. Nevertheless given heredity of some sort as a fact, the author provides a stimulating discussion of the importance of the phenomena in sociological study. It is a useful essay, though it might have appeared at any time since Weismann cleared our ideas as to the non-inheritance of parental experience.

Taken as a whole, the author's propositions are so much more consistent with fact than those advanced by his opponents that we are reluctant to raise objections which can be deferred. But, admitting substantial truth underlying his contentions, we trust that in his future writings he will discriminate more strictly between the various evils against which humanity has to struggle. Alcoholism, effects of slum conditions, and the various infirmities of mind or body which heredity transmits are not really comparable either in their physiological nature or in their sociological consequences. Moreover the remedies appropriate to the various cases are exceedingly diverse. Fuller knowledge of heredity is fast revealing these distinctions, and it is in the extension of such analysis that all hope of progress lies. The disposition to contemplate all classes of adverse influences as similar, and to regard the existence of suffering as a guarantee of progressive adaptation, belongs to an earlier stage of evolutionary thought. With these reservations we welcome *The Principles of Heredity* as a help towards enlightenment. What, as we asked at the outset, will happen when that enlightenment actually comes to pass and the facts of heredity are as commonly known as those of bacteriology, for instance? One thing is certain: mankind will begin to interfere; perhaps not in England, but in some country more ready to break with the past and eager for "national efficiency". Mr Galton has suggested a selection at the top, with State encouragement for families of superlative quality. More probably, and we suspect more effectively, selection will begin by elimination at the bottom. Mr Reid tells us of a pair of semi-insane parents who contributed nine insane children to the burdens of the State. Of similar propagation of criminal instinct and bodily disease illustrations abound. When the meaning of these facts is universally appreciated interference is bound to come. In common parlance, people will not stand it. If a farmer saved all his animals, good and worthless alike, and bred from all, his ruin would be assured. That, nevertheless, is exactly what civilisation is doing for our own species. Unlike the farmer, moreover, we apply every resource of science to the reduction of infant mortality and to the preservation of the unfit, and we have public rates to draw on for that purpose. Before long these considerations will become the insistent daily thoughts of all educated people. Something like a fresh era in civilisation must then begin. Contemporary socialism strives for the elevation of the unfit; that of the future will probably aim at their extinction. Ignorance of the remoter consequences of interference has never long postponed such experiments. When power is discovered man always turns it on. The science of heredity will soon provide power on a stupendous scale; and in some country, at some time, not, perhaps, far distant, that power will be applied to control the composition of a nation. Whether the institution of such control will ultimately be good or bad for that nation or for humanity at large is a separate question. To those who look ahead the interesting point is that the attempt will certainly be made. Of all the predicable things "that men shall do and we in our graves" that may be the one which will most profoundly change the destiny of men.

HUXLEY AND EVOLUTION

Nature, 9 May 1925

From time to time I am asked by students, botanical and
other, Was Huxley a great man? Did he do very much? I have
a clear answer. I say, if you were a zoologist you could not ask
that question, for you would know that Huxley worked over
almost the whole face of zoology, and that so much of modern
classification and terminology is the product of his logic and
"organised common-sense" that if we turn to any text-book
earlier than about 1850, when Huxley's operations were beginning, we feel ourselves in zoological pre-history. It is all very
well to say that anybody who chose to look could see that
starfishes, Holothurians and Medusae should not be classed together and with various other creatures, but neither Lamarck
nor Cuvier did notice that Radiata and Polyps were preposterous medleys. Most of the great groups at one time or another
came under Huxley's attention, and his instinct for order and
his morphological sagacity were so sure that his judgment has
been generally accepted by his successors.

I am aware, however, that on the occasion of this centenary
the services we are to commemorate are not those which he
rendered as a great architect of academic morphology. To the
world, scientific as well as lay, Huxley is chiefly famous as the
champion of evolutionary doctrine, whose vigorous and skilful
advocacy counted for so much in obtaining the favourable
verdict of the public. The opportunity was prodigious. He had a
splendid case. Among his opponents were persons of the highest
consequence, some of whom for this particular contest were
equipped with nothing beyond the complacency of ignorance.
He was, moreover, willing to take pains—a very formidable
qualification in a controversialist. Such papers as Huxley on
Suarez, Huxley *v.* Gladstone in the matter of the Gadarene
swine or the order of vertebrate succession, provided a rare
entertainment, of which the like—to compare small with great
—had scarcely been seen since Bentley's *Phalaris*; though without disrespect to the victors in those decisive engagements, one
may perhaps doubt whether either of them went about their
daily business loaded with quite the weight of extensive and
peculiar learning which upon emergency they produced with
perfect spontaneity to the confusion of their opponents.

Looking back over that critical period, we wonder at the

persistent bad leadership of the opposition. The only weapon by which they might have impeded progress was one they never seem to have thought of using, namely, silence. Had authority contented itself with observing that similar notions had been promulgated not infrequently for nearly a century before without meeting the general approval of naturalists, adding possibly a few soothing and carminative words to the effect that, whether true or not, these technicalities left the fundamentals of revelation undisturbed, but disclaiming any particular interest in the topic, trouble would have been long postponed, perhaps avoided indefinitely.

If that course had been pursued, we professionals would be remembering Huxley as a sound naturalist and an acute observer, though scarcely perhaps on a scale amounting to a celebration. Geneticists certainly are not likely to forget him. Through all his triumphant vindications of the doctrine of descent as a general proposition, he never forgot the weak spot. Again and again he declared it to exist in "the group of phenomena which I mentioned to you under the name of Hybridism, and which I explained to consist in the sterility of the offspring of certain species when crossed with one another" (1863).[1] In the same year he writes to Kingsley: "From the first time I wrote about Darwin's book in the *Times*...until now, it has been obvious to me that this is the weak point of Darwin's doctrine. He *has* shewn that selective breeding is a *vera causa* for morphological species; but he has not yet shewn it a *vera causa* for physiological species. But I entertain little doubt that a carefully devised system of experimentation would produce physiological species by selection, only the feat has not been performed yet".

Nothing that has happened since at all mitigates the seriousness of this criticism. The words quoted above may indeed be used to-day with an even stronger emphasis, though I doubt whether many of those best acquainted with modern genetics are so sanguine as Huxley was, that by the most carefully devised system of experimentation are we in the least likely to produce physiological species by selection. Rather have we come to suspect that no amount of selection or accumulation of such variations as we commonly see contemporaneously occurring can ever culminate in the production of that "complete physiological divergence" to which the term species is critically applicable. With entire candour Huxley reiterated that if this were the necessary and inevitable result of all experiments, the

[1] *Collected Essays*, II, 1893, p. 463.

Darwinian hypothesis would be "shattered". Nothing was to be gained by glozing that difficulty. The grounds of the evolutionary faith are otherwise so solid that no alternative can ever be considered again; but chiefly for the reason so prominently named by Huxley, which modern genetical research has so greatly reinforced, the representations of that process which found such facile acceptance in his time no longer satisfy us.

On another occasion Huxley's admirable scientific judgment came near to rendering a great service, if not to science, at least to Darwin. The manuscript of the Pangenesis chapter, published at the end of *Animals and Plants*, was submitted to him for an opinion (1865). What he then replied we do not know, for the letter is not published among his correspondence, being, I imagine, lost. But its tenor may be inferred from the sentence in Darwin's answer, "I do not doubt your judgment is perfectly just, and I will try to persuade myself not to publish". Huxley unfortunately weakened and replied that he had not at all meant to stop the publication, that he really should not like to take that responsibility, etc. So this curious chapter appeared, revealing that Darwin must have gone through life never apprehending the significance of cell division, and almost without curiosity as to what was then already known of the process by which animals and plants are reproduced. From other passages the modern reader of course would suspect as much, but if Huxley's discretion had prevailed, illusion need not have been totally destroyed.

As we can now see very well, both Darwin and Huxley in a sense mistook the character of their own work. They were assembling materials and laying a foundation, well and truly, be it said, though, like so many of their contemporaries, they imagined they were finishing a permanent edifice. Huxley himself, as he stands in Collier's picture, confidently facing his audience with the skull in his hand, might almost be the model for Max Beerbohm's "The Future—as the XIXth Century saw it". Looking forward, the Victorian type sees his successor, the duplicate of himself, the same features, same proportions, same frock coat, only magnified enormously. In biology at least there were no misgivings in those days, and few attempts to look far behind the obvious. Genetics, the experimental study of developmental mechanics, and, in general, the prosecution of more rigorous analysis, are an independent development, related to what went before about as much as the arch was to the architrave.

Late in life Huxley attacked the Gentians, and after a year's work published his *Notes and Queries* on that natural order.[1] It was considered an admirable discussion, and I can believe it to be so. The whole series of genera are there arranged in a logical order of inter-relationships based on the differentiation of the floral parts in adaptation to fertilisation by insects. To be sure, as he explicitly states, this consideration cannot be supposed to have decided the numerous other features of habit, or of leaf-structure, or the various other anatomical points in which the plants also differ, but he has "little doubt that, with larger knowledge, analogous causes will be found operative in all these cases". The "larger knowledge" to which Huxley is looking forward is to be the same kind of knowledge, only more of it. The knowledge his successors seek is of a wholly different order. No one better than Huxley knew that some day the problems of life must be investigated by the methods of physical science if biological speculation is not to degenerate into a barren debate. That ambition, which in Huxley's day was a pious and impotent fantasy, has become the immanent and informing hope in which all modern evolutionary research is directed. The Gentians well illustrate the change; for I suppose we would resign ourselves to ignorance of the teleological meaning of their floral apparatus if some one would give us an analysis of the mechanical forces by which the flowers of *G. campestris* develop their parts in fours, and demonstrate how they are related to the mechanism by which many closely related species divide their flowers into fives.

Yet if our immediate aims are so distinct, our ultimate purpose is the same. In Huxley we shall always reverence one in the fruits of whose victory for truth and liberty we are still sharing. The direction of public opinion is a most precarious art, demanding imagination and a large knowledge of human nature. Of that art Huxley was an incomparable master; and the fact that thousands are now engaged without hindrance in the prosecution of those researches to which he devoted his whole life, is the direct result of his eloquence and courage. "Other men laboured, and ye are entered into their labours."

[1] *Linnean Journal, Botany*, xxiv, 1887.

APPENDIX

ROUGH LIST OF CONTROVERSIES[1]
IN WHICH W. B. ENGAGED

The alleged Mimicry of *Volucellae*. 1892. *Nature*, XLVI, No. 1199, 585, W. B.; *ibid*. XLVII, No. 1202, 28, POULTON; *ibid*. No. 1204, 77, W. B.; *ibid*. No. 1206, 126, POULTON.

The Origin of the Cultivated *Cineraria*.[2] 1895. *Nature*, LI, No. 1330, 605–607, W. B.; *ibid*. LII, No. 1331, 3, THISELTON-DYER; *ibid*. No. 1332, 29, W. B.; *ibid*. No. 1333, 54, WELDON and W. BOTTING-HEMSLEY; *ibid*. No. 1334, 78, DYER and WELDON; *ibid*. No. 1335, 103, W. B.; *ibid*. No. 1336, 128, DYER.
Also notice of the discussion in *Gard. Chron*. 11 May 1895, p. 588. Also correspondence and MS. notes.

Women's Degrees. Fly sheets. Cambridge University. 1897.

Compulsory Greek. Fly sheets. 1891, 1905. Letter to *Nature*, LXXI, No. 1843, February 1905, 390. Speeches in the Senate House, *Cambridge University Reporter*, 17 December 1904, pp. 392, 393; *ibid*. 7 May 1906, pp. 807, 808.

Mendelian Controversy. 1902 and onwards. "Mendel's laws of Alternative (*sic*) Inheritance in Peas." W. F. R. WELDON. *Biometrika*, I, pt 2, 228–254. "A defence of Mendel's Principles of Heredity." W. BATESON, with a translation of Mendel's original papers on Hybridisation. Cambridge University Press, 1902. "On the Ambiguity of Mendel's Categories." W. F. R. WELDON. *Biometrika*, II, pt 1, 44–55. "Mr Bateson's Revisions of Mendel's Theory of Heredity." W. F. R. WELDON. *Biometrika*, II, pt 3, 1903. Letter from W. B. answering Weldon, returned by *Nature* 19 May 1903; the Editor was "not prepared to continue the discussion on *Mendel's Principles* and therefore returns herewith the papers recently sent to him by Mr Bateson". "A Note on Sweet Peas." HUGH RICHARDSON. *Nature*,

---

[1] From a letter to Major C. C. Hurst (received since going to press): 2. 2. 1907. The term *controversial* is conveniently used by those who are wrong to apply to the persons who correct them. Properly the word is not applicable in such cases. It has nothing to do with points of fact, but merely with opinion. W. B.

[2] *Cineraria Stellata* dates from this controversy from crosses made by Mr Lynch in the Cambridge Botanic Garden, for W. B.

S. *tussilaginis* × (S. *cru*. × (*cru*. × *herit*.))
    ♀   ♀   ♀   ♂

October 1902. W. B. wrote a letter discussing and suggesting the probable explanation, which was returned unpublished by the Editor.

This controversy was mixed and merged in the previously quoted Mendel's Principles of Heredity in Mice.

These published controversies were accompanied by much correspondence, which, with rough drafts of his replies, was carefully docketed and kept; W. B. always intended to write the history of "Mendelism".

Mendel's Principles of Heredity in Mice. 1904. "On the result of crossing Japanese Waltzing with Albino Mice." A. D. DARBISHIRE, *Biometrika*, II, pt 1; II, pt 2; III, pt 1. *Nature*, LXVII, No. 1742, 462–463, W. B.; *ibid.* No. 1744, 512, WELDON; *ibid.* No. 1745, 550, Review of Darbishire's "Mice" in *Biometrika*; *ibid.* No. 1747, 585, W. B.; *ibid.* No. 1748, 610, WELDON; *ibid.* LXVIII, No. 1750, 33, W. B. and WELDON.

W. B.'s letter returned to him by *Nature*. Correspondence with Darbishire.

Homotyposis. 1900–1903. "On the Principle of Homotyposis and its Relation to Heredity, to the Variability of the Individual, and to that of the Race." Professor KARL PEARSON, F.R.S. *Phil. Trans. Royal Society*, CXCVII, 1901. "Heredity, Differentiation, and other Conceptions of Biology: a Consideration of Professor Karl Pearson's paper: 'On the Principle of Homotyposis'." W. BATESON. *Proc. Roy. Soc.* LXIX, 1901. "On the Fundamental Conceptions of Biology." KARL PEARSON, F.R.S. *Biometrika*, I, 1902, 320–344. "Variation and Differentiation in Parts and Brethren." W. BATESON. Privately printed, July 1903. Much correspondence as yet unpublished.

In 1905 Mr C. C. Hurst's paper "On the Inheritance of Coat-colour in Horses" was communicated by W. B. to the Royal Society, and read on 7 December of that year. In face of apparently destructive criticism by Professor Weldon, W. B. withdrew this paper, but, after a short study of Weldon's criticism, he was able to recommunicate it at the meeting held 18 January 1906, when a lively discussion took place in which he took a prominent part.

There was also much controversial correspondence on the occasion of Mr C. C. Hurst's demonstration of the "Inheritance of Eye-colour in Man", on the occasion of the British Association Meeting at Leicester, 1907.

"The Ear of Dionysius." Letters to *The Times Literary Supplement*. 5 April 1917 (E. BRABROOK); 19 April 1917 (E. C. CONSTABLE); 26 April 1917 (E. BRABROOK); 3 May 1917 (W. B.); 10 May 1917 (G. W. BALFOUR); 17 May 1917 (W. B.); 24 May 1917 (E. BRABROOK), (ERNEST S. THOMAS); 31 May 1917 (A student of Psychology).

Acquired characters. *Salamandra maculosa*. *Alytes*. Kammerer's lecture given 21 September 1909, published *Natur*, heft 6, 1910.

Correspondence with Kammerer, 17 July 1910 (unpublished).

1913. See *Problems of Genetics*, pp. 199–212. Chapter on Adaptation.

1919. Letters to *Nature*, CIII, No. 2592, 344, W. B.

1920. Letters from Przibram.

1923. Letters to *Nature*, CXI, No. 2793, 637, Dr KAMMERER; *ibid.* No. 2796, 738, W. B.; *ibid.* No. 2799, 841, Professor MACBRIDE; *ibid.* No. 2800, 878, W. B.; CXII, No. 2807, 237, Professor MACBRIDE (also M. PERKINS); *ibid.* No. 2811, 391, W. B.; *ibid.* No. 2825, 899, PRZIBRAM and W. B.

August 1926. Letters to *Nature*, CXVIII, No. 2962, 209, PRZIBRAM and G. K. NOBLE.

INDEX OF PERSONS

INDEX OF SUBJECTS

Teaching, 36, 59, 432, 437
Thoroughness, 432
Todas, 194
Tolerance, 284
Transferable characters, 392, 396, 406, 408
Trotters and pacers, 202, 415
Tuberculosis, 376
Twins, 206, 212

Unconscious memory, 203, 421
Undulatory Hypothesis, 43
Unit characters, 248, 453
Unit factors, 210, 223
Use and disuse, 220, 221, 420, 454

Variability, 169, 189, 278, 286
Variation, and heredity, 218, 219, 220, 284, 332, 390; by loss of factors, 212, 286, 290, 292, 294;

materials for the study of, 42, 52, 54; nature of, 277, 324; study of, 27, 34

War, 127, 357 et seq., 364, 369; South African, 15
War-makers, 374
Wave lengths, 44, 408
Wheat, 251, 269, 300; field of, 414; thrashable, 308
Wheat rust, 188
Wisley, 82
Women's degrees, 59

Y.M.C.A., 130, 133

Zebra, pattern of hide, 408
Zoological Society, 77, 78
Zoology, Professorship of, 106; Readership in, 112
Zygote, 201

PRINTED
BY

WALTER   LEWIS, M.A.

AT
THE CAMBRIDGE
UNIVERSITY
PRESS